MEASURE AND INTEGRATION
Second Edition

REAL ANALYSIS, T*(16-17: 1), B, L. *Measure and Integration, Second Edition*. M.E. Munroe. A-W, 1971, xii + 290 pp, $12.50. Differs from the popular first edition by the addition of a chapter on functional analysis including the Hahn-Banach, Banach-Steinhaus and closed graph theorems along with a discussion of weak and weak* convergence. T.A.V.

MEASURE AND INTEGRATION
Second Edition

M. E. MUNROE
University of New Hampshire

ADDISON-WESLEY PUBLISHING COMPANY
Reading, Massachusetts · Menlo Park, California · London · Don Mills, Ontario

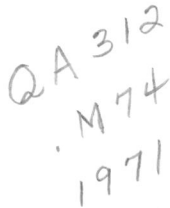

This book is in the
ADDISON-WESLEY SERIES IN MATHEMATICS

Consulting Editor: LYNN H. LOOMIS

AMS 1968 Subject Classification 2801

Copyright © 1971 by Addison-Wesley Publishing Company, Inc.
Philippines copyright 1971 by Addison-Wesley Publishing Company, Inc.

All rights reserved. No part of this publication may be reproduced, stored in a retrieval system, or transmitted, in any form or by any means, electronic, mechanical, photocopying, recording, or otherwise, without the prior written permission of the publisher. Printed in the United States of America. Published simultaneously in Canada. Library of Congress Catalog Card No. 74–127890.

PREFACE TO THE SECOND EDITION

The core of the first edition of this book was devoted to what is commonly called "Carathéodory" measure theory, as contrasted with "Bourbaki" measure theory or "Daniell" integral theory. Without debating the relative merits of these various approaches to a modern theory of the integral, we see no point in changing our basic approach to the subject and, therefore, have made relatively few changes in the central portion of the book. Those who have used the first edition will certainly recognize the chapters on measure (general and specific), measurable functions, integrals, and derivatives. The beginning and the end have undergone some changes.

Chapter 1 of the first edition was written in such a way as to make the book essentially self-contained. In 1953 this seemed realistic because, at that time, the chances were that some of this background material would have to be actively taught as part of a measure theory course. Since then there have appeared a number of adequate texts in undergraduate real analysis, so that today it seems appropriate to summarize this background material in capsule form—definitions and theorems, with proofs and exercises deleted.

The major changes in the present edition come at the end. In the first edition we had a couple of sections designed to inform the student that there is such a thing as functional analysis. In the light of recent recommendations by the Committee on the Undergraduate Program in Mathematics, it now seems desirable to incorporate into the book a genuine introduction to this subject. Accordingly, we have added a new chapter giving the "big three" theorems (Hahn-Banach, Banach-Steinhaus, and closed-graph) together with a fairly thorough discussion of weak and weak* convergence in the standard function spaces.

We want to express our thanks to Addison-Wesley and their consulting editor, Lynn Loomis, for their help and encouragement in the preparation of this revision.

Durham, New Hampshire M. E. M.
September, 1970

PREFACE TO THE FIRST EDITION

This book is based on a course in measure theory as given for the past several years at the University of Illinois. The aim of the book is to present measure theory from the abstract or postulational point of view and yet to do this in such a way that the graduate student as well as the expert will find the work helpful.

The exercises form an important feature of the book. These fall essentially into three classifications: (1) Comparatively trivial problems illustrating and embellishing the material in the text. (2) Theorems whose proofs are modeled after proofs given in the text. (3) Interesting theorems omitted from the text, accompanied by copious hints and outlines, but without complete proofs. Exercises of the first type should be useful as a basis for informal class discussions. Those of the second type should serve for drill and examination purposes. Those of the third type should furnish some special projects for assignment to the students. Also, they should be helpful to the instructor in the preparation of supplementary lectures.

It has been said that real function theory is that branch of mathematics which deals with counterexamples. This "definition" suggests in a facetious way a rather important idea. An essential feature to the spirit of real function theory is that blanket hypotheses (such as analyticity of functions, for example) are to be avoided and that restrictive conditions are to be imposed theorem by theorem as needed. In any such investigation counterexamples obviously play an important part. The point to these remarks is that many of the counterexamples in this book are to be found in the exercises; therefore even the casual reader should not omit them. In order to encourage this procedure in reading the book, we have not been bound by any rigid conventions in the placing of the exercises but have inserted them wherever they seem to fit into the discussion.

The normal prerequisite for a course based on this book would be an introductory course in real function theory. The exact content of such a course is not too important because we have included in Chapter 1 all the specific information from an introductory course that will be needed. However, we should hesitate to recommend this book to a student who has had no experience with "ε, δ" methods. An introduction to modern algebra forms a valuable background because of the mathematical maturity it develops, but we have written this as an analysis book, and there are no specific algebraic prerequisites. We should like to suggest that where it is available an introductory course in point set topology forms an ideal background for this course. Topology not only introduces the student to the fundamental methods of analysis, but it involves many of the specific topics included in our first chapter. In any case, Chapter 1 contains a mixture of topology and real function theory, and we leave it to the individual instructor to edit this to the specific needs and capabilities of his class.

The modern theory of measure and integration is, to our way of thinking, one of the most elegant bodies of mathematical knowledge. The main outline of this theory is developed in the unstarred sections of this book. For pure intellectual enjoyment the student need not go beyond these. However, it would be unfair not to give at least a sampling of the many applications of measure theory to other topics in mathematics. We have tried to do this in the starred sections. We hope that in addition to enhancing the study of measure theory itself, these brief introductions to probability, Banach spaces, orthogonal expansions, etc., will help the student to discover interests which will lead to further graduate study.

The time required for a course based on this book depends on two factors: (1) the preparation of the students and consequent time required to cover the introductory material in Chapter 1, (2) the number of starred sections and other incidental topics that the instructor chooses to put into the course. By cutting the time on each of these features to about one-half that required for thorough coverage, we have given this as a one-semester course at the University of Illinois. We should estimate that the entire book can be covered at a reasonable pace in two quarters. However, it should be useful as a basis for a full-year course, particularly by an instructor who has some pet topics of his own to insert.

The sections of the book are numbered straight through, without regard to chapter numbers. The numbers of theorems, corollaries, etc. carry the section number before the first decimal point. Exercises are labeled with lower case letters within each section, and a reference to an exercise in the current section is merely by the appropriate letter (for example, Ex. *a*). In referring to an exercise outside the current section, we append the section number (for example, Ex. 3.*a*). Each system of postulates is identified by a capital letter indicating the nature of the thing being defined. In order that the reader may find these readily, we have included a separate index of postulate systems.

At the end of each chapter we have listed some suggestions for outside reading. The numbers in these lists refer to the bibliography at the end of the book. For this bibliography we make no pretenses of completeness in any sense. It contains only those works that we have seen fit to include in our reading lists. To the reader who is interested in an exhaustive bibliography we recommend Hahn and Rosenthal [9], where a list of original sources is given at the end of each section.

At the end of each proof we have inserted a horizontal bar in the center of the page. This bar takes the place of a comment such as, "This completes the proof of the theorem." [In the second edition, the symbol ∎ is used to denote the completion of a proof.]

Meyer Jerison and Lowell Schoenfeld used a preliminary version of this book as a text and offered many helpful suggestions on the basis of this experience. Thanks are also due to J. L. Doob, John L. Kelley, and B. J. Pettis for reading the manuscript and suggesting a number of improvements. Thanks are also in order to Evelyn Kinney for help with the proofreading.

November, 1952   M. E. M.

CONTENTS

Chapter 1 Background
1. Sets and functions 1
2. Algebra of sets 5
3. Cardinal numbers 9
4. Pseudo-metric spaces 11
5. Limits and continuity 15
6. Uniform limits 16
7. Function spaces 19
8. General topology 21

Chapter 2 Measure—General Theory
9. Additive classes and Borel sets 25
10. Additive set functions 31
11. Outer measures 42
12. Regular outer measures 50
13. Metric outer measures 57

Chapter 3 Measure—Specific Examples
14. Lebesgue-Stieltjes measures 68
*15. Probability 78
*16. Hausdorff measures 83
*17. Haar measure 86
18. Nonmeasurable sets 93

Chapter 4 Measurable Functions
19. Definitions and basic properties 97
20. Operations on measurable functions 102
21. Approximation theorems 106
*22. Random variables 112

Chapter 5 Integration
23. The integral of a simple function 116
24. Integrable functions 120
25. Elementary properties of the integral 127
26. Additivity of the integral 132
27. Absolute continuity 136
28. Dominated convergence 145

| | 29 | Fubini's theorem | 149 |
| | *30 | Expectation of a random variable | 158 |

Chapter 6 Differentiation
	31	Summary of the problem	167
	32	Vitali coverings	171
	33	Differentiation of additive set functions	176
	34	The Lebesgue decomposition	180
	*35	Metric density and approximate continuity	184
	*36	Differentiation with respect to nets	191

Chapter 7 Convergence Theorems
	37	Uniform and almost everywhere convergence	199
	38	Convergence in measure and in mean	200
	39	Relations among convergence types	206
	40	Convergence of measures and integrals	208
	41	The L_p spaces	215
	42	Cauchy theorems	220
	*43	Orthogonal expansions in Hilbert space	224

Chapter 8 Functional Analysis
	44	Banach spaces	232
	45	The Hahn-Banach theorem	238
	46	Representation of linear functionals	245
	47	Hamel bases	254
	48	Weak and weak* sequential convergence	257
	49	Weak* topologies	262
	50	The closed-graph theorem	268

Bibliography 279

Index of postulates 281

Index of symbols 283

Subject index 287

CHAPTER 1

BACKGROUND

The main business of this book begins in Chapter 2. To embark on that material the reader should have some background knowledge of real analysis. This chapter gives a rapid summary of the background expected. It is written with specific reference to the author's book *Introductory Real Analysis*, Addison-Wesley, 1965, hereafter referred to as *IRA*. Most of the material in this chapter is covered in *IRA* and is therefore presented in a very condensed form here. On occasion we encounter a topic that is introduced in *IRA* but not developed as fully there as will be needed in this book. In these instances we give the additional development in greater detail in this chapter.

1 SETS AND FUNCTIONS

We assume that the notion of *set* is familiar. The notation

$$\alpha \in \Gamma$$

means "α is an element of the set Γ." The *subset* relation may be written

$$\Delta \subset \Gamma \quad \text{or} \quad \Gamma \supset \Delta,$$

meaning Δ is a subset of Γ (if $\alpha \in \Delta$, then $\alpha \in \Gamma$). If $\Delta \subset \Gamma$ and $\Delta \neq \Gamma$, then Δ is a *proper subset* of Γ. We will not use a specific single symbol for the proper subset relation. If $P(\alpha)$ is a proposition concerning α, then $\{\alpha \mid P(\alpha)\}$ means the set of all entities α for which $P(\alpha)$ is true. The *vacuous set* will be denoted by ϕ.

In this book we have adopted certain special notational devices designed to build into the notation an indication of the type of set under discussion. To begin with, Ω will be our standard notation for the *space* under consideration. Frequently, we shall embark on a discussion in which there will appear points of Ω, sets of these points, sets of such point sets, etc. In such discussions we shall use the following terminology and notation:

Type of entity	Terminology	Notation
element of Ω	point	Italic l.c.: x, y, etc.
set of these points	set	Italic caps.: A, B, etc.
class of these sets	class	Script caps.: $\mathscr{A}, \mathscr{B}$, etc.
collection of these classes	collection	German caps.: $\mathfrak{A}, \mathfrak{B}$, etc.

The reader should note that as long as we are dealing with sets of the pure types† mentioned above, the $\in$ relation is possible only between an entity of one type and one of the type immediately following, while the $\subset$ relation is possible only between entities of the same type.

There are many fundamental definitions and theorems concerning sets and their elements which we shall want to apply at various levels in the above outline. Because of the perfectly arbitrary nature of Ω, there is no logical reason why we should not use $A, B, \ldots$ and $x, y, \ldots$ for these sets and their elements. However, we foresee a certain amount of confusion if we prove a theorem with the notation of points and sets and then apply it to sets and classes. Therefore, in fundamental theorems which we intend to apply to various types of entities we shall use Greek caps. (Γ, Δ, etc.) for the sets and Greek l.c. (α, β, etc.) for their elements.

An *ordered pair* is a 2-element set with one of them distinguished as a "first" element. In purely set theoretic terms the ordered pair (α, β) may be defined by saying that

$$(\alpha, \beta) = \{\{\alpha, \beta\}, \{\alpha\}\}.$$

Let Γ and Δ be sets. The *product set* $\Gamma \times \Delta$ is the set of all ordered pairs (α, β) where $\alpha \in \Gamma$ and $\beta \in \Delta$. A *function* from Γ to Δ is a subset f of $\Gamma \times \Delta$ such that every element of Γ appears as a first entry in exactly one element of f. The relation $(\alpha, \beta) \in f$ is usually written

$$f(\alpha) = \beta \quad \text{or} \quad f_\alpha = \beta$$

and β is called the *value* of f at α.

If f is a function from Γ to Δ, then Γ (the set of all first entries in the ordered pairs of f) is called the *domain* of f. The subset of Δ consisting of all second entries in the ordered pairs of f is called the *range* of f. If the range of f is all of Δ, f is said to be from Γ *onto* Δ. Otherwise, we say f is from Γ *to* Δ or from Γ *into* Δ. If each element of the range of f appears as the second entry in exactly one ordered pair of f, then f is called *one-to-one*.

Let f be a function from Γ to Δ and let g be a function from Δ to Ξ. The *composite function* $g \circ f$ is defined as the set of all ordered pairs (α, γ) such that for some $\beta \in \Delta$, $(\alpha, \beta) \in f$ and $(\beta, \gamma) \in g$. This is to say

$$(g \circ f)(\alpha) = g[f(\alpha)].$$

A function f from Γ to Δ generates an *inverse function* f^{-1} from the class of all subsets of Δ to the class of all subsets of Γ. Specifically, if $\Delta' \subset \Delta$,

$$f^{-1}(\Delta') = \{\alpha \mid f(\alpha) \in \Delta'\}.$$

---

† There is no reason why a set should not have as elements some points, some point sets, some classes of point sets, etc. However, we shall have no occasion to consider such sets.

If Δ' consists of the single element β, we shall follow the usual (inconsistent) practice and write $f^{-1}(\beta)$ rather than $f^{-1}(\{\gamma \mid \gamma = \beta\})$. It is readily verified that

$$(g \circ f)^{-1} = f^{-1} \circ g^{-1}.$$

Suppose the domain of f is a product space; that is, let f be a function from $\Gamma \times \Delta$ to Ξ. For each fixed $\alpha \in \Gamma$, f determines a function from Δ to Ξ; namely, the set of all ordered pairs (β, γ) such that $[(\alpha, \beta), \gamma] \in f$. This function will be denoted by $f(\alpha, \ )$. Similarly, $f(\ , \beta)$ will denote the indicated function from Γ to Ξ.

Let f be a function whose domain is Γ_1 and let g be a function whose domain is $\Gamma_2 \supset \Gamma_1$. Furthermore, suppose f and g coincide on Γ_1; that is, for $\alpha \in \Gamma_1$, $f(\alpha) = g(\alpha)$. Then g is called an *extension* of f to Γ_2 and f is called the *restriction* of g to Γ_1. Recalling that a function is a set of ordered pairs, we may note that $f \subset g$ says that f is a restriction of g, or equivalently, that g is an extension of f.

A *sequence* is a function whose domain is the set of positive integers. For the value of the sequence s at the integer i we write s_i rather than $s(i)$. Let s be any sequence and let n be a sequence of positive integers such that $n_i > n_j$ whenever $i > j$. The composite function $s \circ n$ is called a *subsequence* of s. Again, we follow common notation and write s_{n_i} for $(s \circ n)(i)$.

More often than not the functions encountered in this book will have for their ranges subsets of the *real number system*. This system is characterized uniquely by saying that it is a *complete, ordered field*. Details of this may be found in *IRA*, Chapter 2. We summarize here only those items most pertinent to measure theory. A real number b is an *upper bound* for a set E of real numbers if $a \leq b$ for every $a \in E$. If b is an upper bound for E and if also $b \leq c$ for every upper bound c for E, then b is the supremum (or least upper bound) for E, and we write $b = \sup E$. Lower bound and *infimum* (greatest lower bound) are similarly defined by reversing inequalities. The infimum of E is denoted by $\inf E$. A set is *bounded above* (*below*) if it has an upper (lower) bound. The completeness postulate is that every nonvacuous set of real numbers that is bounded above has a supremum. It is easily established that nonvacuous sets bounded below have infima.

Let f be a function from a set Γ to the real number system, and let $\Delta \subset \Gamma$ be defined by the proposition P. That is, let

$$\Delta = \{\alpha \mid P(\alpha)\}.$$

We shall use the following symbols interchangeably:

$$\sup \{f(\alpha) \mid \alpha \in \Delta\} = \sup \{f(\alpha) \mid P(\alpha)\}$$

$$= \sup_{\alpha \in \Delta} f(\alpha) = \sup_{P(\alpha)} f(\alpha).$$

4 BACKGROUND

In the special case in which f is a sequence we shall write

$$\sup_n f_n = \sup \{f_n \mid n = 1, 2, 3, \ldots\}.$$

We shall use similar notation for infima of functions.

The *limit superior* and *limit inferior* of a sequence u of real numbers may be defined as follows:

$$\overline{\lim_n} \, u_n = \inf_k \sup_{n \geq k} u_n;$$

$$\underline{\lim_n} \, u_n = \sup_k \inf_{n \geq k} u_n.$$

A *bounded function* is one whose range is bounded both above and below. Therefore, *every bounded sequence of real numbers has both a limit superior and a limit inferior*.

In measure theory it is convenient to admit "infinite values" for functions. We do this by defining the *extended real number system*, consisting of the real numbers together with the two elements ∞ and $-\infty$. We make the new system an ordered set by postulating that for every real number x,

$$-\infty < x < \infty.$$

The extended system will not be a field, but we connect the new elements with the field operations by postulating that for every real number x,

$$(\pm\infty) + (\pm\infty) = x \pm \infty = \pm\infty + x = \pm\infty,$$

$$(\pm\infty)(\pm\infty) = \infty,$$

$$(\pm\infty)(\mp\infty) = -\infty,$$

$$(\pm\infty)x = x(\pm\infty) = \begin{cases} \pm\infty & \text{if } x > 0, \\ \mp\infty & \text{if } x < 0, \end{cases}$$

$$x/(\pm\infty) = 0.$$

Specifically, we have postulated as algebraic operations in the extended real number system all those which are uniquely defined as operations on limits, and we have left undefined those operations which represent indeterminate forms.

It is easily seen that any set of extended real numbers has both a supremum and an infimum; consequently *every sequence of extended real numbers has a limit superior and a limit inferior*.

It is inherent in the notion of a set that if $\Gamma \neq \phi$, then there exists α such that $\alpha \in \Gamma$. Consequently, if Γ_1 and Γ_2 are two nonvacuous sets, there exist points α_1 and α_2 such that $\alpha_i \in \Gamma_i$ ($i = 1, 2$). Indeed, for any finite† class of nonvacuous sets, we can choose one point from each to form a new set. We shall have occasion to apply this "choosing a point from each" process to infinite classes of sets, and in

---

† For a precise definition of a finite class, see Section 3.

this case it is not obvious (nor, indeed, is it provable from the basic properties of the ∈ relation) that the process is valid. To justify it, we introduce the axiom of choice which specifically postulates the validity of the process. To associate with each set in a class an element "chosen" from it is to define a function over the given class of sets, and it is in these terms that the axiom of choice is usually stated.

Axiom of choice. If $\mathscr{E}$ is any nonvacuous class of nonvacuous sets, then there is a function f defined on $\mathscr{E}$ such that for each $\Gamma \in \mathscr{E}$, $f(\Gamma) \in \Gamma$.

A method of proof involving the axiom of choice is called a *transfinite method*. There are a number of propositions known to be equivalent to the axiom of choice, though in most cases the equivalence is far from obvious. In this book we shall have a few occasions to make a direct application of the axiom of choice and we shall need transfinite methods in a few other instances in which one of the equivalent propositions, known as Zorn's lemma, is more directly applicable.

The setting for Zorn's lemma is a *partially ordered set*. This is a set Γ in which there is defined a binary relation $\leq$ satisfying the conditions:

PO–I. $\alpha \leq \alpha$ for every $\alpha \in \Gamma$.

PO–II. If $\alpha \leq \beta$ and $\beta \leq \alpha$, then $\alpha = \beta$.

PO–III. If $\alpha \leq \beta$ and $\beta \leq \gamma$, then $\alpha \leq \gamma$.

One of the best examples of such a system is obtained by letting Γ be the class of all subsets of a set A and letting $\leq$ mean $\subset$. Note that there may be α and β for which neither $\alpha \leq \beta$ nor $\beta \leq \alpha$. If Γ is a partially ordered set and $\Delta \subset \Gamma$ has the property that either $\alpha \leq \beta$ or $\beta \leq \alpha$ whenever $\alpha \in \Delta$ and $\beta \in \Delta$, then Δ is called a *linearly ordered* subset of Γ. Continuing to let Γ be a partially ordered set and $\Delta \subset \Gamma$, we say that $\alpha \in \Gamma$ is an *upper bound* for Δ if $\beta \leq \alpha$ for all $\beta \in \Delta$; we also say that $\alpha \in \Delta$ is a *maximal element* of Δ if $\beta \in \Delta$ and $\beta \geq \alpha$ implies $\beta = \alpha$. Note that a maximal element may not be an upper bound; informally an upper bound is larger than any other element while a maximal element is merely one for which nothing is larger.

Zorn's Lemma. *If Γ is a partially ordered set such that every linearly ordered subset of Γ has an upper bound in Γ, then Γ has a maximal element.*

2 ALGEBRA OF SETS

If Γ and Δ are two sets, we define the *union of Γ and Δ* (in symbols, $\Gamma \cup \Delta$) as the set of elements belonging to either Γ or Δ (or both). The *intersection of Γ and Δ* (in symbols, $\Gamma \cap \Delta$) is the set of elements belonging to both Γ and Δ. The *complement of Δ with respect to Γ* (in symbols, $\Gamma - \Delta$) is the set of elements that belong to Γ but not to Δ.

In any given discussion, any set Γ appears as a subset of some space. If Γ is a point set, it is a subset of Ω; if it is a class of point sets, it is a subclass of the class of all subsets of Ω, etc. The complement of Γ with respect to the space in which Γ

is embedded is usually called merely the *complement* of Γ and written $-\Gamma$. The reader should note that if $P(\alpha)$ and $Q(\alpha)$ are contradictory propositions (for any α, one or the other is true but not both) and if $\Gamma = \{\alpha \mid P(\alpha)\}$, then $-\Gamma = \{\alpha \mid Q(\alpha)\}$.

The operations of union and intersection can be extended to any number of sets. Let $\mathscr{E}$ be a class of sets Γ; then

$$\bigcup_{\Gamma \in \mathscr{E}} \Gamma = \{\alpha \mid \text{there exists } \Gamma \in \mathscr{E} \text{ such that } \alpha \in \Gamma\},$$

$$\bigcap_{\Gamma \in \mathscr{E}} \Gamma = \{\alpha \mid \text{if } \Gamma \in \mathscr{E}, \text{ then } \alpha \in \Gamma\}.$$

Informally, the *union of a class of sets* is the set of elements belonging to at least one of the given sets; the *intersection of a class of sets* is the set of elements belonging to every one of the given sets.

Theorem 2.1. *If $\mathscr{E}$ is a class of sets and if $\Gamma_0 \in \mathscr{E}$, then*

2.1.1
$$\bigcap_{\Gamma \in \mathscr{E}} \Gamma \subset \Gamma_0 \subset \bigcup_{\Gamma \in \mathscr{E}} \Gamma,$$

2.1.2
$$-\left(\bigcup_{\Gamma \in \mathscr{E}} \Gamma\right) = \bigcap_{\Gamma \in \mathscr{E}} (-\Gamma),$$

2.1.3
$$-\left(\bigcap_{\Gamma \in \mathscr{E}} \Gamma\right) = \bigcup_{\Gamma \in \mathscr{E}} (-\Gamma).$$

Two sets Γ and Δ are *disjoint* if $\Gamma \cap \Delta = \phi$. A class $\mathscr{E}$ is a *class of disjoint sets* provided that if $\Gamma \in \mathscr{E}$ and $\Delta \in \mathscr{E}$ and $\Gamma \neq \Delta$, then $\Gamma \cap \Delta = \phi$. A sequence Γ is a *sequence of disjoint sets* provided that if $m \neq n$, then $\Gamma_m \cap \Gamma_n = \phi$. The reader should note that the range of a sequence of disjoint sets is a class of disjoint sets, but not every sequence defined to a class of disjoint sets is a sequence of disjoint sets.

The notions of union and intersection are defined for sequences of sets as well as classes:

$$\bigcup_{n=k}^{\infty} \Gamma_n = \{\alpha \mid \alpha \in \Gamma_n \text{ for some } n \geq k\};$$

$$\bigcap_{n=k}^{\infty} \Gamma_n = \{\alpha \mid \alpha \in \Gamma_n \text{ for every } n \geq k\}.$$

Unions of sequences of disjoint sets play a vital role in measure theory, and for that reason we shall have many uses for the following formula. It expresses the union of an arbitrary sequence as the union of a sequence of disjoint sets; furthermore, each term of the new sequence is obtained from the terms of the old by a finite number of union and difference operations.

Theorem 2.2. *If Γ is any sequence of sets, then there is a sequence Δ of disjoint sets such that*

$$\bigcup_{n=1}^{\infty} \Gamma_n = \bigcup_{n=1}^{\infty} \Delta_n.$$

Specifically, $\Delta_1 = \Gamma_1$; and for $n > 1$,

$$\Delta_n = \Gamma_n - \bigcup_{k=1}^{n-1} \Gamma_k.$$

Proof. Suppose $\alpha \in \bigcup_{n=1}^{\infty} \Gamma_n$; then for some n, $\alpha \in \Gamma_n$. Let n_0 be the smallest integer such that $\alpha \in \Gamma_{n_0}$; then it follows from the definition of the sequence Δ that $\alpha \in \Delta_{n_0} \subset \bigcup_{n=1}^{\infty} \Delta_n$. Thus,

$$\bigcup_{n=1}^{\infty} \Gamma_n \subset \bigcup_{n=1}^{\infty} \Delta_n.$$

However, if for some n, $\alpha \in \Delta_n$, then $\alpha \in \Gamma_n$; so

$$\bigcup_{n=1}^{\infty} \Gamma_n \supset \bigcup_{n=1}^{\infty} \Delta_n. \quad \blacksquare$$

The *characteristic function of a set* Γ is the real valued function C_Γ defined (over the space in which Γ is embedded) by the formula

$$C_\Gamma(\alpha) = \begin{cases} 1 & \text{for } \alpha \in \Gamma, \\ 0 & \text{for } \alpha \in -\Gamma. \end{cases}$$

Theorem 2.3. *If Γ is a sequence of sets and if*

$$\Gamma^0 = \bigcup_{n=1}^{\infty} \Gamma_n \text{ and } \Gamma_0 = \bigcap_{n=1}^{\infty} \Gamma_n,$$

then for each α,

$$C_{\Gamma^0}(\alpha) = \sup_n C_{\Gamma_n}(\alpha)$$

and

$$C_{\Gamma_0}(\alpha) = \inf_n C_{\Gamma_n}(\alpha).$$

Proof. If $\alpha \in \Gamma^0$, then for some k, $\alpha \in \Gamma_k$; so

$$\sup_n C_{\Gamma_n}(\alpha) = C_{\Gamma_k}(\alpha) = 1.$$

If $\alpha \in -\Gamma^0$, then $\alpha \in -\Gamma_n$ for every n; so $C_{\Gamma_n}(\alpha) = 0$ for every n, and

$$\sup_n C_{\Gamma_n}(\alpha) = 0.$$

This proves the first statement; the other may be proved by a similar argument. $\blacksquare$

Let Γ be a sequence of sets. We define

$$\overline{\lim_n} \Gamma_n = \bigcap_{k=1}^{\infty} \bigcup_{n=k}^{\infty} \Gamma_n;$$

$$\underline{\lim_n} \Gamma_n = \bigcup_{k=1}^{\infty} \bigcap_{n=k}^{\infty} \Gamma_n.$$

These sets are called the *limit superior* and *limit inferior* of the sequence, respectively. The connection between limits superior and inferior for sequences of sets and sequences of numbers is given by 2.3. Specifically, if Γ is a sequence of sets and if $\Gamma^0 = \overline{\lim}_n \Gamma_n$ and $\Gamma_0 = \underline{\lim}_n \Gamma_n$, then for each α,

$$C_{\Gamma^0}(\alpha) = \overline{\lim_n} C_{\Gamma_n}(\alpha) \quad \text{and} \quad C_{\Gamma_0}(\alpha) = \underline{\lim_n} C_{\Gamma_n}(\alpha).$$

It follows from the definitions of union and intersection that $\overline{\lim}_n \Gamma_n$ is the set of points α such that for every positive integer k there exists a positive integer $n \geq k$ such that $\alpha \in \Gamma_n$. Thus, $\overline{\lim}_n \Gamma_n$ consists of those points which belong to Γ_n for an infinite number of values of n. A similar analysis shows that $\underline{\lim}_n \Gamma_n$ is the set of points α such that for some positive integer k, $\alpha \in \Gamma_n$ for all positive integers $n \geq k$. That is, $\underline{\lim}_n \Gamma_n$ consists of those points which belong to Γ_n for all except a finite number of values of n.

Theorem 2.4. *If Γ is any sequence of sets, then*

$$\overline{\lim_n} \Gamma_n \supset \underline{\lim_n} \Gamma_n.$$

Proof. This is most easily seen from the alternative definitions developed above. If the relation $\alpha \in \Gamma_n$ holds for all but a finite number of values of n, then it holds for an infinite number of values of n. Thus, if $\alpha \in \underline{\lim}_n \Gamma_n$, then $\alpha \in \overline{\lim}_n \Gamma_n$. ∎

A sequence of sets is said to be *convergent* if

$$\overline{\lim_n} \Gamma_n = \underline{\lim_n} \Gamma_n.$$

For a convergent sequence, the set which appears as both $\overline{\lim}_n \Gamma_n$ and $\underline{\lim}_n \Gamma_n$ is called

$$\lim_n \Gamma_n.$$

A sequence Γ of sets is called an *expanding sequence* if for each positive integer n,

$$\Gamma_{n+1} \supset \Gamma_n.$$

It is called a *contracting sequence* if for each n,

$$\Gamma_{n+1} \subset \Gamma_n.$$

A *monotone sequence of sets* is one which is either an expanding sequence or a contracting sequence.

Theorem 2.5. *Every monotone sequence of sets is convergent.*

Proof. Suppose Γ is an expanding sequence; then for each k,

$$\bigcap_{n=k}^{\infty} \Gamma_n = \Gamma_k;$$

therefore

$$\varliminf_{n} \Gamma_n = \bigcup_{k=1}^{\infty} \Gamma_k.$$

However, for an expanding sequence, $\bigcup_{n=k}^{\infty} \Gamma_n$ is independent of k; so we may always take $k = 1$, and

$$\varlimsup_{n} \Gamma_n = \bigcap_{k=1}^{\infty} \bigcup_{n=k}^{\infty} \Gamma_n = \bigcup_{n=1}^{\infty} \Gamma_n = \varliminf_{n} \Gamma_n.$$

The contracting case may be treated similarly. In that case,

$$\lim_{n} \Gamma_n = \bigcap_{n=1}^{\infty} \Gamma_n. \quad \blacksquare$$

3 CARDINAL NUMBERS

Let Γ be a set and let $\sim$ be a relation between pairs of elements of Γ satisfying the following postulates:

E–I (reflexivity). $\alpha \sim \alpha$.
E–II (symmetry).   If $\alpha \sim \beta$, then $\beta \sim \alpha$.
E–III (transitivity).   If $\alpha \sim \beta$ and $\beta \sim \gamma$, then $\alpha \sim \gamma$.

Such a relation is called an *equivalence relation*. An equivalence relation in Γ partitions Γ into disjoint sets called *equivalence classes* defined by saying that α and β belong to the same equivalence class if and only if $\alpha \sim \beta$.

Let Γ be a class of sets (in naive set theory it is treacherous to say the class of all sets, though informally this is what we are thinking of) and let $\alpha \sim \beta$ mean that there exists a one-to-one function from α onto β. This is an equivalence relation, and its equivalence classes are called *cardinal numbers*. An equivalence class is aptly described by specifying a representative element of it, and in these terms certain cardinal numbers have generally adopted symbols assigned to them:

The cardinal of ϕ is 0.
The cardinal of the set of integers $1, 2, \ldots, n$ is n.
The cardinal of the set of all positive integers is $\aleph_0$.†
The cardinal of the set of all real numbers is c.

A set whose cardinal is zero or a positive integer is called a *finite set*, and the cardinal of such a set is called a *finite cardinal*. Any other set is called an *infinite set*, and the cardinal of an infinite set is called a *transfinite cardinal*.

---

† The symbol $\aleph$ is the first letter of the Hebrew alphabet. It is read "aleph."

A satisfactory definition of $<$ for cardinals may be obtained by saying that the cardinal of Γ is less than that of Δ if there is a one-to-one mapping of Γ onto some subset of Δ but no such mapping of Δ onto any subset of Γ. The relation $<$ defined in this way linearly orders the cardinals. For anything except finite cardinals, this fact constitutes a rather deep theorem. We omit the proof.†

A set which either is finite or has cardinal $\aleph_0$ is called a *countable set*. Any other set is called a *noncountable set*. Cardinals appear in measure theory primarily because a distinction between countable and noncountable sets is important. The following theorems (proofs in *IRA*, Section 5.3) give the pertinent facts.

Theorem 3.1. *If Γ is countable and if $\Delta \subset \Gamma$, then Δ is countable.*

Corollary 3.1.1. *If Γ is noncountable and if $\Delta \supset \Gamma$, then Δ is noncountable.*

Theorem 3.2. *Any infinite set has a subset with cardinal $\aleph_0$.*

Corollary 3.2.1. *Any infinite set can be mapped one-to-one onto a proper subset of itself.*

Corollary 3.2.2. *$\aleph_0$ is the smallest transfinite cardinal.*

In the light of 3.2.2 it is in order to restate some of our definitions as follows. A set is

finite if it has cardinal less than $\aleph_0$,
infinite if it has cardinal greater than or equal to $\aleph_0$,
countable if it has cardinal less than or equal to $\aleph_0$,
noncountable if it has cardinal greater than $\aleph_0$.

Theorem 3.3. *The union of a countable class of countable sets is a countable set.*

The set of all integers (positive and negative) is the union of two countable sets; therefore it is countable. *Rational numbers* are those of the form p/q where p and q are integers. For a fixed denominator q, the set of distinct rationals of the form p/q is in obvious one-to-one correspondence with a subset of the set of all integers; so by 3.1 each of these sets is countable. The set of all rationals is the union of these sets as q runs through the set of positive integers; thus we have as an application of 3.3 that *the set of all rational numbers is countable*.

The best-known noncountable set is that of all real numbers. To prove that the real number system is noncountable, it suffices (because of 3.1.1) to prove that the real numbers between 0 and 1 form a noncountable set. Each such number has an infinite decimal expansion, and such an expansion is unique except for the fact that
$$.a_1 a_2 \cdots a_n 999 \cdots = .a_1 a_2 \cdots (a_n + 1) 000 \cdots.$$

---

† See Hausdorff [12], pp. 58–61.

Let us agree to use the form with zeros in such cases. Let

$$x_1 = .a_{11}a_{12}a_{13}\cdots$$
$$x_2 = .a_{21}a_{22}a_{23}\cdots$$
$$x_3 = .a_{31}a_{32}a_{33}\cdots$$
$$\cdots$$

be any countable set of such numbers with their decimal expansions. Consider the number $x = .b_1b_2b_3\cdots$ where

$$b_n = \begin{cases} 8 & \text{if } a_{nn} \neq 8, \\ 7 & \text{if } a_{nn} = 8. \end{cases}$$

There are no nines in the expansion of x, so there is no ambiguity, and x is not any one of the numbers x_n. It differs from x_n in the nth decimal place. Yet, $0 < x < 1$, so the arbitrarily chosen countable set does not exhaust the interval. Thus, the interval is noncountable.

It is easy to describe cardinals greater than c. In fact, the following procedure may be used to describe a cardinal greater than any given one. Let Γ be any set and let $\mathscr{E}$ be the class of all subsets of Γ. If the cardinal of Γ is $\aleph$, we say† that the cardinal of $\mathscr{E}$ is $2^\aleph$.

Theorem 3.4. *For any cardinal $\aleph$, $2^\aleph > \aleph$.*

4 PSEUDO-METRIC SPACES

A space Ω is a *pseudo-metric space* if there is a real-valued function ρ on $\Omega \times \Omega$ satisfying the following postulates:

PM–I. $\rho(x, x) = 0$ for every $x \in \Omega$.
PM–II. $\rho(x, y) \leq \rho(z, x) + \rho(z, y)$ for every $x, y, z \in \Omega$.

The function ρ is called the *distance function*; PM–II is called the *triangle postulate*. Distance functions have two other basic properties, frequently given as postulates but provable from PM–I and PM–II.

Theorem 4.1. *If Ω is a pseudo-metric space and if $x, y \in \Omega$, then:*

4.1.1 $\rho(x, y) \geq 0$,

4.1.2 $\rho(x, y) = \rho(y, x)$.

A *metric space* is a pseudo-metric space in which ρ satisfies the additional condition that $\rho(x, y) = 0$ implies $x = y$. In any pseudo-metric space, $\rho(x, y) = 0$ is an equivalence relation; and the set of equivalence classes forms a metric space. Some of the most important spaces appearing in measure theory are pseudo-metric

---

† This notation is inspired by the fact that a set containing n points has 2^n subsets.

only (not metric). There is one school of thought that insists that each of these be replaced by its space of equivalence classes. We find this procedure unduly cumbersome. We also find that with a very few (duly noted) exceptions, pseudo-metric space theory is quite adequate for measure theory.

DEFINITIONS:

Diameter

$$\text{diam}(E) = \sup \{\rho(x, y) \mid x \in E, y \in E\}.$$

Bounded set

$$\text{diam}(E) < \infty.$$

Distance between a point and a set

$$\rho(x, A) = \inf \{\rho(x, y) \mid y \in A\}.$$

Distance between two sets

$$\rho(A, B) = \inf \{\rho(x, y) \mid x \in A, y \in B\}.$$

Neighborhood

$$N(x, \varepsilon) = \{y \mid \rho(x, y) < \varepsilon\}.$$

Deleted neighborhood

$$\underline{N}(x, \varepsilon) = \{y \mid 0 < \rho(x, y) < \varepsilon\}.$$

Point of accumulation

x is a point of accumulation of E if for every $\varepsilon > 0$ there exists $y \in E \cap \underline{N}(x, \varepsilon)$.

Limit of a sequence

$\lim_n x_n = x_0$ if for every $\varepsilon > 0$ there exists n_0 such that $n > n_0$ implies $x_n \in N(x_0, \varepsilon)$.

Cauchy sequence

For every $\varepsilon > 0$ there exists n_0 such that $n > n_0$ and $m > n_0$ imply $\rho(x_m, x_n) < \varepsilon$.

Complete pseudo-metric space

If x is a Cauchy sequence in Ω, there exists $x_0 \in \Omega$ such that $\lim_n x_n = x_0$.

Closure

$x \in \bar{A}$ provided $x \in A$ or x is an accumulation point of A.

Interior

$x \in A^\circ$ provided there exists $\varepsilon > 0$ such that $N(x, \varepsilon) \subset A$.

Closed set

$A = \bar{A}$.

Open set

$A = A^\circ$.

Dense

A is dense in B if $A \subset B \subset \bar{A}$.

Separable space

Ω is separable if there is a countable set that is dense in Ω.

Theorem 4.2. *Given that*

$$\lim_n x_n = a \quad \text{and} \quad \lim_n x_n = b,$$

in a pseudo-metric space it follows that $\rho(a, b) = 0$; thus in a metric space it follows that $a = b$.

The primary difference between metric and pseudo-metric spaces is indicated in 4.2. It involves the sense in which limits are "unique." The following familiar theorems are all valid in pseudo-metric spaces.

Theorem 4.3. *If x_0 is a point of accumulation of E, then there exists a sequence x of points of E such that:*

4.3.1 *if $m \neq n$ then $x_m \neq x_n$;*

4.3.2 $\lim_n x_n = x_0$.

Corollary 4.3.3. *If x_0 is a point of accumulation of E, then for every $\varepsilon > 0$, the set $E \cap N(x_0, \varepsilon)$ is an infinite set.*

Corollary 4.3.4. *If E is a finite set, then it has no points of accumulation.*

Corollary 4.3.5. *If x_0 is a point of accumulation of the range of the sequence x, then there is a subsequence $x_{n()}$ of x which converges to x_0.*

Theorem 4.4. *Neighborhoods are open sets.*

Theorem 4.5. *If A is open, then $-A$ is closed; if A is closed, then $-A$ is open.*

Theorem 4.6. *If $\lim_n x_n = x_0$, then x is a Cauchy sequence.*

Theorem 4.7. *If x is a Cauchy sequence and if for some subsequence*

$$\lim_i x_{n_i} = x_0,$$

then

$$\lim_n x_n = x_0.$$

A set E has the *Bolzano-Weierstrass property* if every infinite subset of E has a point of accumulation in E. It has the *Heine-Borel property* if every open covering of E is reducible to a finite covering.

Theorem 4.8. *In a pseudo-metric space the Bolzano-Weierstrass and Heine-Borel properties are equivalent.*

A *compact set* is one that has the Heine-Borel (Bolzano-Weierstrass) property. Ω is *locally compact* if every $x \in \Omega$ has a neighborhood whose closure is compact.

A set E in a pseudo-metric space is *nowhere dense* if $(\bar{E})^0 = \phi$. A set is of the *first category* (of Cat. I) if it is a countable union of nowhere dense sets. A set that is not of Cat. I is of the *second category* (of Cat. II).

Theorem 4.9 (Baire category theorem). *If Ω is a complete pseudo-metric space, then Ω is of Cat. II.*

Most applications of 4.9 exploit the obvious fact that ϕ is of Cat. I and take the form of applications of the following corollary.

Corollary 4.9.1. *If Ω is complete, then $E \subset \Omega$ is nonvacuous if $-E$ is of Cat. I.*

A much-used example of a nowhere dense set in the unit interval is the *Cantor ternary set* K. Not only is K nowhere dense; it is also *perfect* (every point of K is an accumulation point of K). It should also be noted that K is noncountable. The construction of K proceeds as follows.

Let E_0^1 be an open interval, the middle third of the unit interval $[0, 1]$. Let E_1^1 and E_1^2 be the open middle thirds of the two closed intervals of $[0, 1] - E_0^1$. Extract these, and let $E_2^1, E_2^2, E_2^3, E_2^4$ be the open middle thirds of the remaining four closed intervals, etc. (see Fig. 1). The Cantor ternary set K is defined as

$$K = [0, 1] - \bigcup_{n=0}^{\infty} \bigcup_{k=1}^{2^n} E_n^k.$$

Since each E_n^k is open, the union of all of them is; so K is closed. The distance between adjacent sets E_n^k for $n < n_0$ is $1/3^{n_0}$; therefore given $x \in K$ and $\varepsilon > 0$, $N(x, \varepsilon)$ intersects at least two sets E_n^k. This proves two things: (1) that $\bar{K} - K$

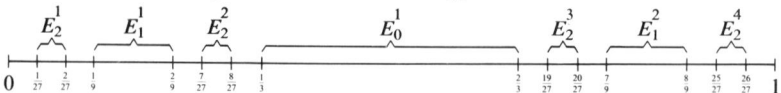

Figure 1

contains no such neighborhood, and thus K is nowhere dense in $[0, 1]$; (2) that $N(x, \varepsilon)$ contains at least two points of K and so contains one different from x; thus x is a point of accumulation of K, and so K is perfect.

5 LIMITS AND CONTINUITY

Let f be a function from a pseudo-metric space X to a pseudo-metric space Z. If x_0 is a point of accumulation of X and if $z \in Z$, then we define

$$\lim_{x \to x_0} f(x) = z$$

to mean that given any $\varepsilon > 0$ there is a $\delta > 0$ such that if

$$x \in \underline{N}(x_0, \delta),$$

then

$$f(x) \in N(z, \varepsilon).$$

An equivalent condition is that for every sequence x of distinct points such that

$$\lim_n x_n = x_0,$$

it follows that

$$\lim_n f(x_n) = z.$$

Theorem 5.1. *If f is a function from a pseudo-metric space X to a pseudo-metric space Z and if*

$$\lim_{x \to x_0} f(x) = z_1 \quad \text{and} \quad \lim_{x \to x_0} f(x) = z_2,$$

then $\rho(z_1, z_2) = 0$.

Note that for this theorem to hold it is essential that limits at x_0 be defined only if x_0 is an accumulation point of X.

On the other hand, we will say that f is *continuous* at x_0 provided that given $\varepsilon > 0$ there is a $\delta > 0$ such that if

$$x \in N(x_0, \delta),$$

then

$$f(x) \in N[f(x_0), \varepsilon].$$

If f is continuous at each point of its domain, we say that f is a *continuous function*. Frequently we want to characterize continuity of f at x_0 by the condition

$$\lim_{x \to x_0} f(x) = f(x_0). \tag{1}$$

If x_0 is an isolated point of the domain, then (1) is not equivalent to the definition of continuity just given. It helps to adopt two different interpretations of the "lim" symbol; $x = x_0$ is specifically excluded unless we are talking about continuity, and then $x = x_0$ is specifically included.

Theorem 5.2. *If f is defined from a pseudo-metric space X to a pseudo-metric space Z, if x_0 is a point of accumulation of X and if Z is complete, then a necessary and sufficient condition that there exists a point $z \in Z$ such that*

$$\lim_{x \to x_0} f(x) = z$$

is that for every sequence x of distinct points with

$$\lim_n x_n = x_0$$

the sequence $f(x_n)$ be a Cauchy sequence.

Theorem 5.3. *A function f from a pseudo-metric space X to a pseudo-metric space Z is continuous if and only if for every open set $G \subset Z$, $f^{-1}(G)$ is an open set in X.*

Corollary 5.3.1. *A function f from a pseudo-metric space X to a pseudo-metric space Z is continuous if and only if for every closed set $F \subset Z$, $f^{-1}(F)$ is a closed set in X.*

6 UNIFORM LIMITS

Let f be a function from $X \times Y$ to Z; let X and Z be pseudo-metric spaces; let x_0 be an accumulation point of X; and let g be a function from Y to Z. We say that

$$\lim_{x \to x_0} f(x, y) = g(y) \text{ pointwise on } Y$$

provided that given $\varepsilon > 0$ and given $y \in Y$ there exists $\delta > 0$ such that

$$x \in \underline{N}(x_0, \delta) \quad \text{implies} \quad f(x, y) \in N[g(y), \varepsilon].$$

We say that

$$\lim_{x \to x_0} f(x, y) = g(y) \text{ uniformly on } Y$$

provided that given $\varepsilon > 0$ there exists $\delta > 0$ such that

$$x \in \underline{N}(x_0, \delta) \quad \text{implies} \quad f(x, y) \in N[g(y), \varepsilon]$$

for every $y \in Y$. The distinction is in the order of choice of δ and y. In pointwise convergence, y is given first and δ may depend on it. In uniform convergence, δ must be chosen first and may not depend on y.

Let X and Z be pseudo-metric, f from X to Z, and $Y \subset X$. Define ϕ on $X \times Y$ by

$$\phi(x, y) = f(x) - f(y).$$

Then, f is continuous on Y if
$$\lim_{x \to y} \phi(x, y) = 0 \text{ pointwise on } Y;$$
f is *uniformly continuous* on Y if
$$\lim_{x \to y} \phi(x, y) = 0 \text{ uniformly on } Y.$$

Theorem 6.1. *Let X and Z be pseudo-metric and let Y be a compact subset of X. A function from X to Z that is continuous on Y is uniformly continuous on Y.*

This theorem is usually given only for the case $Y = X$. The generalization stated here (and proved in *IRA*) is very useful in proving a classical theorem in integration theory (see 24.4).

Let X, Y, and Z all be pseudo-metric spaces and f a function from $X \times Y$ to Z. The statement that
$$\lim_{x \to x_0} f(x, y) = g(y)$$
uniformly on some deleted neighborhood of y_0 means that there exists $\eta > 0$ such that for every $\varepsilon > 0$ there is a $\delta > 0$ such that
$$y \in \underline{N}(y_0, \eta) \quad \text{and} \quad x \in \underline{N}(x_0, \delta)$$
imply
$$f(x, y) \in N[g(y), \varepsilon].$$

If we reverse the order of choice of ε and η here (let the size of the neighborhood of y_0 depend on ε), we define a very useful concept known as *subuniform convergence*. Specifically,
$$\lim_{x \to x_0} f(x, y) = g(y) \text{ subuniformly at } y_0$$
provided that given $\varepsilon > 0$ there exist $\eta > 0$ and $\delta > 0$ such that
$$y \in \underline{N}(y_0, \eta) \quad \text{and} \quad x \in \underline{N}(x_0, \delta)$$
imply
$$f(x, y) \in N[g(y), \varepsilon].$$

6.2. Consider the statements that
$$\lim_{x \to x_0} f(x, y) = g(y):$$

6.2.1 *uniformly on* $Y - \{y_0\}$,

6.2.2 *uniformly on some deleted neighborhood of* y_0,

6.2.3 *subuniformly at* y_0.

Because of the logical structure of the definitions, 6.2.1 implies 6.2.2 implies 6.2.3. In general these implications are not reversible but there is one important case in which they are. Let I be the set of positive integers and denote by $\bar{I}$ the set I plus a point called ∞. Define distance by $\rho(m, n) = |(1/m) - (1/n)|$, $\rho(n, \infty) = 1/n$. If the space Y is $\bar{I}$, we are talking about sequences of functions on X; so we turn to the usual notation $f_n(x)$. Since a neighborhood of ∞ in $\bar{I}$ contains all but a finite number of points of I, we have the following.

Theorem 6.3. *If*
$$\lim_{x \to x_0} f_n(x) = a_n$$
subuniformly at ∞ in $\bar{I}$, then
$$\lim_{x \to x_0} f_n(x) = a_n$$
uniformly on I.

The two major theorems on subuniform convergence are the following. The first of these (6.4) usually appears with a third conclusion concerning the existence of a "double limit" and is often called the "double limit theorem." So far as major applications in real analysis are concerned, double limits are superfluous and the two conclusions listed here are the important ones. (For a complete discussion, see *IRA*, Chapter 6.)

Theorem 6.4 (Moore-Osgood). *Let X, Y, and Z be pseudo-metric spaces with Z complete and let f be a function from $X \times Y$ to Z such that:*

6.4.1 $\lim_{y \to y_0} f(x, y) = f(x, y_0)$ *pointwise on* $X - \{x_0\}$,

6.4.2 $\lim_{x \to x_0} f(x, y) = f(x_0, y)$ *pointwise on* $Y - \{y_0\}$,

and

6.4.3 $\lim_{y \to y_0} f(x, y) = f(x, y_0)$ *subuniformly at x_0.*

Then

6.4.4 $\lim_{x \to x_0} \lim_{y \to y_0} f(x, y) = \lim_{y \to y_0} \lim_{x \to x_0} f(x, y)$,

and

6.4.5 $\lim_{x \to x_0} f(x, y) = f(x_0, y)$ *subuniformly at y_0.*

For our purposes the important case is that in which $Y = \bar{I}$ and $y_0 = \infty$. Then 6.4.1 and 6.4.3 say that $f_n \to f_\infty$ pointwise on $X - \{x_0\}$ and subuniformly at x_0; 6.4.2 says that each f_n is continuous at x_0. It follows by a simple computation from the conclusion 6.4.4 that f_∞ is continuous at x_0. Conclusion 6.4.5 says (in the light of 6.3) that
$$\lim_{x \to x_0} f_n(x) = f_n(x_0) \text{ uniformly on } I.$$

This is (by definition) the condition that the functions f_n be *equicontinuous* at x_0.

Theorem 6.5 (Osgood). *Let f be a pointwise convergent sequence of continuous functions from a pseudo-metric space X to a complete pseudo-metric space Z. Then, the points of X at which the sequence f does not converge subuniformly form a set of Cat. I.*

Corollary 6.5.1. *If, in 6.5, X is also complete, then there exists $x_0 \in X$ such that the limit function is continuous at x_0 and the functions f_n are equicontinuous at x_0.*

7 FUNCTION SPACES

The following definitions of distance make the indicated sets of functions into metric spaces:

R_n (Euclidean n-space): the set of all n-tuples of real numbers; i.e. the set of all functions on $\{1, 2, \ldots, n\}$.

$$\rho(x, y) = \sum_{i=1}^{n} (|x_i - y_i|^2)^{1/2}.$$

l_2 (Hilbert space): the set of all sequences x of real numbers such that

$$\sum_{i=1}^{\infty} x_i^2$$

is convergent.

$$\rho(x, y) = \sum_{i=1}^{\infty} (|x_i - y_i|^2)^{1/2}.$$

c_0: the set of all sequences x of real numbers such that $\lim_i x_i = 0$.

$$\rho(x, y) = \sup_i |x_i - y_i|.$$

c: the set of all convergent sequences of real numbers.

$$\rho(x, y) = \sup_i |x_i - y_i|.$$

m: the set of all bounded sequences of real numbers.

$$\rho(x, y) = \sup_i |x_i - y_i|.$$

s: the set of all sequences of real numbers.

$$\rho(x, y) = \sum_{i=1}^{\infty} \frac{|x_i - y_i|}{2^i (1 + |x_i - y_i|)}.$$

C: the set of all continuous real-valued functions on $[0, 1]$.

$$\rho(x, y) = \sup_{0 < t < 1} |x(t) - y(t)|.$$

M: the set of all bounded real-valued functions on $[0, 1]$.

$$\rho(x, y) = \sup_{0 < t < 1} |x(t) - y(t)|.$$

Each of these spaces is complete. Of these spaces, only R_n is locally compact. The spaces R_n, l_2, c_0, c, s, and C are separable; m and M are nonseparable. Convergence in Hilbert space will be studied later (Chapter 7). For each of the spaces in which distance is defined by a supremum, convergence means uniform convergence of the functions. This is to say, for example, that $\rho(x_n, x_0) \to 0$ in C if and only if $x_n \to x_0$ uniformly on $[0, 1]$. In R_n, convergence means pointwise convergence, but since the domain is finite this is equivalent to uniform convergence.

Of more importance is the fact that convergence in s is equivalent to pointwise convergence. This is to say that

$$\lim_n \rho(x^n, x^0) = 0$$

in s if and only if for each i,

$$\lim_n x_i^n = x_i^0.$$

This is proved in *IRA* (Chapter 5, Theorem 12). Virtually the same argument may be used to prove the following generalization which will be used on occasion in this book.

Theorem 7.1. *Let a and b be double sequences (functions on $I \times I$) and let c be a sequence such that*

$$0 \le a_i^n \le b_i^n \le c_i$$

for each i and each n. Then

$$\lim_n \sum_{i=1}^{\infty} \frac{a_i^n}{2^i(1 + b_i^n)} = 0$$

if and only if

$$\lim_n a_i^n = 0$$

for each i.

If we set $a_i^n = b_i^n$, we get the theorem that convergence in s is pointwise convergence. There are other spaces in which this same device is used to define distance but with $a \ne b$ satisfying 7.1. It is useful to know that these, too, are pointwise convergence spaces.

Finally, we note that each of the spaces mentioned here is a metric space, not just pseudo-metric. The pseudo-metric spaces which are not full metric spaces appear when measure theoretic devices are used to define distance.

8 GENERAL TOPOLOGY

By and large the "prerequisites" for this book have been summarized in the preceding seven sections. In general, topology here will be metric topology. Ideas and theorems from this section will be used only in Sections 17 and 49.

A topological space is usually described by taking a set and designating certain subsets as "open." It can equally well be done with neighborhoods, and this approach seems preferable here. Suppose that for each $x \in \Omega$ there is a class $\mathcal{N}_x$ of subsets of Ω such that (i) for each $N \in \mathcal{N}_x$, $x \in N$ and (ii) if $N_1 \in \mathcal{N}_x$ and $N_2 \in \mathcal{N}_x$, there exists $N_3 \in \mathcal{N}_x$ such that $N_3 \subset N_1 \cap N_2$. The class $\mathcal{N}_x$ is called a neighborhood system for x and members of $\mathcal{N}_x$ are called *neighborhoods* of x. If $N \in \mathcal{N}_x$, then $\underline{N} = N - \{x\}$ is a *deleted neighborhood* of x. Point of accumulation, closure, interior, closed set, and open set are now defined just as in metric spaces. The first item of metric space theory that breaks down is 4.3. In general topology accumulation points are not necessarily limits of sequences; to get a comparable theorem we must turn from sequences to nets.

Let D be a partially ordered set with the added property that if $d_1 \in D$ and $d_2 \in D$ there exists $d_3 \in D$ such that $d_1 \leq d_3$ and $d_2 \leq d_3$. (Informally, given two points of D, there is a larger one.) Then D is called a *directed set*. A *net* is a function whose domain is a directed set. The positive integers with the usual ordering form a directed set; so a sequence is a special example of a net. Let x be a net in a topological space;

$$\lim_\alpha x_\alpha = x_0$$

provided that for every $N \in \mathcal{N}_{x_0}$ there exists α_0 such that

$$\alpha \geq \alpha_0 \quad \text{implies} \quad x_\alpha \in N.$$

Theorem 8.1. *If x_0 is an accumulation point of A, there exists a net x in $A - \{x_0\}$ such that $\lim_\alpha x_\alpha = x_0$.*

Theorem 8.2. *If $\bar{A}$ is the set of all limits of nets in A, then $\bar{\bar{A}} = \bar{A}$.*

If the space is pseudo-metric, 8.1 and 8.2 hold with nets specialized to sequences. In general, this specialization is not permissible. In Section 49 we give an example of the failure of 8.2 for sequences.

A topological space Ω is a T_1-space if given $x \neq y$ there exists $N \in \mathcal{N}_x$ such that $y \notin N$. Equivalently, a T_1-space is one in which all singleton sets are closed. If a pseudo-metric space is T_1 then it is metric. In general, it takes more than the T_1-property to achieve the major advantage of metric spaces over pseudo-metric ones. A topological space is a *Hausdorff space* (T_2-space) provided that given $x \neq y$ there exist $N_1 \in \mathcal{N}_x$ and $N_2 \in \mathcal{N}_y$ such that $N_1 \cap N_2 = \phi$.

Theorem 8.3. *In a Hausdorff space limits of nets are unique.*

An alternative characterization of an ordered pair is that it is a function on the set $\{1, 2\}$. In these terms $A \times B$ is the set of all functions x on $\{1, 2\}$ such that

$x_1 \in A$ and $x_2 \in B$. Generalizing this idea, let A be any set and for each $a \in A$ let X_a be a topological space; then

$$\underset{a \in A}{\times} X_a$$

is the set of all functions x on A such that $x_a \in X_a$ for each $a \in A$. The *product topology* is described by saying that if y is a net in $\times_{a \in A} X_a$, then $\lim_\alpha y^\alpha = x$ if and only if $\lim_\alpha y_a^\alpha = x_a$ for every $a \in A$. This is to say that convergence of nets in the product topology is equivalent to pointwise convergence on the index set A of the corresponding net of functions. A neighborhood of x in $\times_{a \in A} X_a$ assumes the form

$$\{y \mid y_a \in N_a, N_a \in \mathcal{N}_{x_a} \text{ in } X_a, a \in F\},$$

where F is a finite subset of A. Informally, y is close to x in the product space if y_a is close to x_a for a finite number of indices a.

In many important examples of product spaces the spaces X_a are the same for all a. In this case, the usual notation for

$$\underset{a \in A}{\times} X \quad \text{is} \quad X^A;$$

this is the space of all functions from A to X. In the product topology convergence in X^A is pointwise convergence on A. In examples that we shall be concerned with X will be the real number system R. In the product topology of R^A a neighborhood of x assumes the form

$$\{y \mid |y_a - x_a| < \varepsilon_a, a \in F, F \text{ a finite subset of } A\}.$$

To put it another way, in R^A a neighborhood of x is determined by a finite set $\{a_1, a_2, \ldots, a_n\} \subset A$ and a set $\{\varepsilon_1, \varepsilon_2, \ldots, \varepsilon_n\}$ of positive numbers; y belongs to this neighborhood if and only if

$$|y_{a_i} - x_{a_i}| < \varepsilon_i \quad (i = 1, 2, \ldots, n).$$

Note that if R is the real number system and I is the set of positive integers, then R^I with the product topology is (see 7.1 and surrounding comments) precisely the space s. Thus in s a neighborhood of the zero sequence assumes the form

$$\{x \mid |x_{n_i}| < \varepsilon_i; \quad i = 1, 2, \ldots, k\}.$$

Now s is a metric space, but many product spaces (particularly those overcountable index sets) cannot be represented as metric spaces.

In general, the Heine-Borel and Bolzano-Weierstrass properties are not equivalent in a topological space. The H-B property always implies the B-W, and in general topology a *compact set* is defined as one for which every open covering is reducible to a finite covering. One of the major milestones in the development of general topology was the appearance of the following theorem, whose proof (omitted here—see Kelley [16]) relies heavily on transfinite methods.

Theorem 8.4 (Tychonoff). *Any product of compact topological spaces is compact in the product topology.*

Compactness can also be characterized by the *finite intersection property*. Let X be compact and let F be a class of closed subsets of X with the property that every finite subclass of F has a nonvacuous intersection; then the entire family F has a nonvacuous intersection. This is merely the contrapositive of the Heine-Borel property stated in terms of the complementary sets.

A closed subset of a compact set is always compact, and in a Hausdorff space compact sets are always closed. Let C_1 and C_2 be disjoint compact sets in a Hausdorff space; then there exist disjoint open sets G_1 and G_2 such that $G_1 \supset C_1$ and $G_2 \supset C_2$.

REFERENCES FOR FURTHER STUDY

On the algebra of sets:
   Halmos [11]

On metric spaces:
   Goffman [8]
   Kelley [16]
   Rudin [26]

On transfinite induction and cardinals:
   Goffman [8]

On uniform limits:
   Goffman [8]
   Munroe [21]

On general topology:
   Kelley [16]

CHAPTER 2

MEASURE—GENERAL THEORY

The notion of a measure function might well be introduced by considering the problem of formulating a mathematical description of a mass distribution. To give a complete description of a mass distribution, we should list the amount of mass to be found in each portion of the space concerned. That is, we should associate a number (the mass) with each element of a certain class of point sets in the space. In mathematical terminology, we should define a real valued function whose domain is a class of point sets.

This is the first thing to be noted about a measure function. It is a function of sets, not a function of points.

However, not every function of sets will suffice to describe our intuitive picture of a mass distribution. The function μ for which $\mu(E) = 1$ for every $E \subset \Omega$ would describe a situation in which every set had unit mass. Now, mass was never like this. If $\mu(E)$ equals the mass of E, we expect that for disjoint sets E_1 and E_2,

$$\mu(E_1 \cup E_2) = \mu(E_1) + \mu(E_2).$$

This much our intuition demands of a mass function. The great contribution of Lebesgue and other early-twentieth-century mathematicians was to observe that a lot more can be done with a measure function μ if it has the property that for every sequence E of disjoint sets in its domain,

$$\mu\left(\bigcup_{n=1}^{\infty} E_n\right) = \sum_{n=1}^{\infty} \mu(E_n).$$

This is the second thing to be noted about a measure function. It satisfies an additivity condition.

These are the two key properties of a measure function. By adding a few details it is easy to give a precise axiomatic definition and to prove from these axioms enough properties of measure functions to form the basis for a theory of integration. Indeed, we accomplish this in one section of the present chapter (Section 10). However, we know of only one example of any practical importance of a function which is obviously a measure function. So, before turning to the study of integration, we shall take a look at some of the examples of measure functions which require a more elaborate presentation. Our method of attack on this project is to develop in this chapter a general method for the construction of

9 ADDITIVE CLASSES AND BOREL SETS

measure functions. Then, in Chapter 3, we apply this method to a number of specific examples.

9 ADDITIVE CLASSES AND BOREL SETS

We look at measure functions themselves in Section 10. Before that we want to examine some of the properties that will be demanded of the domain of a measure function.

Let Ω be any space, and let $\mathscr{A}$ be a class of subsets of Ω. We say that $\mathscr{A}$ is a *finitely additive class of sets* if it satisfies the following postulates:

c–I. $\phi \in \mathscr{A}$.

c–II.   If $A \in \mathscr{A}$ and $B \in \mathscr{A}$, then $A - B \in \mathscr{A}$.

c–III.   If $A \in \mathscr{A}$ and $B \in \mathscr{A}$, then $A \cup B \in \mathscr{A}$.

We say that $\mathscr{A}$ is a *completely additive class of sets* if it satisfies the stronger postulates:

C–I. $\phi \in \mathscr{A}$.

C–II.   If $A \in \mathscr{A}$, then $-A \in \mathscr{A}$.

C–III.   If A is any sequence of sets from $\mathscr{A}$, then

$$\bigcup_{n=1}^{\infty} A_n \in \mathscr{A}.$$

The reader should be warned that neither this pair of postulate systems nor this terminology is in universal use. The most common variant in the postulate systems is to replace C–II by c–II. As far as other terminology is concerned, a ring, field, algebra, or weakly additive class is a system satisfying c–III and some other conditions; a σ-ring, σ-field, σ-algebra, Borel field, additive class, or countably additive class is a system satisfying C–III and some other conditions. Each time the reader turns to a new author on measure theory, he is advised to check the definitions of any of these terms that appear.

In any space Ω there are two obvious examples of completely additive classes: namely, the class of all subsets of Ω and the class consisting of the two sets Ω and ϕ.

Completely additive classes will appear throughout our discussion of measure theory. For the present we want only the following fundamental theorems concerning them.

Theorem 9.1. *Let $\mathscr{A}$ be a completely additive class of sets. Then,*

9.1.1 $\Omega \in \mathscr{A}$;

9.1.2 *if A is a sequence of sets from $\mathscr{A}$, then*

$$\bigcap_{n=1}^{\infty} A_n \in \mathscr{A};$$

9.1.3 if A is a sequence of sets from $\mathscr{A}$, then

$$\overline{\lim_n} A_n \in \mathscr{A} \quad \text{and} \quad \underline{\lim_n} A_n \in \mathscr{A};$$

9.1.4 $\mathscr{A}$ is a finitely additive class of sets.

Proof. 9.1.1 follows from C–I and C–II. 9.1.2 follows from C–II, C–III, and 2.1.2. 9.1.3 follows from C–III and 9.1.2. To prove 9.1.4, we note first that if $A_n = \phi$ for $n > 2$, then C–III reduces to c–III. Using this and C–II, we have that the intersection of two sets from $\mathscr{A}$ is a set in $\mathscr{A}$; c–II then follows from the fact that $A - B = A \cap (-B)$. ∎

Let $\mathscr{E}$ be any class of subsets of Ω. A completely additive class $\mathscr{A}$ is called the *minimal completely additive class containing* $\mathscr{E}$, provided $\mathscr{A} \supset \mathscr{E}$ and $\mathscr{A} \subset \mathscr{B}$ whenever $\mathscr{B}$ is completely additive and $\mathscr{B} \supset \mathscr{E}$. Clearly, such a minimal class is unique. If each of the classes $\mathscr{A}$ and $\mathscr{B}$ is a minimal completely additive class containing $\mathscr{E}$, then $\mathscr{A} \subset \mathscr{B}$ because $\mathscr{A}$ is minimal, and $\mathscr{A} \supset \mathscr{B}$ because $\mathscr{B}$ is minimal. The interesting thing is that such a minimal class always exists.

Theorem 9.2. *If $\mathscr{E}$ is any class of subsets of Ω, then there is a minimal completely additive class $\mathscr{A}$ containing $\mathscr{E}$.*

Proof. Let $\mathfrak{A}$ be the collection of all completely additive classes $\mathscr{B}$ of subsets of Ω such that $\mathscr{B} \supset \mathscr{E}$. $\mathfrak{A}$ is not vacuous because the class of all subsets of Ω is an element. Let

$$\mathscr{A} = \bigcap_{\mathscr{B} \in \mathfrak{A}} \mathscr{B}.$$

$\mathscr{A}$ is completely additive. We have $\phi \in \mathscr{B}$ for each $\mathscr{B} \in \mathfrak{A}$, and so $\phi \in \mathscr{A}$. If $A \in \mathscr{A}$, then $A \in \mathscr{B}$ for each $\mathscr{B} \in \mathfrak{A}$; so $-A \in \mathscr{B}$ for each $\mathscr{B} \in \mathfrak{A}$, and it follows that $-A \in \mathscr{A}$. If $A_n \in \mathscr{A}$ for each positive integer n, then for each $\mathscr{B} \in \mathfrak{A}$, $A_n \in \mathscr{B}$ for each n, and so $\bigcup_{n=1}^{\infty} A_n \in \mathscr{B}$; thus $\bigcup_{n=1}^{\infty} A_n \in \mathscr{A}$. Since, by definition, $\mathscr{E} \subset \mathscr{B}$ for each $\mathscr{B} \in \mathfrak{A}$, it follows that $\mathscr{E} \subset \mathscr{A}$. Finally, it follows from 2.1.1 that $\mathscr{A}$ is minimal. ∎

EXERCISES

a. Let Ω be any space, and let $\mathscr{A}$ consist of the countable subsets of Ω and those whose complements are countable. Show that $\mathscr{A}$ is a completely additive class.

b. Let Ω be any infinite space, and let $\mathscr{A}$ consist of the finite subsets of Ω and those whose complements are finite. Show that $\mathscr{A}$ is finitely, but not completely, additive.

c. Let us define a half-open interval in R_1 as a set of the form $(a, b]$—always open on the left, closed on the right. Show that the class consisting of ϕ and all sets of the form $\bigcup_{k=1}^{n} I_k$, where the I_k are disjoint half-open intervals, is a finitely additive class.

d. Show that the class of Exercise c is not completely additive. In particular, C–II and C–III both fail.

e. Let Ω be a noncountable space, and let $\mathscr{A}$ be the class of all countable subsets of Ω. Show that $\mathscr{A}$ satisfies c–II and C–III but not C–II.

f. Let $\mathscr{E}$ be any class of subsets of Ω. Show that there is a minimal finitely additive class containing $\mathscr{E}$.

g. Let $\mathscr{A}$ be a finitely additive class, and let $A_k \in \mathscr{A}$ for $k = 1, 2, \ldots, n$. Show that $\bigcup_{k=1}^{n} A_k \in \mathscr{A}$.

h. Show that the class of all nowhere dense subsets of Ω is a finitely additive class.

i. Show that the class consisting of those subsets of Ω which are of Cat. I and those whose complements are of Cat. I is a completely additive class.

Let $\mathscr{E}$ be any class of subsets of Ω. We define†

$$\mathscr{E}_\sigma = \left\{ \bigcup_{n=1}^{\infty} E_n \mid E_n \in \mathscr{E}; n = 1, 2, 3, \ldots \right\},$$

$$\mathscr{E}_\delta = \left\{ \bigcap_{n=1}^{\infty} E_n \mid E_n \in \mathscr{E}; n = 1, 2, 3, \ldots \right\},$$

That is, $\mathscr{E}_\sigma$ is the class of all sets which are countable unions of sets from $\mathscr{E}$; $\mathscr{E}_\delta$ is the class of all sets which are countable intersections of sets from $\mathscr{E}$. By taking a sequence each of whose terms is the same set, we can express any set in $\mathscr{E}$ as a member of either $\mathscr{E}_\sigma$ or $\mathscr{E}_\delta$; that is,

$$\mathscr{E} \subset \mathscr{E}_\sigma \quad \text{and} \quad \mathscr{E} \subset \mathscr{E}_\delta.$$

It follows from 3.3 that a double union (or intersection),

$$\bigcup_{n=1}^{\infty} \bigcup_{k=1}^{\infty} E_{nk} \left[\text{or} \bigcap_{n=1}^{\infty} \bigcap_{k=1}^{\infty} E_{nk} \right],$$

may be rearranged as a simple, countable union (or intersection). Thus, we always have

$$(\mathscr{E}_\sigma)_\sigma = \mathscr{E}_\sigma \quad \text{and} \quad (\mathscr{E}_\delta)_\delta = \mathscr{E}_\delta.$$

However, these operators may be applied alternately and (in many cases) continue to produce new classes of sets. The usual notation is to write the subscripts in the order in which the operations are performed. For example, for

$$\{[(\mathscr{E}_\sigma)_\delta]_\sigma\}_\delta \quad \text{we write} \quad \mathscr{E}_{\sigma\delta\sigma\delta}.$$

We shall use the σ and δ operators on classes of sets in several different connections. However, they are particularly important in connection with the classes

---

† The symbols σ and δ used in this way seem to be Greek abbreviations for the German words *Summe* (sum) and *Durchschnitt* (intersection).

of open and closed sets in a pseudo-metric space. Accordingly, let Ω be a pseudo-metric space. We shall denote by $\mathscr{F}$ the class of all closed subsets of Ω and by $\mathscr{G}$ the class of all open sets of Ω. For obvious reasons, F will be a standard symbol for a closed set and G a standard symbol for an open set.† First, we note the following:

> **Theorem 9.3.** *The union of a finite number of closed sets is closed; the intersection of a finite number of open sets is open. The intersection of any (countable or noncountable) class of closed sets is closed; the union of any class of open sets is open.*

Proof. Let us prove the first and last of these four statements. The other two then follow because of 4.5 and 2.1.2.

Let $F_1, F_2, \ldots, F_n$ be closed sets, and let x be a point of accumulation of $\bigcup_{k=1}^{n} F_k$. By 4.3 there is a sequence of distinct points of $\bigcup_{k=1}^{n} F_k$ converging to x. Since there are only a finite number of sets, there is at least one of them, F_{k_0}, which contains an infinite number of points from this sequence. However, the subsequence with range contained in F_{k_0} also converges to x; and since the points of the sequence are distinct, it follows that x is a point of accumulation of F_{k_0}. Since F_{k_0} is closed, we have

$$x \in F_{k_0} \subset \bigcup_{k=1}^{n} F_k,$$

and the first statement is proved.

Let $\mathscr{E}$ be any class of open sets, and let

$$x \in \bigcup_{G \in \mathscr{E}} G;$$

then there is some set $G_0 \in \mathscr{E}$ such that $x \in G_0$. Since G_0 is open, there is an $\varepsilon > 0$ such that

$$N(x, \varepsilon) \subset G_0 \subset \bigcup_{G \in \mathscr{E}} G.$$

That is, x is an interior point of $\bigcup_{G \in \mathscr{E}} G$. ∎

It thus appears that

$$\mathscr{F}_\delta = \mathscr{F} \quad \text{and} \quad \mathscr{G}_\sigma = \mathscr{G};$$

however, the class $\mathscr{F}_\sigma$ introduces some new sets. For example, a one-point set is closed; so the set of rationals is an $\mathscr{F}_\sigma$ set in R_1, but it is neither open nor closed. It follows from 4.5 and 2.1.2 that if $E \in \mathscr{F}_\sigma$, then $-E \in \mathscr{G}_\delta$; thus the set of irrationals is a $\mathscr{G}_\delta$ set in R_1 which is neither open nor closed.

---

† The use of F and G for closed and open sets, respectively, is quite general among mathematicians. Presumably, these are abbreviations of the French *fermé* (closed) and the German *Gebiet* (region).

Theorem 9.4. $\mathscr{F} \subset \mathscr{F}_\sigma$; $\quad \mathscr{G} \subset \mathscr{G}_\delta$; $\quad \mathscr{F} \subset \mathscr{G}_\delta$; $\quad \mathscr{G} \subset \mathscr{F}_\sigma$.

Proof. The first two statements are obvious. To prove the third, suppose $F \in \mathscr{F}$. For each positive integer n, let

$$G_n = \bigcup_{x \in F} N(x, 1/n).$$

Each set $\mathscr{G}_n$ is a union of open sets and therefore open because of 9.3. Now, $F \subset G_n$ for each n, therefore

$$F \subset \bigcap_{n=1}^{\infty} G_n.$$

Let $y \in \bigcap_{n=1}^{\infty} G_n$; then for each n, there is a point $x \in F$ such that $y \in N(x, 1/n)$. Thus, y is either a point of F or an accumulation point of F. Since F is closed, $y \in F$ in any case. Therefore,

$$F = \bigcap_{n=1}^{\infty} G_n.$$

Thus, $F \in G_\delta$; the last statement follows from this by 4.5 and 2.1.2. ∎

Figure 2

Continuing to apply the σ and δ operators, we obtain two sequences of classes which constitute the *Borel classification of sets*. There is an important set of inclusion relations among these classes that can be outlined as shown in Fig. 2. This is easier to formulate specifically if we adopt the following notation:

$$\mathscr{F}^0 = \mathscr{F}; \quad \mathscr{F}^1 = \mathscr{F}_\sigma; \quad \mathscr{F}^2 = \mathscr{F}_{\sigma\delta}; \quad \mathscr{F}^3 = \mathscr{F}_{\sigma\delta\sigma}; \quad \cdots$$

$$\mathscr{G}^0 = \mathscr{G}; \quad \mathscr{G}^1 = \mathscr{G}_\delta; \quad \mathscr{G}^2 = \mathscr{G}_{\delta\sigma}; \quad \mathscr{G}^3 = \mathscr{G}_{\delta\sigma\delta}; \quad \cdots$$

Theorem 9.5. *For each positive integer n,*

$$\mathscr{F}^{n-1} \subset \mathscr{F}^n; \quad \mathscr{G}^{n-1} \subset \mathscr{G}^n; \quad \mathscr{F}^{n-1} \subset \mathscr{G}^n; \quad \mathscr{G}^{n-1} \subset \mathscr{F}^n.$$

Proof. This can be proved by induction, and 9.4 covers the case $n = 1$; all we have to do is prove the general inductive step. The first two statements are just as obvious in the general case as they were for $n = 1$; so we pass on to the proof of the third. Let us assume that $\mathscr{F}^{n-1} \subset \mathscr{G}^n$ and consider the case in which n is odd. If $H \in \mathscr{F}^n$, then

$$H = \bigcup_{n=1}^{\infty} A_n,$$

where

$$A_n \in \mathscr{F}^{n-1} \subset \mathscr{G}^n;$$

but for odd n, the union of sets in $\mathscr{G}^n$ is a set in $\mathscr{G}^{n+1}$; therefore $H \in \mathscr{G}^{n+1}$. The proofs for even n and for the relation $\mathscr{G}^{n-1} \subset \mathscr{F}^n$ are completely analogous. ∎

We define the class $\mathscr{B}$ of *Borel sets* as the minimal completely additive class containing $\mathscr{F}$. Clearly, for each positive integer n,

$$\mathscr{F}^n \subset \mathscr{B} \quad \text{and} \quad \mathscr{G}^n \subset \mathscr{B}.$$

We shall call a set of class $\mathscr{F}^n$ or $\mathscr{G}^n$ a *Borel set of order n*. There are† Borel sets which are not of finite order. Such a set might be of the form

$$H = \bigcup_{n=1}^{\infty} A_n,$$

where for each n, $A_n \in \mathscr{F}^n$ but $A_n \in -\mathscr{F}^{n-1}$.

EXERCISES

j. Show that if $\mathscr{A}$ is a completely additive class, then $\mathscr{A}_\sigma = \mathscr{A}_\delta = \mathscr{A}$.

k. Let $G_n \in \mathscr{G}$ for each positive integer n; show that

$$\overline{\lim_n} \, G_n \in \mathscr{G}_\delta \quad \text{and} \quad \underline{\lim_n} \, G_n \in \mathscr{G}_{\delta\sigma}.$$

l. Show that in Exercise *k* the first result cannot be improved to say that

$$\overline{\lim_n} \, G_n \in \mathscr{G}.$$

m. Show that if F is closed and $-F$ is dense in Ω, then F is nowhere dense.

n. Show that if $H \in \mathscr{F}_\sigma$ and $-H$ is dense in Ω, then H is of Cat. I.

o. Show from Exercise *n* that the set of irrationals is not of class $\mathscr{F}_\sigma$.

p. Show that

$$\mathscr{F}_\sigma \not\subset \mathscr{G}_\delta; \quad \mathscr{G}_\delta \not\subset \mathscr{F}_\sigma; \quad \mathscr{F}_\sigma \neq \mathscr{F}_{\sigma\delta}; \quad \mathscr{G}_\delta \neq \mathscr{G}_{\delta\sigma}.$$

q. Show that in Exercise *k* the result cannot be improved to say that

$$\underline{\lim_n} \, G_n \in \mathscr{G}_\delta.$$

---

† See Hausdorff [12], p. 182.

r. Formulate Borel classification characterizations for the limits superior and inferior of a sequence of closed sets, and show that they cannot be improved.

s. If $\mathscr{A}$ is a finitely additive class, does it follow that $\mathscr{A}_\sigma$ is completely additive?

t. Let $\mathscr{A}$ satisfy C–I, C–II, and c–III. Does it now follow that $\mathscr{A}_\sigma$ is completely additive?

10 ADDITIVE SET FUNCTIONS

In line with our first comment in the introduction to this chapter, we shall be concerned here with extended real valued functions whose domains are classes of point sets. To help distinguish these from functions whose domains are point sets, we shall use Latin letters f, g, h for functions defined over sets of points and Greek letters μ, σ, τ for functions defined over classes of sets. For example:

Type	Symbol for function	Symbol for function value
Point function	f	$f(x)$
Set function	μ	$\mu(E)$

We shall say that an extended real valued function σ is a *completely additive set function* provided it satisfies the following postulates:

A–I. The domain of σ is a completely additive class $\mathscr{A}$ of sets.

A–II. If E is a sequence of disjoint sets from $\mathscr{A}$, then

$$\sum_{n=1}^{\infty} \sigma(E_n)$$

is defined in the extended real number system and

$$\sigma\left(\bigcup_{n=1}^{\infty} E_n\right) = \sum_{n=1}^{\infty} \sigma(E_n).$$

A–III. $\sigma(\phi) = 0$.

Some authors either delete postulate A–I or replace it by a weaker one. In that case, A–II must be modified to say that the equation holds whenever $\bigcup_{n=1}^{\infty} E_n \in \mathscr{A}$. We have included A–I because it seems to us that a completely additive class is the natural domain for a function satisfying A–II. Also, our definition serves to simplify a few proofs, and it detracts nothing from the overall results in the theory of integration.

The requirement in A–II that $\sum_{n=1}^{\infty} \sigma(E_n)$ be defined in the extended real number system is best clarified by an example. Let Ω be the set of positive integers, and let u be a sequence of real numbers. For each set E of positive integers, let us define

10.1. $$\sigma(E) = \sum_{n \in E} u_n.$$

Let us also define sequences p and q such that

$$p_n = \begin{cases} u_n & \text{if } u_n \geq 0, \\ 0 & \text{if } u_n < 0, \end{cases}$$

$$q_n = \begin{cases} u_n & \text{if } u_n \leq 0, \\ 0 & \text{if } u_n > 0. \end{cases}$$

We now distinguish three cases:

(1) $\sum p_n$ and $\sum q_n$ are both convergent.
(2) One of these series is convergent, and the other is divergent.
(3) Both series are divergent.

In case (1), $\sum u_n$ is absolutely convergent, so every subseries is convergent. In case (2), every subseries is either convergent or divergent to $\pm \infty$; divergence to ∞ is possible only if $\sum p_n = \infty$, while divergence to $-\infty$ is possible only if $\sum q_n = -\infty$. In case (3), there are always subseries whose partial sums oscillate. Thus, in cases (1) and (2), 10.1 defines a completely additive set function, while in case (3) it does not.

EXERCISES

a. Show that if Ω is a countable space and σ is a completely additive set function whose domain is the class of all subsets of Ω, then σ must be of the form 10.1.

b. Show that if σ is a completely additive set function on the class of all subsets of a countable space, then a necessary and sufficient condition for σ to be identically zero is that it vanish on every set consisting of a single point.

c. Let Ω be any space, and let $\sigma(E)$ equal the number of points in E for finite sets with $\sigma(E) = \infty$ for infinite sets. Show that σ is a completely additive set function.

If the numbers u_n in the above example are all nonnegative (that is, if $u_n \equiv p_n$) and if, in addition,

$$\sum_{n=1}^{\infty} p_n = 1,$$

then we say that 10.1 describes a *discrete probability distribution*. Roughly, the picture is this: Suppose the possible results of an experiment form a countable set. We associate with result number n a number $p_n \geq 0$ which is (to the best of our knowledge) proportional to the likelihood of occurrence of this result. It is customary to choose the scale of measurement of these likelihoods (probabilities) so that $\sum p_n = 1$. Then an *event* can be defined as a set of positive integers. The physical significance of the event E is that the result of the experiment be one of those corresponding to an integer in the set E. Then, 10.1 can be used to define the *probability of an event*:

$$\mu(E) = \sum_{n \in E} p_n.$$

Clearly, the probability distribution μ defined in this way is a completely additive set function whose domain is the class of all sets of positive integers.

If an experiment has only a finite set of possible results, we can still represent the situation by this model. We set $p_n = 0$ for each value of n that does not represent a possible result.

EXERCISES

d. An elementary "definition of probability" is: If an experiment can produce n different results, all equally likely, then the probability that the result will be one of r specified results is r/n. Construct a completely additive set function on the class of all sets of positive integers which will describe any situation covered by this definition.

e. If two dice are thrown, the possible totals range from 2 to 12, inclusive. Construct a completely additive set function in a space of 11 points which will give the probability distribution for events involving these totals.

f. The probability of throwing a 7 with two dice is 1/6; that of throwing anything else is 5/6. Assume that on n successive throws the probability of a specified sequence of results is the product of the probabilities of the individual results. Let the point n represent the results, "A 7 is thrown for the first time on the nth throw." Let ∞ represent the result, "A 7 is never thrown." Construct the indicated discrete probability distribution on the countable space consisting of the positive integers and ∞.

g. In the light of Exercise *f* discuss the relation between "impossible events" and "probability zero."

If we modify the postulates for a completely additive set function to read as follows, we define a *finitely additive set function*:

a–I. The domain of σ is a finitely additive class $\mathscr{A}$ of sets.

a–II. If E_1 and E_2 are disjoint sets in $\mathscr{A}$, then

$$\sigma(E_1) + \sigma(E_2)$$

is defined in the extended real number system, and

$$\sigma(E_1 \cup E_2) = \sigma(E_1) + \sigma(E_2).$$

a–III. $\sigma(\phi) = 0$.

As noted in Section 9, every completely additive class is finitely additive, so A–I implies a–I. If in A–II we set $E_n = \phi$ for $n > 2$, then by using† A–III we reduce A–II to a–II. Thus, every completely additive set function is finitely additive. We shall use the phrase *additive*‡ *set function* to mean one which is at least finitely additive.

---

† The reader should note that without A–III it is not obvious that A–II can be reduced to a–II. See Exercise *l*.

‡ The reader should be warned that some authors use the word additive in connection with set functions to mean completely additive.

Theorem 10.2. Let σ be an additive set function defined on $\mathscr{A}$, and let $A \in \mathscr{A}$ and $B \in \mathscr{A}$; then

10.2.1 If $A \supset B$ and $\sigma(B)$ is finite, then
$$\sigma(A - B) = \sigma(A) - \sigma(B);$$

10.2.2 If $A \supset B$ and $\sigma(B)$ is infinite, then
$$\sigma(A) = \sigma(B);$$

10.2.3 If $A \supset B$ and $\sigma(A)$ is finite, then $\sigma(B)$ is finite;

10.2.4 If $\sigma(A) = +\infty$, then $\sigma(B) \neq -\infty$.

Proof. If $A \supset B$, then
$$A = B \cup (A - B);$$
so in this case
$$\sigma(A) = \sigma(B) + \sigma(A - B).$$

If $\sigma(B)$ is finite, we can subtract it from each side of this equation and thus prove 10.2.1.

If $\sigma(B)$ is infinite and $\sigma(B) + \sigma(A - B)$ is defined, then this sum must be equal to $\sigma(B)$. This proves 10.2.2; 10.2.3 follows from 10.2.2.

Finally, suppose $\sigma(A) = +\infty$ and $\sigma(B) = -\infty$; then we must have
$$\sigma(A) = \sigma(A \cap B) + \sigma(A - B) = +\infty,$$
$$\sigma(B) = \sigma(B \cap A) + \sigma(B - A) = -\infty.$$

If $\sigma(A \cap B)$ is infinite, one of these results is impossible, so we conclude that
$$\sigma(A - B) = +\infty,$$
$$\sigma(B - A) = -\infty.$$

However, $A - B$ and $B - A$ are disjoint, so this contradicts a–II. ∎

EXERCISES

h. Extend a–II by induction to give the result that for any finite class $E_1, E_2, \ldots, E_n$ of sets from $\mathscr{A}$,
$$\sigma\left(\bigcup_{k=1}^{n} E_k\right) = \sum_{k=1}^{n} \sigma(E_k).$$

i. Let $\sum u_n$ be a convergent series. Define σ by 10.1 for finite sets of positive integers, and let $\sigma(E) = \infty$ if E is an infinite set of positive integers. Show that σ is finitely but not completely additive.

j. Replace A–III by the following postulate: There exists a set $A \in \mathscr{A}$ such that $\sigma(A)$ is finite. Using this and A–II, derive a–II.

k. From a–II and the postulate of Exercise j, derive A–III.

l. Using Exercise j, show that A–II alone (without A–III) implies a–II.

m. Given that σ satisfies A–I and A–II, the imposition of A–III rules out only one example. What is it?

The following limit theorem gives an important continuity property of completely additive set functions.

Theorem 10.3. *If σ is a completely additive set function on $\mathscr{A}$, and if E is an expanding sequence of sets from $\mathscr{A}$, then*

$$\lim_n \sigma(E_n) = \sigma(\lim_n E_n).$$

Proof. Let us recall that for an expanding sequence of sets,

$$\lim_n E_n = \bigcup_{n=1}^{\infty} E_n.$$

To shorten the notation, let $E_0 = \phi$; then it is easily seen that for an expanding sequence,

$$\lim_n E_n = \bigcup_{n=1}^{\infty} E_n = \bigcup_{n=1}^{\infty} (E_n - E_{n-1}).$$

The sets $E_n - E_{n-1}$ are disjoint; so

$$\sigma(\lim_n E_n) = \sum_{n=1}^{\infty} \sigma(E_n - E_{n-1}) = \lim_k \sum_{n=1}^{k} \sigma(E_n - E_{n-1})$$

$$= \lim_k \sigma\left[\bigcup_{n=1}^{k} (E_n - E_{n-1})\right] = \lim_k \sigma(E_k). \quad \blacksquare$$

Corollary 10.3.1. *If σ is a completely additive set function defined on $\mathscr{A}$ and if E is a contracting sequence of sets from $\mathscr{A}$ with $\sigma(E_n)$ finite for some n, then*

$$\lim_n \sigma(E_n) = \sigma(\lim_n E_n).$$

Proof. Suppose $\sigma(E_{n_0})$ is finite. Since

$$\lim_n E_n = \bigcap_{n=1}^{\infty} E_n \subset E_{n_0},$$

it follows from 10.2.3 that $\sigma(\lim_n E_n)$ is finite. Thus, from 10.2.1, we have

$$\sigma[\lim_n (E_{n_0} - E_n)] = \sigma(E_{n_0} - \lim_n E_n)$$

$$= \sigma(E_{n_0}) - \sigma(\lim_n E_n). \quad (1)$$

However, by 10.3,

$$\sigma[\lim_n (E_{n_0} - E_n)] = \lim_n \sigma(E_{n_0} - E_n). \quad (2)$$

Finally, it follows from 10.2.3 that $\sigma(E_n)$ is finite for $n > n_0$; so on combining (1) and (2) and using 10.2.1 again, we have

$$\sigma(E_{n_0}) - \sigma(\lim_n E_n) = \lim_n \sigma(E_{n_0} - E_n)$$

$$= \lim_n [\sigma(E_{n_0}) - \sigma(E_n)] = \sigma(E_{n_0}) - \lim_n \sigma(E_n). \blacksquare$$

EXERCISES

n. Show that the hypothesis of complete additivity in 10.3 cannot be replaced by one of finite additivity. [*Hint:* Use the example of Exercise *i.*]

o. Let Ω be the set of positive integers; let σ be the function of Exercise *c*; let $E_n = \{k \mid k \geq n\}$. Use this example to show that the finiteness condition in 10.3.1 is essential.

p. Prove the following partial converse to 10.3: If σ is additive on a completely additive class $\mathscr{A}$ and if for every expanding sequence E of sets from $\mathscr{A}$, $\lim_n \sigma(E_n) = \sigma(\lim E_n)$, then σ is completely additive.

A set function σ is called *nondecreasing* if

$$\sigma(A) \geq \sigma(B) \quad \text{whenever} \quad A \supset B,$$

nonincreasing if

$$\sigma(A) \leq \sigma(B) \quad \text{whenever} \quad A \supset B,$$

and *monotone* if it is either nondecreasing or nonincreasing. These notions are defined for any set functions, but they are particularly interesting in the case of additive set functions. Since ϕ is a subset of every set, it follows from a–III that if σ is additive and nondecreasing (nonincreasing), then it is everywhere nonnegative (nonpositive). What is even more interesting is that for additive functions this implication goes the other way. Suppose $A \supset B$; then if σ is additive,

$$\sigma(A) = \sigma(B) + \sigma(A - B) \begin{cases} \geq \sigma(B) & \text{if } \sigma(A - B) \geq 0, \\ \leq \sigma(B) & \text{if } \sigma(A - B) \leq 0. \end{cases}$$

Thus, if σ is additive and everywhere nonnegative (nonpositive), then it is nondecreasing (nonincreasing).

If σ is additive with domain $\mathscr{A}$, then for each $E \in \mathscr{A}$, we define the *upper variation of σ on E*:

$$\overline{V}(\sigma, E) = \sup \{\sigma(A) \mid A \in \mathscr{A}; A \subset E\}.$$

Similarly, we define the *lower variation of σ on E*:

$$\underline{V}(\sigma, E) = \inf \{\sigma(A) \mid A \in \mathscr{A}; A \subset E\}.$$

Since we always have $\phi \in \mathscr{A}$ and $\phi \subset E$, it follows from a–III that

$$\underline{V}(\sigma, E) \leq 0 \leq \overline{V}(\sigma, E).$$

We also define the *total variation of σ on E*:

$$V(\sigma, E) = |\overline{V}(\sigma, E)| + |\underline{V}(\sigma, E)| = \overline{V}(\sigma, E) - \underline{V}(\sigma, E).$$

Theorem 10.4. *If σ is a completely additive set function on $\mathscr{A}$, then $\overline{V}(\sigma, \ )$ and $\underline{V}(\sigma, \ )$ are monotone, completely additive set functions on $\mathscr{A}$.*

Proof. Let us give the proof for $\overline{V}(\sigma, \ )$. Since we have already observed that $\overline{V}(\sigma, E) \geq 0$ for every $E \in \mathscr{A}$, monotonicity follows as soon as we prove additivity. Postulate A–I is obviously satisfied; the domain of $\overline{V}(\sigma, \ )$ is the same as that of σ. As for A–II, there is no trouble about sums being defined when the numbers involved are all nonnegative; so we note that A–III is satisfied because

$$\overline{V}(\sigma, \phi) = \sigma(\phi) = 0,$$

and turn to the proof that if E is a sequence of disjoint sets from $\mathscr{A}$, then

$$\overline{V}\left(\sigma, \bigcup_{n=1}^{\infty} E_n\right) = \sum_{n=1}^{\infty} \overline{V}(\sigma, E_n).$$

We take an arbitrary set $A \in \mathscr{A}$ such that $A \subset \bigcup_{n=1}^{\infty} E_n$; then

$$\sigma(A) = \sum_{n=1}^{\infty} \sigma(A \cap E_n) \leq \sum_{n=1}^{\infty} \overline{V}(\sigma, E_n).$$

Since this is true for every such A, we have

$$\overline{V}\left(\sigma, \bigcup_{n=1}^{\infty} E_n\right) \leq \sum_{n=1}^{\infty} \overline{V}(\sigma, E_n). \tag{1}$$

If for any n, $\overline{V}(\sigma, E_n) = \infty$, then there are sets $B \subset E_n \subset \bigcup_{n=1}^{\infty} E_n$ for which $\sigma(B)$ is arbitrarily large; so

$$\overline{V}\left(\sigma, \bigcup_{n=1}^{\infty} E_n\right) = \infty,$$

and the equality in (1) must hold. If $\overline{V}(\sigma, E_n)$ is always finite, then given $\varepsilon > 0$, there is for each n a set $A_n \subset E_n$ such that

$$\sigma(A_n) \geq \overline{V}(\sigma, E_n) - \frac{\varepsilon}{2^n}.$$

Thus

$$\overline{V}\left(\sigma, \bigcup_{n=1}^{\infty} E_n\right) \geq \sigma\left(\bigcup_{n=1}^{\infty} A_n\right) = \sum_{n=1}^{\infty} \sigma(A_n)$$

$$\geq \sum_{n=1}^{\infty} \left[\overline{V}(\sigma, E_n) - \frac{\varepsilon}{2^n}\right] = \sum_{n=1}^{\infty} \overline{V}(\sigma, E_n) - \varepsilon.$$

Since this holds for every $\varepsilon > 0$, it follows that

$$\overline{V}\left(\sigma, \bigcup_{n=1}^{\infty} E_n\right) \geq \sum_{n=1}^{\infty} \overline{V}(\sigma, E_n).$$

This, together with (1), gives the desired equality. ∎

Theorem 10.5. *If σ is completely additive and $\overline{V}(\sigma, E) = \infty$, then $\sigma(E) = \infty$; if σ is completely additive and $\underline{V}(\sigma, E) = -\infty$, then $\sigma(E) = -\infty$.*

Proof. Let us prove the first statement. The proof of the other is completely analogous.

First, we note that, because of 10.2.2, in order to prove that $\sigma(E) = \infty$ we need only find a subset A of E such that $\sigma(A) = \infty$.

Suppose $\overline{V}(\sigma, E) = \infty$; then there is a set $E_1 \subset E$ such that

$$\sigma(E_1) > 1.$$

Because of the additivity of $\overline{V}(\sigma, \ )$, it follows that either $\overline{V}(\sigma, E_1) = \infty$ or $\overline{V}(\sigma, E - E_1) = \infty$. Let A_1 be either E_1 or $E - E_1$ just so $\overline{V}(\sigma, A_1) = \infty$.

For each positive integer n, let $E_n \subset A_{n-1}$ be such that

$$\sigma(E_n) > n.$$

Let A_n be either E_n or $A_{n-1} - E_n$ just so $\overline{V}(\sigma, A_n) = \infty$. There are now two cases:

(1) For an infinite number of values of n,

$$A_n = A_{n-1} - E_n.$$

(2) For all $n \geq n_0$,

$$A_n = E_n.$$

In case (1) we have a sequence E_{n_k} of disjoint sets; so

$$\sigma\left(\bigcup_{k=1}^{\infty} E_{n_k}\right) = \sum_{k=1}^{\infty} \sigma(E_{n_k}) \geq \sum_{k=1}^{\infty} n_k = \infty.$$

Since $\bigcup_{k=1}^{\infty} E_{n_k} \subset E$, this proves the theorem for case (1).

In case (2) we have a contracting sequence of sets

$$E_{n_0}, \quad E_{n_0+1}, \quad E_{n_0+2}, \quad \ldots.$$

If $\sigma(E_{n_0}) = \infty$, the theorem is proved. If not, 10.3.1 applies, and

$$\sigma(\lim_n E_n) = \lim_n \sigma(E_n) \geq \lim_n n = \infty.$$

Since $\lim_n E_n = \bigcap_{n=n_0}^{\infty} E_n \subset E$, we again have the desired result. ∎

Theorem 10.6 (Jordan decomposition theorem). *Every completely additive set function is the sum of two monotone, completely additive set functions; specifically, for each $E \in \mathscr{A}$,*

$$\sigma(E) = \overline{V}(\sigma, E) + \underline{V}(\sigma, E).$$

Proof. Because of 10.5, we cannot have simultaneously $\overline{V}(\sigma, E) = \infty$ and $\underline{V}(\sigma, E) = -\infty$; thus for $\sigma(E) = \pm\infty$, the result is obvious.

If $\sigma(E)$ is finite, let $A \in \mathscr{A}$ and $A \subset E$; then

$$\sigma(A) = \sigma(E) - \sigma(E - A) \begin{cases} \leq \sigma(E) - \underline{V}(\sigma, E), \\ \geq \sigma(E) - \overline{V}(\sigma, E). \end{cases}$$

Since this is true for every such A,

$$\overline{V}(\sigma, E) \leq \sigma(E) - \underline{V}(\sigma, E),$$
$$\underline{V}(\sigma, E) \geq \sigma(E) - \overline{V}(\sigma, E).$$

Since everything is finite, we can transpose in these inequalities and get

$$\overline{V}(\sigma, E) + \underline{V}(\sigma, E) \leq \sigma(E) \leq \overline{V}(\sigma, E) + \underline{V}(\sigma, E). \quad \blacksquare$$

EXERCISES

q. Let f be an everywhere finite real valued function whose domain is the real number system. Define a function τ_0 on the half-open intervals by the relation

$$\tau_0\{(a, b]\} = f(b) - f(a).$$

Let τ be defined on the class of all sets R which can be expressed as the union of a finite number of disjoint half-open intervals by the relation

$$\tau(R) = \sum_{k=1}^{n} \tau_0(I_k) \quad \text{where} \quad R = \bigcup_{k=1}^{n} I_k.$$

Show that τ is uniquely defined and is an extension of τ_0. Show also that τ is finitely additive.

r. Show that the upper and lower variations of a finitely additive set function are finitely additive set functions.

s. Let f be defined by

$$f(x) = \begin{cases} 1/x & \text{for } x \neq 0, \\ 0 & \text{for } x = 0; \end{cases}$$

let τ be defined from f as in Exercise q. Use this example to show that 10.5 does not apply to finitely additive functions.

t. State and prove a Jordan decomposition theorem for finitely additive set functions. Note: In the light of Exercise s, an additional hypothesis will be called for.

u. A function f of a real variable is said to be *of bounded variation* if there exists a finite number M such that for every finite set $(a_1, b_1], (a_2, b_2], \ldots, (a_n, b_n]$ of disjoint intervals,

$$\sum_{k=1}^{n} |f(b_k) - f(a_k)| \leq M.$$

Show from Exercise t that any function of bounded variation is the sum of two monotone functions of bounded variation. *Hint*:

$$f(x) = \begin{cases} f(0) + \overline{V}\{\tau, (0, x]\} + \underline{V}\{\tau, (0, x]\} & \text{for } x \geq 0, \\ f(0) - \overline{V}\{\tau, (x, 0]\} - \underline{V}\{\tau, (x, 0]\} & \text{for } x \leq 0. \end{cases}$$

v. Is the function of Exercise s the sum of two monotone functions?

w. For 10.4 and 10.5 we gave a representative half of the proof in each case. Write out the other half of each of these proofs.

x. Define upper and lower variations for an arbitrary set function (not necessarily additive), and show that they are monotone though not necessarily of constant sign.

We now define a *measure function* as a nonnegative (therefore nondecreasing), completely additive set function. Having arrived at 10.6, we see that the study of completely additive set functions in general can be reduced to the study of measure functions. This is an important step, because measure functions are much easier to handle than more general completely additive set functions. To a large extent this ease in manipulation for measure functions stems from three important inequalities. The inequality defining monotonicity is one. The other two are given by the following theorems.

Theorem 10.7. *If μ is a measure function on $\mathscr{A}$ and if E is any sequence of sets from $\mathscr{A}$, then†*

$$\mu\left(\bigcup_{n=1}^{\infty} E_n\right) \leq \sum_{n=1}^{\infty} \mu(E_n).$$

Proof. By 2.2 we can replace the sequence E by a sequence A of disjoint sets having the same union and such that $A_n \subset E_n$ for each n. Thus

$$\mu\left(\bigcup_{n=1}^{\infty} E_n\right) = \mu\left(\bigcup_{n=1}^{\infty} A_n\right) = \sum_{n=1}^{\infty} \mu(A_n) \leq \sum_{n=1}^{\infty} \mu(E_n). \blacksquare$$

Theorem 10.8. *If μ is a measure function on $\mathscr{A}$ and if E is any sequence of sets from $\mathscr{A}$, then*

$$\mu(\varliminf_n E_n) \leq \varliminf_n \mu(E_n).$$

Proof. For each n, let

$$A_n = \bigcap_{k=n}^{\infty} E_k;$$

then $A_n \subset E_n$, so

$$\mu(A_n) \leq \mu(E_n).$$

Thus

$$\varliminf_n \mu(A_n) \leq \varliminf_n \mu(E_n).$$

However, A is an expanding sequence; so 10.3 applies, and we have

$$\varliminf_n \mu(E_n) \geq \varliminf_n \mu(A_n) = \mu(\lim_n A_n) = \mu\left(\bigcup_{n=1}^{\infty} A_n\right) = \mu(\varliminf_n E_n). \blacksquare$$

---

† The property of μ described by this inequality is called *subadditivity* or, frequently, *countable subadditivity* to distinguish it from the property that for any two sets, $\mu(E_1 \cup E_2) \leq \mu(E_1) + \mu(E_2)$.

Corollary 10.8.1. *If μ is a measure function on $\mathscr{A}$ and if E is any sequence of sets from $\mathscr{A}$ for which $\mu(\bigcup_{n=1}^{\infty} E_n) < \infty$, then*

$$\mu(\overline{\lim_n} E_n) \geq \overline{\lim_n} \mu(E_n).$$

Proof. This is proved from 10.8 by taking complements with respect to $\bigcup_{n=1}^{\infty} E_n$. ∎

Corollary 10.8.2. *If μ is a measure function on $\mathscr{A}$ and if E is a convergent sequence of sets from $\mathscr{A}$ with $\mu(\bigcup_{n=1}^{\infty} E_n) < \infty$, then*

$$\mu(\lim_n E_n) = \lim_n \mu(E_n).$$

Proof. To show this, we apply 10.8 and 10.8.1:

$$\mu(\lim_n E_n) = \mu(\underline{\lim_n} E_n) \leq \underline{\lim_n} \mu(E_n) \leq \overline{\lim_n} \mu(E_n) \leq \mu(\overline{\lim_n} E_n) = \mu(\lim_n E_n). \ \blacksquare$$

Corollary 10.8.3. *If σ is any completely additive set function on $\mathscr{A}$ and if E is a convergent sequence of sets from $\mathscr{A}$ with $\sigma(\bigcup_{n=1}^{\infty} E_n)$ finite, then*

$$\sigma(\lim_n E_n) = \lim_n \sigma(E_n).$$

Proof. It follows at once from 10.5 that if $\sigma(\bigcup_{n=1}^{\infty} E_n)$ is finite, then the measure functions of the Jordan decomposition of σ satisfy the conditions of 10.8.2. Thus the result follows at once from 10.6 and 10.8.2. ∎

EXERCISES

y. Write out the details of the proof of 10.8.1.

z. Show by an example that the hypothesis

$$\mu\left(\bigcup_{n=1}^{\infty} E_n\right) < \infty \quad \text{or} \quad \sigma\left(\bigcup_{n=1}^{\infty} E_n\right) \text{ finite,}$$

as the case may be, is essential in 10.8.1, 10.8.2, and 10.8.3. [*Hint:* See Exercise *o*.]

Theorems 10.3 and 10.8 and their corollaries constitute a standard set of limit theorems for completely additive set functions and measure functions. It should be noted that some limit theorems hold unrestrictedly; others require finiteness; still others require nonnegativity. In general, these limit theorems are not valid for finitely additive set functions. Indeed, a rough picture of the situation is that finite additivity plus a limit theorem yields complete additivity. There are various ways to make this rough observation precise. One appears in Exercise *p* above; a slightly different version for later reference is as follows. We say that an additive set function σ is *continuous at* ϕ if for every contracting sequence E of sets in the domain of σ such that $\lim_n E_n = \phi$ we have $\lim_n \sigma(E_n) = 0$.

Theorem 10.9. *If σ is finitely additive on a completely additive class and if σ is continuous at ϕ, then σ is completely additive.*

Proof. Let E be a sequence of disjoint sets. For each n,

$$\sigma\left(\bigcup_{k=1}^{\infty} E_k\right) = \sum_{k=1}^{n-1} \sigma(E_k) + \sigma\left(\bigcup_{k=n}^{\infty} E_k\right).$$

As $n \to \infty$, $\bigcup_{k=n}^{\infty} E_k$ tends monotonically to ϕ, and the result follows. ∎

11 OUTER MEASURES

If in 10.1 we stipulate that $u_n \geq 0$ for each n, then 10.1 indicates a general method for the construction of measure functions in countable spaces. In other spaces the general problem is not quite so simple. An important difference between the discrete case 10.1 and the general case is that in the discrete case the domain of the measure function is the class of all subsets of Ω, while in most other cases of practical importance the domain of the measure function is some smaller completely additive class.

Our approach to the general construction of measure functions will be to give a postulational description of what is called an outer measure function. In general, this will not be a measure function, but it will have for its domain the class of all subsets of Ω, and we shall prove that a suitable restriction of an outer measure function to a smaller domain always yields a measure function. We then proceed to describe a general method for the construction of outer measure functions.

Let μ^* be an extended real valued function whose domain is the class of all subsets of Ω. We say that μ^* is an *outer measure function* (or, frequently, merely an *outer measure*) if it satisfies the following postulates:†

M–I. μ^* is nondecreasing.

M–II. For any sequence E of subsets of Ω,

$$\mu^*\left(\bigcup_{n=1}^{\infty} E_n\right) \leq \sum_{n=1}^{\infty} \mu^*(E_n).$$

M–III. $\mu^*(\phi) = 0$.

EXERCISES

a. Show that if μ is a measure function whose domain is the class of all subsets of Ω, then μ is an outer measure.

b. Show that the equations $\mu^*(\phi) = 0$, $\mu^*(E) = 1$ if $E \neq \phi$ define a function μ^* which is always an outer measure; however, if Ω contains more than one point, μ^* is not a measure function.

c. Show that an outer measure function is always nonnegative.

---

† These, together with M–IV (Section 13), are essentially the postulates of Carathéodory [4]. For this reason, outer measures are frequently called *Carathéodory outer measures*.

d. Give an example of a nonnegative set function satisfying M–II and M–III but not M–I. Compare this with the situation for additive set functions.

e. Exercises 10.*j* and 10.*k* show that postulate A–III may be replaced by the postulate stated in Exercise 10.*j*. Show by an example that M–III may not be so replaced.

Let μ^* be an outer measure function. We say that a set E is *measurable* with respect to μ^* if for every $A \subset \Omega$,

$$\mu^*(A) = \mu^*(A \cap E) + \mu^*(A - E).$$

Measurability is not an intrinsic property of sets; it depends on the outer measure function. There may well be two outer measures for the same space such that a given set E is measurable with respect to one but not the other. The most accurate terminology would be "E is μ^*-measurable"; however, when (as in most cases) there is no possibility of ambiguity, we shall merely say "E is measurable," meaning measurable with respect to the outer measure under consideration.

It might be in order to say a word here about the intuitive picture of measurability. Postulate M–II does not, in general, give even finite additivity for an outer measure function. However, the usual picture is that if two disjoint sets are not too badly intertwined, the outer measure will add properly for them and their union. Thus, a nonmeasurable set is one which is so snarled up with its complement that additivity of the outer measure fails somewhere across its boundary. A measurable set is one which is "smooth" enough so that it breaks no set in such a way as to destroy additivity for the outer measure.

Finally, we might note that in writing down the formula that defines measurability of E, we start, not with $\mu^*(E)$, but with $\mu^*(A)$, where A is an arbitrary "test set." Measurability of E has nothing to do with the outer measure of E itself. It depends on what E does to the outer measure of other sets.

Theorem 11.1. *A necessary and sufficient condition that E be measurable is that for every $A \subset \Omega$ for which $\mu^*(A) < \infty$,*

$$\mu^*(A) \geq \mu^*(A \cap E) + \mu^*(A - E).$$

Proof. This inequality, together with M–II, gives the equality which defines measurability of E. If $\mu^*(A) = \infty$, the above inequality automatically holds; so it suffices to test with sets of finite outer measure. This proves sufficiency of the condition. Necessity follows from the fact that the defining equation is a special case of the above inequality. ∎

EXERCISES

f. Let Ω be any space. Let $\mu^*(E)$ equal the number of points in E for finite sets, and let $\mu^*(E) = \infty$ for infinite sets. Show that μ^* is an outer measure, and determine the class of measurable sets.

g. Let Ω be any space. Let $\mu^*(\phi) = 0$ and $\mu^*(E) = 1$ for $E \neq \phi$. Show that μ^* is an outer measure, and determine the class of measurable sets.

h. Let Ω be any space. Let $\mu^*(\phi) = 0$, $\mu^*(\Omega) = 2$ and $\mu^*(E) = 1$ for all other sets. Show that μ^* is an outer measure, and determine the class of measurable sets.

i. Let Ω be a noncountable space. Let $\mu^*(E) = 0$ if E is countable, $\mu^*(E) = 1$ if E is noncountable. Show that μ^* is an outer measure, and determine the class of measurable sets.

j. Let Ω be a complete metric space. Let $\mu^*(E) = 0$ if E is of Cat. I, $\mu^*(E) = 1$ if E is of Cat. II. Show that μ^* is an outer measure, and determine the class of measurable sets.

k. Construct an example to show that if M–III is removed from the definition of an outer measure, there may be no measurable sets.

l. Let μ^* be an outer measure, and let H be a μ^*-measurable set. Let μ_0^* be the restriction of μ^* to the subsets of H. Show that μ_0^* is an outer measure in the space H such that the μ_0^*-measurable sets are precisely the μ^*-measurable subsets of H.

m. Suppose in Exercise l that H is not μ^*-measurable. Show that μ_0^* is still an outer measure. Show, further, that if $E \subset H$ and E is μ^*-measurable, then E is μ_0^*-measurable. Show, finally, that there are μ_0^*-measurable sets which are not μ^*-measurable.

n. Prove that if an outer measure is finitely additive, then it is completely additive. *Hint:* If E is a sequence of disjoint sets, then for each n,

$$\mu^*\left(\bigcup_{k=1}^{\infty} E_k\right) \geq \mu^*\left(\bigcup_{k=1}^{n} E_k\right) = \sum_{k=1}^{n} \mu^*(E_k).$$

We turn now to the key theorem on outer measures.

Theorem 11.2. *Let μ^* be an outer measure function, and let $\mathcal{M}$ be the class of μ^*-measurable sets. Then, $\mathcal{M}$ is a completely additive class, and the restriction of μ^* to $\mathcal{M}$ is a measure function.*

Proof. The proof of this fundamental theorem is easier to follow if we break it up into several steps.

11.2.1 *If $\mu^*(E) = 0$, then E is measurable; thus, in particular, ϕ is measurable.*

For any set A, $A \cap E \subset E$ and $A - E \subset A$; thus, using M–I and the hypothesis that $\mu^*(E) = 0$, we have

$$\mu^*(A \cap E) + \mu^*(A - E) \leq \mu^*(E) + \mu^*(A) = \mu^*(A),$$

so E is measurable by 11.1. Measurability of ϕ follows from M–III.

11.2.2 *If E is measurable, then $-E$ is measurable.*

Since $A \cap E = A - (-E)$, the equation defining measurability of E is the same as that defining measurability of $-E$.

11.2.3 *Any finite union of measurable sets is measurable.*

First, we prove this for two sets. Given measurable sets E_1 and E_2 and an arbitrary set A, we have

$$\mu^*(A) = \mu^*(A \cap E_1) + \mu^*(A - E_1) \tag{1}$$

because E_1 is measurable. Testing measurability of E_2 by the set $A - E_1$, we have

$$\mu^*(A - E_1) = \mu^*[(A - E_1) \cap E_2] + \mu^*[(A - E_1) - E_2] \\ = \mu^*[(A - E_1) \cap E_2] + \mu^*[A - (E_1 \cup E_2)]. \qquad (2)$$

However, $[(A - E_1) \cap E_2] \cup (A \cap E_1) = A \cap (E_1 \cup E_2)$; so we substitute (2) into (1) and use M–II to see that

$$\mu^*(A) = \mu^*(A \cap E_1) + \mu^*[(A - E_1) \cap E_2] + \mu^*[A - (E_1 \cup E_2)] \\ \geq \mu^*[A \cap (E_1 \cup E_2)] + \mu^*[A - (E_1 \cup E_2)].$$

Measurability of $E_1 \cup E_2$ follows from 11.1.

We complete the proof of 11.2.3 by an induction on the number of sets involved. Suppose every union of n measurable sets is measurable; then

$$\bigcup_{k=1}^{n+1} E_k = \left(\bigcup_{k=1}^{n} E_k\right) \cup E_{n+1}$$

expresses the union of $n + 1$ measurable sets as the union of two measurable sets.

11.2.4 *If E is a sequence of disjoint measurable sets and if, for each n,*

$$S_n = \bigcup_{k=1}^{n} E_k,$$

then for each n and for any set A,

$$\mu^*(A \cap S_n) = \sum_{k=1}^{n} \mu^*(A \cap E_k).$$

We prove this by induction on n. For $n = 1$, it is trivial. Assuming the result for n, we recall 11.2.3 and use $A \cap S_{n+1}$ as a test set for measurability of S_n:

$$\mu^*(A \cap S_{n+1}) = \mu^*(A \cap S_{n+1} \cap S_n) + \mu^*(A \cap S_{n+1} - S_n) \\ = \mu^*(A \cap S_n) + \mu^*(A \cap E_{n+1}) \\ = \sum_{k=1}^{n} \mu^*(A \cap E_k) + \mu^*(A \cap E_{n+1}) \\ = \sum_{k=1}^{n+1} \mu^*(A \cap E_k).$$

11.2.5 *If E is a sequence of disjoint measurable sets and if*

$$S = \bigcup_{k=1}^{\infty} E_k,$$

then for any set A,

$$\mu^*(A \cap S) = \sum_{k=1}^{\infty} \mu^*(A \cap E_k).$$

Recalling the notation and result of 11.2.4, we see that for each n, $A \cap S \supset A \cap S_n$; so for each n,

$$\mu^*(A \cap S) \geq \mu^*(A \cap S_n) = \sum_{k=1}^{n} \mu^*(A \cap E_k).$$

Letting $n \to \infty$, we have

$$\mu^*(A \cap S) \geq \sum_{k=1}^{\infty} \mu^*(A \cap E_k),$$

and the reverse inequality follows from M–II.

11.2.6 *Any countable union of disjoint measurable sets is measurable.*

Again we employ the notation of 11.2.4 and 11.2.5. Let A be any set. We note that S_n is measurable by 11.2.3; then we use 11.2.4; finally we note that $-S_n \supset -S$. These observations give us that for each n,

$$\mu^*(A) = \mu^*(A \cap S_n) + \mu^*(A - S_n)$$

$$= \sum_{k=1}^{n} \mu^*(A \cap E_k) + \mu^*(A - S_n) \geq \sum_{k=1}^{n} \mu^*(A \cap E_k) + \mu^*(A - S).$$

Letting $n \to \infty$, and then using M–II, we have

$$\mu^*(A) \geq \sum_{k=1}^{\infty} \mu^*(A \cap E_k) + \mu^*(A - S) \geq \mu^*(A \cap S) + \mu^*(A - S),$$

so S is measurable by 11.1.

11.2.7 *Any countable union of measurable sets is measurable.*

By 11.2.3 any finite union of measurable sets is measurable, and by 11.2.2 the complement of a measurable set is measurable. Since

$$A - B = -[(-A) \cup B],$$

it follows that the difference of two measurable sets is measurable. Therefore, the formula of 2.2 expresses the union of an arbitrary sequence of measurable sets as the union of a sequence of disjoint measurable sets, and the result follows from 11.2.6.

The proof of 11.2 is now complete. Postulates C–I, C–II, and C–III for $\mathcal{M}$ are given by 11.2.1, 11.2.2, and 11.2.7, respectively. This gives A–I for the restriction of μ^* to $\mathcal{M}$. We obtain A–II by setting $A = S$ in 11.2.5, and A–III comes directly from M–III. Since μ^* is nondecreasing, its restriction to $\mathcal{M}$ is nondecreasing; so this restriction is a measure function. ∎

Theorem 11.2 indicates the role played by an outer measure in the construction of a measure function. The next problem would seem to be that of constructing outer measure functions.

Let $\mathscr{C}$ be a class of subsets of Ω. We say that $\mathscr{C}$ is a *sequential covering class* if for every $A \subset \Omega$ there is a sequence E of sets from $\mathscr{C}$ such that

$$A \subset \bigcup_{n=1}^{\infty} E_n.$$

In order that finite unions may appear as special cases of infinite unions, we shall always assume that $\phi \in \mathscr{C}$.

The covering class itself need not be countable. In fact, one of the most important examples of a sequential covering class is the set of all open intervals on the real line. Clearly, this is a noncountable class of sets, but every set on the real line is contained in some countable union of open intervals.

Method I—Construction of outer measures. Take a sequential covering class $\mathscr{C}$ and a nonnegative, extended real valued function τ defined on $\mathscr{C}$ such that $\tau(\phi) = 0$. For each $A \subset \Omega$, let

$$\mu^*(A) = \inf \left\{ \sum_{n=1}^{\infty} \tau(E_n) \,\middle|\, E_n \in \mathscr{C};\ \bigcup_{n=1}^{\infty} E_n \supset A \right\}.$$

We shall prove below that this procedure always yields an outer measure. We have called it Method I because in Section 13 we want to introduce a slightly more complicated procedure which we shall call Method II. In a trivial sense, Method I is completely general. That is, given any μ^*, let $\mathscr{C}$ be the class of all subsets of Ω, and let $\tau = \mu^*$; then Method I gives μ^* right back again. However, the practical point of view is that we want to work from smaller classes $\mathscr{C}$ and simpler functions τ. The excuse for Method II is that for certain $\mathscr{C}, \tau$ combinations, Method I does not give an outer measure with all the additional properties we want.

Theorem 11.3. *For any sequential covering class $\mathscr{C}$ and any nonnegative function τ defined on $\mathscr{C}$ for which $\tau(\phi) = 0$, Method I yields an outer measure.*

Proof. Postulate M–III follows from the fact that $\tau(\phi) = 0$. Since τ is nonnegative, it is clear that $\mu^*(A) \geq \mu^*(\phi)$ for every $A \subset \Omega$. As for monotonicity where two nonvacuous sets are involved, suppose $A \subset B$; given $\varepsilon > 0$, there is a sequence E of sets from $\mathscr{C}$ such that

$$\bigcup_{n=1}^{\infty} E_n \supset B \supset A$$

and

$$\sum_{n=1}^{\infty} \tau(E_n) \leq \mu^*(B) + \varepsilon.$$

However, since $A \subset \bigcup_{n=1}^{\infty} E_n$, it follows that

$$\mu^*(A) \leq \sum_{n=1}^{\infty} \tau(E_n) \leq \mu^*(B) + \varepsilon.$$

Since this is true for every $\varepsilon > 0$, $\mu^*(A) \leq \mu^*(B)$, and M–I holds.

To check M–II, let A be any sequence of sets, and let ε be any positive number. For each positive integer n there is a sequence E_n of sets from $\mathscr{C}$ such that

$$\bigcup_{k=1}^{\infty} E_{nk} \supset A_n$$

and

$$\sum_{k=1}^{\infty} \tau(E_{nk}) \leq \mu^*(A_n) + \frac{\varepsilon}{2^n}.$$

Now

$$\bigcup_{n=1}^{\infty} \bigcup_{k=1}^{\infty} E_{nk} \supset \bigcup_{n=1}^{\infty} A_n;$$

so

$$\mu^*\left(\bigcup_{n=1}^{\infty} A_n\right) \leq \sum_{n=1}^{\infty} \sum_{k=1}^{\infty} \tau(E_{nk}) \leq \sum_{n=1}^{\infty} \left[\mu^*(A_n) + \frac{\varepsilon}{2^n}\right] = \sum_{n=1}^{\infty} \mu^*(A_n) + \varepsilon.$$

Since this is true for every $\varepsilon > 0$, M–II follows. ∎

Theorem 11.4. *Let $\mathscr{C}$ be a sequential covering class and let τ_1 and τ_2 be nonnegative functions, each defined on $\mathscr{C}$, vanishing at ϕ, and yielding, by Method I, outer measures μ_1^* and μ_2^*, respectively. If $\mu_1^*(E) = \mu_2^*(E)$ for every $E \in \mathscr{C}$, then μ_1^* and μ_2^* are identical.*

Proof. Since any set $E \in \mathscr{C}$ covers itself, it is obvious that for $E \in \mathscr{C}$,

$$\mu_i^*(E) \leq \tau_i(E) \qquad (i = 1, 2).$$

Let A be any set and ε any positive number. There exists a sequence E of sets from $\mathscr{C}$ such that $\bigcup_{n=1}^{\infty} E_n \supset A$ and

$$\mu_1^*(A) + \varepsilon \geq \sum_{n=1}^{\infty} \tau_1(E_n) \geq \sum_{n=1}^{\infty} \mu_1^*(E_n) = \sum_{n=1}^{\infty} \mu_2^*(E_n) \geq \mu_2^*\left(\bigcup_{n=1}^{\infty} E_n\right) \geq \mu_2^*(A).$$

Since this holds for every $\varepsilon > 0$, we have $\mu_1^*(A) \geq \mu_2^*(A)$. On interchanging the subscripts 1 and 2, we have the reverse inequality. ∎

We now have the machinery set up for the construction of measure functions in any space. Let us review this procedure briefly and, in the course of this review, codify certain notation that we want to adopt as standard from here on.

In our space Ω we choose:

$\mathscr{C}$—a sequential covering class,

τ—a nonnegative function defined on $\mathscr{C}$ and such that $\tau(\phi) = 0$.

The application of Method I yields:

μ^*—an outer measure.

This outer measure determines:

$\mathcal{M}$—the class of measurable sets.

Finally, the restriction of μ^* to $\mathcal{M}$ is:

μ—a measure function.

To apply this machinery to what is probably the most important specific example, let Ω be the real line, and let $\mathcal{C}$ be the class consisting of ϕ and all open intervals. For each interval I, let $\tau(I)$ be defined as the length of I. The outer measure μ^* is then called *Lebesgue outer measure*, the sets of $\mathcal{M}$ are called *Lebesgue measurable*, and μ is called *Lebesgue measure*.

We could turn to other examples of classes $\mathcal{C}$ and functions τ and thereby list a number of important measure functions, but a study of such specific examples will be more profitable after we learn a little more about outer measures in general. For example, we know that the Lebesgue measurable sets form a completely additive class, but what familiar sets (if any) belong to that class? We started the construction of Lebesgue measure by assigning a natural weight to each of the open intervals, but after we wander through the maze of Method I and the definition of measurability, are these intervals themselves measurable sets?

Instead of trying to answer such questions directly for a specific measure, we shall, in the next two sections, develop such answers in terms of properties of $\mathcal{C}$ and τ. Then, the treatment of specific measures is relatively simple. We specify $\mathcal{C}$ and τ, and the measure is defined. We check on certain properties of $\mathcal{C}$ and τ, and many useful properties of the measure are determined.

EXERCISES

o. Let Ω be any space. Let $\mathcal{C}$ consist of Ω and ϕ, together with all one-point sets. Let $\tau(\Omega) = \infty$, $\tau(\phi) = 0$, $\tau(E) = 1$ otherwise. Describe the outer measure obtained by Method I.

p. Let Ω be any space. Let $\mathcal{C}$ consist of the two sets Ω and ϕ, and let $\tau(\Omega) = 1$. Describe μ^*.

q. Let Ω be any space. Let $\mathcal{C}$ consist of all the proper subsets of Ω. Let $\tau(E) = 1$ for every nonvacuous $E \in \mathcal{C}$. Describe μ^*.

r. Let Ω be a noncountable space. Let $\mathcal{C}$ be defined as in Exercise o. Let $\tau(\Omega) = 1$, $\tau(E) = 0$ otherwise. Describe μ^*.

s. Describe $\mathcal{C}$ and τ so that Method I will give the outer measure of Exercise j.

t. In 10.1 there is a general description of measure in a countable space. Formulate this description in terms of a class $\mathcal{C}$ and a function τ.

u. Prove that if μ^* is constructed by Method I, then it satisfies M–I, because $\inf A \leq \inf B$ if $B \subset A \subset R_1$.

v. Show that if μ^* comes from $\mathscr{C}$ and τ by Method I and if $E \in \mathscr{C}$, then $\mu^*(E) \leq \tau(E)$. Construct examples in which the inequality holds.

w. Prove that the inequality in Exercise v is always an equality in the case of Lebesgue outer measure; that is, the Lebesgue outer measure of an open interval is its length.

x. Show that the Lebesgue outer measure of any countable set on the real line is zero; hence, in this case $\mu^*\{[a, b]\} = \mu^*\{(a, b)\}$. [*Hint*: For a sequence x, cover x_n with an interval of length $\varepsilon/2^n$.]

12 REGULAR OUTER MEASURES

An outer measure μ^* is called *regular* if for every $A \subset \Omega$ there is a measurable set $E \supset A$ such that
$$\mu(E) = \mu^*(A).$$
We shall refer to E as a *measurable cover* for A. In these terms, an outer measure is regular if it determines measurable sets in such a way that every set has a measurable cover.

EXERCISES

a. In Exercises 11.*f* through 11.*j* several outer measures are described. Which of these are regular outer measures?

d. In Exercise 11.*l* assume that μ^* is regular. Show that μ_0^* is regular.

c. Same as Exercise *b* for the situation in Ex. 11.*m*.

d. Show that in Exercise 11.*m*, μ_0^* may be regular even though μ^* is not.

e. Show that if μ^* is regular and A is any sequence of sets, then there exist measurable covers E_i for A_i such that $\bigcup_{i=1}^{\infty} E_i$ is a measurable cover for $\bigcup_{i=1}^{\infty} A_i$.

Limit theorems for measure functions are given by 10.3, 10.8, 10.8.1, and 10.8.2. In general, these results do not apply to outer measure functions (except, of course, on measurable sets). However, for regular outer measures we can get the first two of these results on arbitrary sets.

Theorem 12.1. *If μ^* is a regular outer measure and if A is any sequence of sets, then*
$$\mu^*(\varliminf_n A_n) \leq \varliminf_n \mu^*(A_n).$$

Proof. For each n, let E_n be a measurable cover for A_n. Then,
$$\varliminf_n A_n \subset \varliminf_n E_n,$$
and $\varliminf_n E_n$ is measurable; so we have from 10.8 that
$$\mu^*(\varliminf_n A_n) \leq \mu(\varliminf_n E_n) \leq \varliminf_n \mu(E_n) = \varliminf_n \mu^*(A_n). \blacksquare$$

Corollary 12.1.1. *If μ^* is a regular outer measure and A is an expanding sequence of sets, then*

$$\mu^*(\lim_n A_n) = \lim_n \mu^*(A_n).$$

Proof. For an expanding sequence,

$$\lim_n A_n = \bigcup_{n=1}^{\infty} A_n \supset A_n$$

for every n; therefore

$$\mu^*(\lim_n A_n) \geq \mu^*(A_n)$$

for every n, and it follows that

$$\mu^*(\lim_n A_n) \geq \lim_n \mu^*(A_n).$$

The reverse inequality is given by 12.1. ∎

EXERCISES

f. Show that 10.8.1 and 10.8.2 cannot be generalized even to regular outer measures. [*Hint:* Let Ω be the set of positive integers; use the outer measure of Exercise 11.*g*; let $A_n = \{k \mid k \geq n\}$.]

g. Show that 12.1 and 12.1.1 cannot be generalized to arbitrary outer measures. [*Hint:* Let Ω be the set of positive integers; use the outer measure of Exercise 11.*h*; let $A_n = \{k \mid k \leq n\}$.]

h. Let Ω be the set of positive integers. If E contains n points ($n = 0, 1, 2, 3, \ldots$), let $\mu^*(E) = n/(n+1)$. If E is infinite, let $\mu^*(E) = 1$. Show that μ^* is a non-regular outer measure, but the limit theorem for expanding sequences of arbitrary sets still applies. Thus, regularity is not a necessary condition in 12.1.1.

i. Show that if μ^* is an outer measure for which the result in 12.1.1 holds, then the result in 12.1 also holds for μ^*.

j. Show that regularity is not a necessary condition in 12.1.

In many discussions† of Lebesgue measure the outer measure is defined as we indicated at the end of Section 11. Then, for bounded sets A, an *inner measure* μ_* is defined by the relation

$$\mu_*(A) = b - a - \mu^*[(a, b) - A],$$

where (a, b) is an interval covering A. A set E is then called measurable if

$$\mu_*(E) = \mu^*(E).$$

---
† Note, in particular, the original discussion of Lebesgue [19].

Fundamentally, this amounts to taking a space $\Omega = (a, b)$ of finite measure and saying that E is measurable if

$$\mu^*(E) + \mu^*(-E) = \mu^*(\Omega).$$

This is our measurability equation with Ω used as the test set. We require, however, that the equation

$$\mu^*(A) = \mu^*(A \cap E) + \mu^*(A - E)$$

hold for all test sets A. Formally, then, our requirement is much stronger. It is not too surprising that the inner measure approach is equivalent to ours in the special case of Lebesgue measure, but it is interesting to note that the two definitions of measurability are equivalent for all regular outer measures for which $\mu(\Omega)$ is finite.

Theorem 12.2. *If μ^* is a regular outer measure for which $\mu^*(\Omega) < \infty$, then a necessary and sufficient condition that E be measurable is that*

$$\mu^*(\Omega) = \mu^*(E) + \mu^*(-E).$$

Proof. Necessity is obvious. If we take $A = \Omega$ in the definition of measurability, we get the condition stated here.

To prove sufficiency, let A be any set, and let B be a measurable cover for A. From measurability of B, we have

$$\mu^*(E) = \mu^*(E \cap B) + \mu^*(E - B),$$
$$\mu^*(-E) = \mu^*(B - E) + \mu^*(-E - B);$$

thus

$$\mu(\Omega) = \mu^*(E) + \mu^*(-E)$$
$$= \mu^*(E \cap B) + \mu^*(E - B) + \mu^*(B - E) + \mu^*(-E - B)$$
$$\geq \mu(B) + \mu(-B) = \mu(\Omega).$$

Therefore

$$\mu^*(B \cap E) + \mu^*(B - E) + \mu^*(E - B) + \mu^*(-E - B) = \mu(B) + \mu(-B).$$

Noting that all quantities are finite, we subtract from this equation the inequality

$$\mu^*(E - B) + \mu^*(-E - B) \geq \mu(-B);$$

this gives the result

$$\mu^*(B \cap E) + \mu^*(B - E) \leq \mu(B).$$

Now, $A \cap E \subset B \cap E$ and $A - E \subset B - E$; so we have

$$\mu^*(A \cap E) + \mu^*(A - E) \leq \mu^*(B \cap E) + \mu^*(B - E) \leq \mu(B) = \mu^*(A),$$

and E is measurable by 11.1. ∎

EXERCISES

k. Prove the following corollary to 12.2: If μ^* is a regular outer measure and if there is a measurable set $H \supset E$ such that $\mu(H) < \infty$ and $\mu(H) = \mu^*(E) + \mu^*(H - E)$, then E is measurable.

l. Use the outer measure of Exercise 11.h to show that 12.2 need not hold for nonregular outer measures.

m. Show that 12.2 holds for the nonregular outer measure of Exercise g.

n. Modify the outer measure of Exercise 11.h by setting $\mu^*(\Omega) = \frac{3}{2}$. Show that 12.1 still fails, but 12.2 now holds.

o. Construct an example in which 12.1 holds but 12.2 fails.

p. Show that if μ^* is regular and finite, then every set A has a measurable subset E such that $\mu(E) = \mu_*(A)$. We call E a *measurable kernel* for A.

q. Let μ^* be regular and finite and let $\nu^* = (\mu^* + \mu_*)/2$. Show that ν^* is an outer measure. *Hint:* Verification of M–II is nontrivial. The following outline suggests the major steps. (i) It suffices to show that $\nu^*(A_0) \le \sum_{n=1}^{\infty} \nu^*(A_n)$ where A_n are disjoint and $A_0 = \bigcup_{n=1}^{\infty} A_n$. (ii) Let E_n be a measurable cover for A_n with (Exercise e) $E_0 = \bigcup_{n=1}^{\infty} E_n$ a measurable cover for A_0; replace $\bigcup E_n$ by a disjoint union $\bigcup (E_n - B_n)$ (see 2.2) and let $B_0 = \bigcup_{n=1}^{\infty} (B_n \cap E_n)$. (iii) $\mu(B_0) \le \sum_{n=1}^{\infty} \mu(B_n \cap E_n) = \sum_{n=1}^{\infty} \mu^*(A_n) - \mu^*(A_0)$. (iv) Let F_n be a measurable kernel for A_n ($n = 0, 1, 2, \ldots$); $C_n = \bigcup_{k \ne n} E_k$; $G_n = F_n \cup (F_0 - C_n)$; $G_0 = \bigcup_{n=1}^{\infty} G_n$; $R = F_0 - G_0$. (v) G_n is a measurable kernel for A_n; $R \cup G_0$ is a measurable kernel for A_0; so $\mu(R) = \mu_*(A_0) - \sum_{n=1}^{\infty} \mu_*(A_n)$. (vi) $R \subset \bigcap_{n=1}^{\infty} C_n \subset B_0$, and the result follows from (iii) and (v).

r. Show that if ν^* in Exercise q is regular, then every set is ν^*-measurable. Note: We shall point out later that if μ^* is Lebesgue outer measure, then ν^* has some nonmeasurable sets and therefore is not regular. See Exercise 18.c.

Returning to Method I for the construction of outer measures, we now look for conditions under which we get a regular outer measure.

Theorem 12.3. *If μ^* is constructed from $\mathscr{C}$ and τ by Method I, then given any set A, there is a set $E_{00} \in \mathscr{C}_{\sigma\delta}$ such that $E_{00} \supset A$ and $\mu^*(E_{00}) = \mu^*(A)$.*

Proof. Let A be any set. For each positive integer n, there is a sequence E_n of sets from $\mathscr{C}$ such that

$$\bigcup_{k=1}^{\infty} E_{nk} \supset A$$

and

$$\sum_{k=1}^{\infty} \tau(E_{nk}) \le \mu^*(A) + \frac{1}{n}.$$

We now set

$$E_{n0} = \bigcup_{k=1}^{\infty} E_{nk};$$

then for each n, $E_{n0} \in \mathscr{C}_\sigma$. Remembering that $\mu^*(B) \leq \tau(B)$ for every $B \in \mathscr{C}$, we have
$$\mu^*(E_{n0}) \leq \sum_{k=1}^\infty \mu^*(E_{nk}) \leq \sum_{k=1}^\infty \tau(E_{nk}) \leq \mu^*(A) + \frac{1}{n}.$$
Setting
$$E_{00} = \bigcap_{n=1}^\infty E_{n0},$$
we see that $E_{00} \in \mathscr{C}_{\sigma\delta}$ and that $E_{00} \supset A$; therefore
$$\mu^*(E_{00}) \geq \mu^*(A).$$
However,
$$\mu^*(E_{00}) \leq \mu^*(E_{n0}) \leq \mu^*(A) + 1/n$$
for every n; so
$$\mu^*(E_{00}) \leq \mu^*(A). \blacksquare$$

Corollary 12.3.1. *If μ^* is constructed by Method I and if $\mathscr{C}$ consists of μ^*-measurable sets, then μ^* is regular.*

Proof. Since $\mathscr{M}$ is completely additive, it follows that if $\mathscr{C} \subset \mathscr{M}$, then $\mathscr{C}_{\sigma\delta} \subset \mathscr{M}$. Thus, the set E_{00} of 12.3 serves as a measurable cover for A. $\blacksquare$

As a criterion for regularity, 12.3.1 seems a little awkward. True, it involves a property of $\mathscr{C}$, but that property of $\mathscr{C}$ is determined by μ^*. It appears that an examination of $\mathscr{C}$ and τ alone is not sufficient to determine whether or not μ^* is regular. This is not as serious a drawback as it seems. We shall find that for metric outer measures (Section 13) all Borel sets are automatically measurable. Therefore, in those cases we can check the hypotheses of 12.3.1 by looking at the topological properties of the sets in $\mathscr{C}$.

We have approached the subject of measure through that of outer measure. The question naturally arises whether or not all measures come from outer measures by the procedure we have outlined. Though this is not strictly the case, in a certain sense it is "almost" so. To make this statement more precise, let us turn to the following considerations.

Let $\mathscr{A}$ be a completely additive class and let μ be a measure function whose domain is $\mathscr{A}$. Let
$$\mathfrak{z} = \{Z \mid Z \subset A \in \mathscr{A}; \mu(A) = 0\};$$
that is, $\mathfrak{z}$ is the class of all subsets of sets of measure zero. Now, if $\mathscr{E}$ is the class of all sets of the form $Z \cup A$, where $Z \in \mathfrak{z}$ and $A \in \mathscr{A}$, it is easily checked that $\mathscr{E}$ is a completely additive class containing $\mathscr{A}$. Furthermore, there is an obvious extension of μ to $\mathscr{E}$ defined by
$$\bar{\mu}(Z \cup A) = \mu(A).$$
Clearly, $\bar{\mu}$ is a measure. It is called the *completion* of μ. It is easily seen that $\bar{\bar{\mu}} = \bar{\mu}$. If $\mu = \bar{\mu}$, we say that μ is a *complete measure*. Thus, a complete measure is one whose domain contains all subsets of sets of measure zero.

There are important examples of measures which are not complete. Probably the outstanding one is that of *Borel measure*. We shall show (see 13.2.1 and Exercise 13.*i*) that every Borel set on the real line is Lebesgue measurable. On the other hand, there are (see Section 19) Lebesgue measurable sets of measure zero which are not Borel sets. It follows from 12.3 that any such set is contained in a Borel set of measure zero. Thus, Borel measure—defined as the restriction of Lebesgue measure to the class $\mathscr{B}$—is not a complete measure.

From 11.2.1, we see that any measure generated by an outer measure is complete. The interesting thing is that for an important class of spaces every complete measure comes not only from an outer measure but from a regular outer measure. This result, along with others of interest, is contained in the following theorem.

Theorem 12.4. *Let $\mathscr{C}$ be the domain of a measure function μ; let $\tau = \mu$; let μ_0^* be constructed from $\mathscr{C}$ and τ by Method I.*

12.4.1 *μ_0^* is an extension of μ.*

12.4.2 *$\mathscr{M}_0 \supset \mathscr{C}$; hence μ_0^* is regular.*

12.4.3 *If $A \in \mathscr{M}_0$ and $A = \bigcup_{n=1}^{\infty} A_n$, where $\mu_0^*(A_n) < \infty$ for each n, then $A = E \cup Z$ where $E \in \mathscr{C}$ and $Z \subset B \in \mathscr{C}$ with $\mu(B) = 0$.*

12.4.4 *If $\Omega = \bigcup_{n=1}^{\infty} A_n$ where $\mu_0^*(A_n) < \infty$ for each n, then μ_0 is the completion of μ.*

12.4.5 *If μ is the restriction of a regular outer measure μ^* to its class of measurable sets, then $\mu^* = \mu_0^*$.*

Proof. Since, in this case, $\mathscr{C}$ is a completely additive class, if a set A is covered by a countable subclass of $\mathscr{C}$, it is covered by a single set from $\mathscr{C}$—the union of this subclass. Furthermore, because of 10.7, the use of this single covering set does not increase the function for which we are taking an infimum. So, in this proof, we may modify Method I to say

$$\mu_0^*(A) = \inf \{\mu(E) \mid E \in \mathscr{C}; E \supset A\}.$$

If $E \in \mathscr{C}$, it is its own covering; so $\mu_0^*(E) \leq \mu(E)$. However, if $B \supset E$ and $B \in \mathscr{C}$, then $\mu(B) \geq \mu(E)$; so $\mu_0^*(E) \geq \mu(E)$. Thus, μ and μ_0^* coincide on $\mathscr{C}$, and 12.4.1 is proved.

Suppose $E \in \mathscr{C}$; let A be any set; let ε be any positive number. There is a set $B \in \mathscr{C}$ such that $B \supset A$ and

$$\mu(B) \leq \mu_0^*(A) + \varepsilon.$$

Thus

$$\mu_0^*(A \cap E) + \mu_0^*(A - E) \leq \mu_0^*(B \cap E) + \mu_0^*(B - E)$$
$$= \mu(B \cap E) + \mu(B - E) = \mu(B) \leq \mu_0^*(A) + \varepsilon.$$

Since this holds for every $\varepsilon > 0$,
$$\mu_0^*(A \cap E) + \mu_0^*(A - E) \leq \mu_0^*(A),$$
and E is measurable by 11.1. So, $\mathcal{M}_0 \supset \mathcal{C}$; regularity of μ_0^* now follows from 12.3.1.

In proving 12.4.3, let us first assume that $\mu_0^*(A) < \infty$. Since $\mathcal{C}$ is completely additive, $\mathcal{C}_{\sigma\delta} = \mathcal{C}$; so by 12.3, A has a measurable cover $C \in \mathcal{C}$. Now, μ_0^* is additive on $\mathcal{M}_0$; so if $A \in \mathcal{M}_0$ and $\mu_0^*(A) < \infty$, we have
$$\mu_0^*(C - A) = \mu_0^*(C) - \mu_0^*(A) = 0.$$
We now choose a measurable cover $H \in \mathcal{C}$ for the set $C - A$ and a measurable cover $B \in \mathcal{C}$ for the set $A \cap H$. Since
$$\mu_0^*(A \cap H) \leq \mu(H) = \mu_0^*(C - A) = 0,$$
we have
$$\mu(B) = 0.$$
It is easily checked that
$$A = (C - H) \cup (A \cap H);$$
so we set $E = C - H \in \mathcal{C}$ and $Z = A \cap H \subset B$, and 12.4.3 is proved for the case $\mu_0^*(A) < \infty$.

To complete the proof of 12.4.3, we apply what we have just proved to the measurable sets (of finite measure) $A \cap D_n$, where for each n, D_n is a measurable cover for A_n. This gives us
$$A = \bigcup_{n=1}^{\infty} A_n = \bigcup_{n=1}^{\infty} (A \cap D_n) = \bigcup_{n=1}^{\infty} (E_n \cup Z_n) = \left(\bigcup_{n=1}^{\infty} E_n\right) \cup \left(\bigcup_{n=1}^{\infty} Z_n\right),$$
where for each n, $E_n \in \mathcal{C}$ and $Z_n \subset B_n \in \mathcal{C}$ with $\mu(B) = 0$. Setting
$$E = \bigcup_{n=1}^{\infty} E_n, \quad Z = \bigcup_{n=1}^{\infty} Z_n, \quad B = \bigcup_{n=1}^{\infty} B_n,$$
we have the required decomposition of A.

If Ω is the union of a countable number of sets for each of which μ_0^* is finite, then every set A satisfies the restrictive condition in 12.4.3; thus it follows from 12.4.3 that the domain of μ_0 is that of the completion of μ. That μ_0 has the right values follows at once from the fact that it is monotone. Thus, 12.4.4 is established.

To prove 12.4.5, let A be any set, and let E be a μ^*-measurable cover for A; then
$$\mu_0^*(A) \leq \mu(E) = \mu^*(A).$$
However, if $A \subset B \in \mathcal{C}$, $\mu^*(A) \leq \mu(B)$; so
$$\mu^*(A) \leq \inf \mu(B) = \mu_0^*(A). \quad \blacksquare$$

EXERCISES

s. Let Ω be a noncountable space; let $\mathscr{C}$ be the class of all sets which either are countable or have countable complements; for finite sets $E \in \mathscr{C}$ let $\mu(E)$ equal the number of points in E; for infinite sets $E \in \mathscr{C}$ let $\mu(E) = \infty$. Show that μ_0 as constructed in 12.4 is not the completion of μ.

t. Show that the measure μ of Exercise s is a complete measure which is not the restriction to its measurable sets of any outer measure.

u. Show that the requirement of regularity for μ^* cannot be dropped in 12.4.5. [*Hint:* Let μ^* be the outer measure of Exercise 11.*h*.]

v. Show that if the requirement of regularity for μ^* is dropped in 12.4.5, the result is $\mu^* \leq \mu_0^*$.

w. Assuming that Borel sets on the real line are Lebesgue measurable, show that Lebesgue measure is the completion of Borel measure.

13 METRIC OUTER MEASURES

If Ω is a metric space, it is desirable to have some connection between the topological properties (openness, closure, etc.) of sets and the property of measurability. This is achieved in a very satisfactory manner by adding another postulate to the definition of outer measure. We say that μ^* is a *metric outer measure* if in addition to M–I, M–II, and M–III it satisfies:

M–IV. If $\rho(A, B) > 0$, then $\mu^*(A \cup B) = \mu^*(A) + \mu^*(B)$.

Our first use of M–IV will be to show that for metric outer measures all Borel sets are measurable. This will conquer once and for all a rather difficult point in the study of any specific example—namely, the determination of a large and useful class of measurable sets. For the rather trivial examples of outer measures we have encountered so far (Lebesgue measure excepted), the classes of measurable sets can be determined by inspection; but for the more important examples of Chapter 3 this is no longer feasible. Instead, we prove here that M–IV gives measurability for Borel sets; then we have only to check M–IV for each example.

The first step in proving measurability for Borel sets is to prove a limit theorem for an expanding sequence constructed in a certain way from an open set. If the outer measure is regular, the limit theorem holds for all expanding sequences; but we do not want to make use of this fact here, because we want to use the end result of this argument to show that certain outer measures are regular.

Lemma 13.1 (Carathéodory). *Let μ^* be a metric outer measure; let G be an open set; let A_0 be any set such that $A_0 \subset G$; for each positive integer n, let*

$$A_n = \{x \mid x \in A_0; \rho(x, -G) \geq 1/n\}.$$

Then

$$\lim_n \mu^*(A_n) = \mu^*(A_0).$$

Proof. Since A is an expanding sequence with $A_n \subset A_0$ for each n, we have only to prove that

$$\varlimsup_n \mu^*(A_n) \geq \mu^*(A_0). \tag{1}$$

The existence of the limit and the reverse inequality follow from the monotonicity of μ^*.

Since $A_0 \supset A_n$ for each n,

$$A_0 \supset \bigcup_{n=1}^{\infty} A_n.$$

Since G is open, each point of A_0 is an interior point of G and so must be a point of A_n for sufficiently large n. That is,

$$A_0 \subset \bigcup_{n=1}^{\infty} A_n.$$

Thus,

$$A_0 = \bigcup_{n=1}^{\infty} A_n. \tag{2}$$

For each n, let us define

$$D_n = A_{n+1} - A_n.$$

It follows from (2) that for each n,

$$A_0 = A_{2n} \cup \left(\bigcup_{k=2n}^{\infty} D_k \right) = A_{2n} \cup \left(\bigcup_{k=n}^{\infty} D_{2k} \right) \cup \left(\bigcup_{k=n}^{\infty} D_{2k+1} \right).$$

Therefore,

$$\mu^*(A_0) \leq \mu^*(A_{2n}) + \sum_{k=n}^{\infty} \mu^*(D_{2k}) + \sum_{k=n}^{\infty} \mu^*(D_{2k+1}).$$

If both of these last two sums tend to zero as $n \to \infty$, we have

$$\mu^*(A_0) \leq \varlimsup_n \mu^*(A_{2n}) \leq \varlimsup_n \mu^*(A_n),$$

as required. Otherwise, at least one of the series

$$\sum \mu^*(D_{2k}), \quad \sum \mu^*(D_{2k+1})$$

is divergent. Let us suppose it is the first of these. The proof for the other case is completely analogous.

From the definition of the sets A_n, we have

$$\rho(D_{2k}, D_{2k+2}) \geq \frac{1}{2k+1} - \frac{1}{2k+2} > 0$$

for each k. Since (for $n > 1$)

$$A_{2n} \supset \bigcup_{k=1}^{n-1} D_{2k},$$

we have, using M–IV, that
$$\mu^*(A_{2n}) \geq \mu^*\left(\bigcup_{k=1}^{n-1} D_{2k}\right) = \sum_{k=1}^{n-1} \mu^*(D_{2k}).$$

Since $\sum \mu^*(D_{2k})$ is assumed divergent, we now have
$$\varlimsup_n \mu^*(A_n) \geq \varlimsup_n \mu^*(A_{2n}) = \infty,$$

and the inequality (1) must hold. ∎

Theorem 13.2. *If μ^* is a metric outer measure, then every closed set is measurable.*

Proof. Let F be a closed set, and let A be any set. Then, $A - F$ is contained in the open set $-F$; so by 13.1 there is a sequence E of sets such that
$$\rho(E_n, F) \geq 1/n \tag{1}$$
for each n, and
$$\lim_n \mu^*(E_n) = \mu^*(A - F). \tag{2}$$

Using (1) and M–IV, we have for each n,
$$\mu^*(A) \geq \mu^*[(A \cap F) \cup E_n] = \mu^*(A \cap F) + \mu^*(E_n).$$

Letting $n \to \infty$ and applying (2), we have
$$\mu^*(A) \geq \mu^*(A \cap F) + \mu^*(A - F). \quad \blacksquare$$

Corollary 13.2.1. *If μ^* is a metric outer measure, then every Borel set is measurable.*[†]

Proof. The class $\mathcal{M}$ of measurable sets is completely additive, and by 13.2, $\mathcal{M} \supset \mathcal{F}$. Therefore, by the definition of the class $\mathcal{B}$ of Borel sets, $\mathcal{M} \supset \mathcal{B}$. ∎

EXERCISES

a. Prove the converse to 13.2.1: If every Borel set is μ^*-measurable, then μ^* is a metric outer measure. *Hint:* Work with the open sets. If $\rho(E_1, E_2) > 0$, let $G \supset E_1$ and $G \cap E_2 = \phi$. Write the condition for measurability of G using $E_1 \cup E_2$ as the test set.

b. Show that in Exercise 12.q, ν^* satisfies M–IV if μ^* does. Note: This will furnish an example of a nonregular metric outer measure. See Exercise 18.c.

c. Let Ω be the real number system. Of the outer measures described in Exercises 11.f through 11.j which are metric outer measures?

d. Let Ω be the real number system, and let x be the sequence of points defined by the relation $x_n = 1/n$. For each A, let $\mu^*(A)$ be the number of points of the sequence x which are contained in A. Show that μ^* is a regular metric outer measure.

[†] In general, not every measurable set is a Borel set. See Section 19.

In order to study the construction of metric outer measures, we need to consider what might be called fine-mesh covering classes. Suppose $\mathscr{C}$ is a sequential covering class in a metric space. Let us define for each positive integer n the class

$$\mathscr{C}_n = \{E \mid E \in \mathscr{C}; d(E) \leq 1/n\}.$$

Let us also assume that $\phi \in \mathscr{C}_n$ for each n. For the construction of metric outer measures we shall be interested in the cases in which $\mathscr{C}_n$ is a sequential covering class for each n. If this is the case, and if $\tau \geq 0$ is defined on $\mathscr{C}$ and vanishes at ϕ, we can apply Method I, using $\mathscr{C}_n$ and τ, to define an outer measure μ_n^*. Since, for each n,

$$\mathscr{C}_{n+1} \subset \mathscr{C}_n,$$

it follows that for each $A \subset \Omega$ and each n,

$$\mu_{n+1}^*(A) \geq \mu_n^*(A);$$

so for each A, $\mu_n^*(A)$ tends to a (finite or infinite) limit as $n \to \infty$. These considerations lead us to the following procedure.

Method II—Construction of outer measures. Take a sequential covering class $\mathscr{C}$ such that for each n, $\mathscr{C}_n$ is a sequential covering class. Take a nonnegative, extended real valued function τ, defined on $\mathscr{C}$ and vanishing at ϕ. For each n, use Method I to construct an outer measure μ_n^* from $\mathscr{C}_n$ and τ. Finally, define μ_0^* by the relation

$$\mu_0^*(A) = \lim_n \mu_n^*(A).$$

Theorem 13.3. *Any function μ_0^* constructed by Method II is a metric outer measure.*

Proof. Postulates M–I and M–III apply to μ_n^* for each n. By a simple passage to the limit, it follows that they apply to μ_0^*. To check M–II, let E be any sequence of sets; from postulate M–II for μ_n^* and the monotonicity of μ^* as a function of n, it follows that for each n,

$$\mu_n^*\left(\bigcup_{k=1}^{\infty} E_k\right) \leq \sum_{k=1}^{\infty} \mu_n^*(E_k) \leq \sum_{k=1}^{\infty} \mu_0^*(E_k).$$

Therefore

$$\mu_0^*\left(\bigcup_{k=1}^{\infty} E_k\right) = \lim_n \mu_n^*\left(\bigcup_{k=1}^{\infty} E_k\right) \leq \sum_{k=1}^{\infty} \mu_0^*(E_k).$$

As for M–IV, suppose $\rho(A, B) > 0$; then there is an integer n_0 such that for every $n > n_0$,

$$\rho(A, B) > \frac{1}{n}. \tag{1}$$

Given $\varepsilon > 0$, there is for each n a sequence E_n of sets from $\mathscr{C}_n$ such that

$$\bigcup_{k=1}^{\infty} E_{nk} \supset A \cup B$$

and

$$\sum_{k=1}^{\infty} \tau(E_{nk}) \leq \mu_n^*(A \cup B) + \varepsilon.$$

However, if $n > n_0$, it follows from (1) that no set E_{nk} contains points of both A and B; therefore part of the sequence covers A, and the rest covers B. Thus, for $n > n_0$,

$$\mu_n^*(A) + \mu_n^*(B) \leq \sum_{k=1}^{\infty} \tau(E_{nk}) \leq \mu_n^*(A \cup B) + \varepsilon.$$

Since this is true for every $\varepsilon > 0$,

$$\mu_n^*(A) + \mu_n^*(B) \leq \mu_n^*(A \cup B).$$

Since this holds for every $n > n_0$, we let $n \to \infty$, and

$$\mu_0^*(A) + \mu_0^*(B) \leq \mu_0^*(A \cup B).$$

This result, together with M–II, gives M–IV. ∎

Let us say that a sequence τ of set functions on some common domain $\mathscr{D}$ is *convergent* if $\lim_n \tau_n(E)$ exists in the extended number system for each $E \in \mathscr{D}$. Method II involves a convergent sequence of outer measures, and we see there that the limit function is also an outer measure. If the functions of the convergent sequence are measures, is the limit function also a measure? The following exercises indicate some answers to this question, and a more extensive discussion appears in Section 40.

EXERCISES

e. Let Ω be the set of positive integers; let u be a sequence of numbers such that $0 \leq u_k \leq 1$ for each k and such that $\sum_{k=1}^{\infty} u_k < \infty$; for each n let v_n be the sequence defined by

$$v_{nk} = \begin{cases} u_k & \text{for } k \leq n, \\ 1 & \text{for } k > n; \end{cases}$$

let μ be the sequence of measures defined by

$$\mu_n(E) = \sum_{k \in E} v_{nk}.$$

Show that μ is a convergent (indeed, monotone) sequence of measures for which the limit function is not even an outer measure.

f. Prove that for a nondecreasing sequence of measures on a common domain the limit function is a measure. [*Hint:* Use Exercise 11.n.]

Theorem 13.4. *If μ_0^* is constructed from $\mathscr{C}$ and τ by Method II, then for every $A \subset \Omega$ there is a set $E_{00} \in \mathscr{C}_{\sigma\delta}$ such that $E_{00} \supset A$ and $\mu^*(E_{00}) = \mu_0^*(A)$.*

Proof. Let A be any set, and for each n let E_n be a sequence of sets from $\mathscr{C}_n \subset \mathscr{C}$ such that

$$\bigcup_{k=1}^{\infty} E_{nk} \supset A$$

and

$$\sum_{k=1}^{\infty} \tau(E_{nk}) \leq \mu_n^*(A) + \frac{1}{n}.$$

Let

$$E_{n0} = \bigcup_{k=1}^{\infty} E_{nk} \in \mathscr{C}_\sigma;$$

then $E_{n0} \supset A$, and

$$\mu_n^*(E_{n0}) \leq \sum_{k=1}^{\infty} \tau(E_{nk}) \leq \mu_n^*(A) + \frac{1}{n}.$$

Now, let

$$E_{00} = \bigcap_{n=1}^{\infty} E_{n0} \in \mathscr{C}_{\sigma\delta};$$

then $E_{00} \supset A$, and for each n,

$$\mu_n^*(E_{00}) \leq \mu_n^*(E_{n0}) \leq \mu_n^*(A) + \frac{1}{n}.$$

Letting $n \to \infty$, we have

$$\mu_0^*(E_{00}) \leq \mu_0^*(A).$$

Since $E_{00} \supset A$, the reverse inequality follows from M–I. ∎

Thus, we have for outer measures constructed by Method II the same result as we obtained in 12.3 for those constructed by Method I. However, in the case of Method II we always get metric outer measures; so 13.2.1 gives us an independent criterion for measurability, and we can say that if in Method II the sets of $\mathscr{C}$ are Borel sets, then μ_0^* is regular. If we get a regular outer measure in this manner, the measurable covers may all be taken from $\mathscr{C}_{\sigma\delta}$. Now, if $\mathscr{C} \subset \mathscr{G}$, then $\mathscr{C}_{\sigma\delta} \subset \mathscr{G}_{\sigma\delta} = \mathscr{G}_\delta$; so if the sets of $\mathscr{C}$ are open sets, the class order of the measurable covers is reduced by one. This is one reason for the extensive use of open sets in the construction of outer measures. Pursuing this line of thought a little further, we have the following results.

Theorem 13.5. *If μ^* is a metric outer measure constructed from $\mathscr{C}$ and τ by either Method I or Method II, if $\mathscr{C}$ consists of open sets, and if $\Omega = \bigcup_{n=1}^{\infty} A_n$ where $\mu^*(A_n) < \infty$ for each n, then each of the following conditions is necessary and sufficient for measurability of a set E:*

13.5.1 *There exists a set $H \in \mathscr{G}_\delta$ such that $H \supset E$ and $\mu^*(H - E) = 0$.*

13.5.2 *There exists a set $K \in \mathscr{F}_\sigma$ such that $E \supset K$ and $\mu^*(E - K) = 0$.*

Proof. Sets of measure zero, $\mathscr{F}_\sigma$ sets and $\mathscr{G}_\delta$ sets are all measurable; furthermore, the union and difference of two measurable sets are measurable, so the sufficiency of each condition is obvious.

To prove necessity, let E be measurable, and let us first suppose $\mu(E) < \infty$. By 13.4 (or 12.3, as the case may be) there is a set $H \in \mathscr{G}_\delta$ such that $H \supset E$ and $\mu(H) = \mu(E)$; so by 10.2.1, $\mu(H - E) = 0$. Thus, 13.5.1 is necessary in case $\mu(E) < \infty$.

Next we show that 13.5.2 is also necessary in case $\mu(E) < \infty$. In order to do this, we must first consider the special case in which $E = H \in \mathscr{G}_\delta$. Let us write

$$H = \bigcap_{n=1}^{\infty} G_n,$$

where $G_n \in \mathscr{G}$ for each n. For each n, we can write (see 9.4)

$$G_n = \bigcup_{k=1}^{\infty} F_{nk},$$

where for each n, F_n is an expanding sequence of closed sets. It follows from 13.2.1 and 13.4 (or 12.3) that μ^* is regular; so by 12.1.1, we have for each n,

$$\mu(H) = \mu(H \cap G_n) = \lim_k \mu(H \cap F_{nk}).$$

Thus, for each n and each positive integer i, there is a positive integer k_{ni} such that

$$\mu[H - (H \cap F_{nk_{ni}})] = \mu(H) - \mu(H \cap F_{nk_{ni}}) \leq \frac{1}{i2^n}.$$

Setting

$$F^{(i)} = \bigcap_{n=1}^{\infty} F_{nk_{ni}} \in \mathscr{F},$$

we have $F^{(i)} \subset G_n$ for every n; therefore $F^{(i)} \subset H$, and

$$\mu(H) - \mu(F^{(i)}) = \mu(H - F^{(i)})$$
$$= \mu\left\{\bigcup_{n=1}^{\infty} [H - (H \cap F_{nk_{ni}})]\right\} \leq \sum_{n=1}^{\infty} \frac{1}{i2^n} = \frac{1}{i}.$$

Setting

$$B = \bigcup_{i=1}^{\infty} F^{(i)} \in \mathscr{F}_\sigma,$$

we have $B \subset H$ and $\mu(H - B) = \mu(H) - \mu(B) \leq 1/i$ for every i, whence

$$\mu(H - B) = 0,$$

as required.

With this special case established, let E be any measurable set with $\mu(E) < \infty$. Let $H \in \mathscr{G}_\delta$ be a measurable cover for E; let $D \in \mathscr{G}_\delta$ be a measurable cover for

$H - E$; let $B \in \mathscr{F}_\sigma$ be the subset of H constructed above. Setting
$$K = B - D = B \cap (-D) \in \mathscr{F}_\sigma,$$
we have $K \subset E$, and
$$\mu(E - K) = \mu(E) - \mu(K) = \mu(H) - \mu(B - D)$$
$$= \mu(H) - \mu(B) + \mu(B \cap D) \leq \mu(H) - \mu(B) + \mu(D) = 0.$$
Thus, each condition is necessary if $\mu(E) < \infty$.

In the general case, we let C_n be a measurable cover for A_n and set $E_n = E \cap C_n$; then each set E_n fits the case already proved. If K_n is the $\mathscr{F}_\sigma$ set corresponding to E_n, we let
$$K = \bigcup_{n=1}^{\infty} K_n \in \mathscr{F}_\sigma,$$
and we have
$$\mu(E - K) \leq \sum_{n=1}^{\infty} \mu(E_n - K_n) = 0.$$
Thus 13.5.2 is necessary in all cases. Finally, if E is measurable, we have 13.5.2 for $-E$, and this is 13.5.1 for E. ∎

If there is a set $H \in \mathscr{G}_\delta$ such that $H \supset E$ and $\mu^*(H - E) = 0$, we should suspect that for every $\varepsilon > 0$ there would be an open set $G \supset E$ such that $\mu^*(G - E) < \varepsilon$. This, however, is not the case in general. It may happen (as in Section 16) that there are no open sets of finite measure. Nevertheless, we can obtain the following result.

Theorem 13.6. *If μ^* is a metric outer measure, if $\Omega = \bigcup_{n=1}^{\infty} B_n$ where, for each n, $B_n \in \mathscr{G}$ and $\mu(B) < \infty$, and if $E \subset H \in \mathscr{G}_\delta$ with $\mu^*(H - E) = 0$, then given $\varepsilon > 0$, there is an open set $G \supset E$ such that $\mu^*(G - E) < \varepsilon$.*

Proof. For each n, let $E_n = E \cap B_n$ and $H_n = H \cap B_n$. Then, $H_n \in \mathscr{G}_\delta$; so
$$H_n = \bigcap_{k=1}^{\infty} G'_{nk},$$
where $G'_{nk} \in \mathscr{G}$. If we let
$$G_{nk} = \bigcap_{i=1}^{k} G'_{ni} \cap B_n,$$
we have, for each n, a contracting sequence G_n with $G_{nk} \in \mathscr{G}$, $G_{nk} \subset B_n$ and $H_n = \lim_k G_{nk}$; so, the G_{nk} being measurable, 10.8.2 applies, and
$$\mu(H_n) = \lim_k \mu(G_{nk}).$$
Since these are measurable sets of finite measure,
$$\lim_k \mu(G_{nk} - H_n) = 0.$$

Thus, given $\varepsilon > 0$, we have for each n an index k_n such that

$$\mu(G_{nk_n} - H_n) < \frac{\varepsilon}{2^n}.$$

Letting

$$G = \bigcup_{n=1}^{\infty} G_{nk_n} \in \mathcal{G},$$

we have

$$\mu(G - H) \leq \sum_{n=1}^{\infty} \mu(G_{nk_n} - H_n) < \varepsilon;$$

so

$$\mu^*(G - E) \leq \mu(G - H) + \mu^*(H - E) < \varepsilon. \quad \blacksquare$$

Theorem 13.7. *If in 13.5, the sets A_n are open sets, then each of the following conditions† is necessary and sufficient for a set E to be measurable:*

13.7.1 *Given $\varepsilon > 0$, there exists an open set $G \supset E$ such that $\mu^*(G - E) < \varepsilon$.*

13.7.2 *Given $\varepsilon > 0$, there exists a closed set $F \subset E$ such that $\mu^*(E - F) < \varepsilon$.*

Proof. If we prove that 13.7.1 is necessary and sufficient, the result follows for 13.7.2 by taking complements.

Suppose E is measurable; then 13.5.1 applies, and by 13.6, 13.7.1 applies, so 13.7.1 is necessary.

Assuming 13.7.1, let $E \subset G_n \in \mathcal{G}$ with $\mu^*(G_n - E) < 1/n$, and let

$$H = \bigcap_{n=1}^{\infty} G_n;$$

then $E \subset H \in \mathcal{G}_\delta$, and

$$\mu^*(H - E) \leq \mu^*(G_n - E) < \frac{1}{n}$$

for each n. That is, $\mu^*(H - E) = 0$. This is condition 13.5.1; so E is measurable. $\blacksquare$

EXERCISES

g. Let $\mu_n^*(n = 1, 2, 3, \ldots)$ and μ_0^* be a set of outer measures involved in Method II. Let $\mathcal{M}_n(n = 1, 2, 3, \ldots)$ and $\mathcal{M}_0$ be the corresponding classes of measurable sets. Show that

$$\mathcal{M}_0 \supset \varlimsup_n \mathcal{M}_n.$$

h. Prove that if μ_0^* is constructed by Method II, if all the sets of $\mathcal{C}$ are Borel sets, and if Ω is the union of a countable class of sets for each of which μ_0^* is finite, then every measurable set is the union of a Borel set and a subset of a Borel set of measure zero.

---

† Sometimes a measure with the property that every set E in its domain satisfies 13.7.1 and 13.7.2 is called a *regular measure*. This should not be confused with a regular outer measure.

i. Show that the restriction $\Omega = \bigcup_{n=1}^{\infty} A_n$, where $\mu_0^*(A_n) < \infty$, is essential in Exercise h. *Hint*: Let Ω be the real number system; let $\mathscr{C}$ be the class of one-point sets and open intervals; let $\tau = 1$ on the one-point sets, $\tau = \infty$ on the open intervals. Assuming that there are non-Borel sets, show that the result in Exercise *h* fails. For validation of this assumption, see Section 18.

j. Show that the example of Exercise *d* satisfies the hypotheses of 13.5, but the result in 13.6 does not hold. Thus, it is essential in 13.7 that the A_n be open sets; not even $\mathscr{G}_\delta$ sets will suffice.

There is one final consideration in connection with the construction of outer measures that warrants our attention. The reasons for wanting an outer measure to satisfy M–IV are obvious, but it would seem that in order to achieve that end we must employ Method II. It would be desirable to have a condition under which the simpler Method I yielded metric outer measures. This occurs, of course, when Method II reduces to Method I, and a criterion for this phenomenon is given by the following theorem.

Theorem 13.8. *Let $\mathscr{C}$ be a sequential covering class such that for each n, $\mathscr{C}_n$ is a sequential covering class; let τ be a nonnegative function on $\mathscr{C}$ such that $\tau(\phi) = 0$ and such that, given any $E_0 \in \mathscr{C}$, any $\varepsilon > 0$, and any positive integer n, there exists a sequence E of sets from $\mathscr{C}_n$ such that $\bigcup_{k=1}^{\infty} E_k \supset E_0$ and*

$$\sum_{k=1}^{\infty} \tau(E_k) \leq \tau(E_0) + \varepsilon.$$

Then, Method I yields a metric outer measure.

Proof. Let us adopt the following notation: μ^* will be the outer measure coming from $\mathscr{C}$ and τ by Method I; μ_n^* will be the outer measure coming from $\mathscr{C}_n$ and τ by Method I; μ_0^* will be the outer measure constructed by Method II. Since $\mathscr{C} \supset \mathscr{C}_n$, we have that for every n,

$$\mu^* \leq \mu_n^* \leq \mu_0^*.$$

We shall show that, subject to the present hypothesis, $\mu_n^* \leq \mu^*$ for every n; thus $\mu_n^* \equiv \mu^*$, and it follows that $\mu_0^* = \lim_n \mu_n^* = \mu^*$. Since μ_0^* is known to be a metric outer measure, this will prove the theorem.

Given any set A and given $\varepsilon > 0$, there is a sequence E of sets from $\mathscr{C}$ such that $\bigcup_{i=1}^{\infty} E_i \supset A$ and

$$\sum_{i=1}^{\infty} \tau(E_i) \leq \mu^*(A) + \frac{\varepsilon}{2}.$$

However, according to our present hypothesis, given any n, each set E_i is covered by a sequence B_i of sets from $\mathscr{C}_n$ such that

$$\sum_{k=1}^{\infty} \tau(B_{ik}) \leq \tau(E_i) + \frac{\varepsilon}{2^{i+1}};$$

so (since the double sequence B covers A),

$$\mu_n^*(A) \le \sum_{i=1}^{\infty} \sum_{k=1}^{\infty} \tau(B_{ik}) \le \sum_{i=1}^{\infty} \tau(E_i) + \frac{\varepsilon}{2} \le \mu^*(A) + \varepsilon.$$

Since ε is arbitrary, $\mu_n^*(A) \le \mu^*(A)$. ∎

EXERCISES

k. Use 13.8 to show that Lebesgue outer measure is a metric outer measure. Show then that Lebesgue outer measure is regular and has all the properties listed in 13.5 and 13.7.

l. Show that the Cantor ternary set (Section 4) has Lebesgue measure zero. *Hint:* Compute the measure of its complement with respect to $[0, 1]$.

m. By changing the lengths of the extracted intervals, construct a nowhere dense perfect set having Lebesgue measure $\frac{1}{2}$.

n. Generalize Exercise *m*. Show that for every $\varepsilon > 0$, there exists a perfect set which is nowhere dense in $[0, 1]$ and has Lebesgue measure greater than $1 - \varepsilon$.

o. Consider a union of sets from Exercise *n* to show that there is a set of Lebesgue measure zero which is of Cat. II in $[0, 1]$.

p. Show that the sets of Lebesgue measure zero and their complements form a completely additive class.

q. Show that the union of two completely additive classes need not be an additive class. [*Hint:* Use the class of Exercise *p* and that of Exercise 9.*i*.]

REFERENCES FOR FURTHER STUDY

On measures and outer measures in general:
  Carathéodory [4], [5]
  Hahn and Rosenthal [9]
  Halmos [11]
  Hewitt and Stromberg [13]
  von Neumann [23]
  Saks [27]

On Lebesgue measure in R_1:
  Hobson [14]
  Lebesgue [19]
  Natanson [22]
  Royden [25]
  Titchmarsh [28]
  de la Vallée Poussin [29]

CHAPTER 3

MEASURE—SPECIFIC EXAMPLES

By way of illustrating various points in the theory, we have given in Chapter 2 (particularly in the exercises) a number of examples of measure functions. Each of these is interesting for first one reason and then another, but only two of them are of any real practical importance. The discrete probability measure described in Section 10 is certainly one of the important examples. It has a rather simple structure, however, and its properties should be fairly obvious. Lebesgue measure on the real line, described at the end of Section 11, is probably the most-used example of a completely additive measure. It will appear again in this chapter as a special case of a Lebesgue-Stieltjes measure, but the emphasis here is on the more general case. If, at this stage, the reader does not have a fair picture of Lebesgue measure and its properties, he should return to Exercises 11.w, 11.x, and 13.k through 13.q.

In this chapter we want to describe some of the more complicated examples of outer measures that have been found useful in various investigations. In each case, we shall base our description on the general theory developed in Chapter 2. That is, we shall describe a sequential covering class $\mathscr{C}$ and a nonnegative function τ and then call for either Method I or Method II. The important properties of these measures will then follow from the appropriate theorems in Chapter 2.

14 LEBESGUE-STIELTJES MEASURES

Lebesgue measure is described by weighting the intervals on the real line according to their lengths. In Exercise 10.q we derived from a point function f a weighting of the intervals according to something other than their lengths. This is the basic idea behind the construction of a Lebesgue-Stieltjes outer measure. Let us proceed to the details of this construction.

Let f be an everywhere finite, nondecreasing, real valued function whose domain is the real line R_1. In addition, let f be right continuous at every point; that is,

$$\lim_{x \to y^+} f(x) = f(y).$$

Such one-sided limits always exist for monotone functions, so this last restriction serves only to fix the function values at the points of discontinuity. It does not materially restrict the class of functions under consideration.

For each half-open interval $(a, b]$, we define
$$\tau_0\{(a, b]\} = f(b) - f(a).$$
Next, we define a function τ on the class consisting of ϕ and the open intervals by setting $\tau(\phi) = 0$ and
$$\tau\{(a, b)\} = \tau_0\{(a, b]\}.$$
Letting $\mathscr{C}$ be the class consisting of ϕ and the open intervals, and letting τ be the function just defined, we employ Method I to construct an outer measure which we shall designate by μ_f^*. We say that μ_f^* is the *Lebesgue-Stieltjes outer measure induced by f* and that f is a† *distribution function* for μ_f^*.

Since we have used Method I, we must check the hypotheses of 13.8 to show that μ_f^* is a metric outer measure. Accordingly, let (a, b) be any open interval, and let n be any positive integer. We form a partition of (a, b) by points x_i such that
$$a = x_0 < x_1 < \cdots < x_m = b$$
and
$$x_i - x_{i-1} < 1/n \quad (i = 1, 2, \ldots, m).$$
Now,
$$\bigcup_{i=1}^{m} (x_{i-1}, x_i] = (a, b] \supset (a, b),$$
and
$$\sum_{i=1}^{m} \tau_0\{(x_{i-1}, x_i]\} = f(x_m) - f(x_0) = \tau\{(a, b)\}. \tag{1}$$
The open intervals (x_{i-1}, x_i) do not cover (a, b), but we can extend each of them to the right to a point x_i' so that
$$\bigcup_{i=1}^{m} (x_{i-1}, x_i') \supset (a, b),$$
and still have
$$x_i' - x_{i-1} \leq 1/n \quad (i = 1, 2, \ldots, m).$$
Furthermore, using right continuity of f, given $\varepsilon > 0$, we can choose the points x_i' so that for each i $(i = 1, 2, \ldots, m)$
$$\tau\{(x_{i-1}, x_i')\} = f(x_i') - f(x_{i-1}) < f(x_i) - f(x_{i-1}) + \frac{\varepsilon}{m}$$
$$= \tau_0\{(x_{i-1}, x_i]\} + \frac{\varepsilon}{m}.$$
Then, we have from (1) that
$$\sum_{i=1}^{m} \tau\{(x_{i-1}, x_i')\} < \sum_{i=1}^{m} \left[\tau_0\{(x_{i-1}, x_i]\} + \frac{\varepsilon}{m}\right] = \tau\{(a, b)\} + \varepsilon.$$

---

† We say "a," not "the," because f + constant is also a distribution function for μ_f^*.

This is the criterion called for in 13.8, so μ_f^* is a metric outer measure. Thus, open intervals are measurable, and so μ_f^* is regular. Finally, every open interval has finite measure, and R_1 can be covered by a countable class of open intervals. Hence all the results of 13.5 and 13.7 apply to μ_f^*.

We want to use the measure function μ_f in later developments, but we might pause here to see something of the significance of the distribution function f. Suppose a half-open interval $(a, b]$ is covered by open intervals (a_i, b_i); that is,

$$(a, b] \subset \bigcup_{i=1}^{\infty} (a_i, b_i).$$

By right continuity of f, given $\varepsilon > 0$, there exists $\delta > 0$ such that $a + \delta < b$ and

$$f(a + \delta) < f(a) + \varepsilon.$$

By the Heine-Borel theorem, the closed interval $[a + \delta, b]$ is covered by a finite set $(a_{i_1}, b_{i_1}), \ldots, (a_{i_m}, b_{i_m})$ of the given intervals (a_i, b_i). Clearly, for this finite covering

$$a_{i_1} < a + \delta,$$
$$b_{i_m} > b,$$

and the intervals may be so numbered that for each k ($k = 1, 2, \ldots, m$),

$$a_{i_{k+1}} < b_{i_k}.$$

Since f is nondecreasing, we have

$$\sum_{i=1}^{\infty} \tau\{(a_i, b_i)\} \geq \sum_{k=1}^{m} \tau\{(a_{i_k}, b_{i_k})\} = \sum_{k=1}^{m} [f(b_{i_k}) - f(a_{i_k})]$$
$$= f(b_{i_m}) - f(a_{i_1}) + \sum_{k=1}^{m-1} [f(b_{i_k}) - f(a_{i_{k+1}})]$$
$$\geq f(b) - f(a + \delta) > f(b) - f(a) - \varepsilon.$$

Since this holds for every countable covering of $(a, b]$ by open intervals and for every $\varepsilon > 0$, it follows that

$$\mu_f\{(a, b]\} \geq f(b) - f(a).$$

Because of the right continuity of f, given $\varepsilon > 0$, there is a $\delta > 0$ such that

$$(a, b + \delta) \supset (a, b]$$

and

$$\tau\{(a, b + \delta)\} = f(b + \delta) - f(a) < f(b) - f(a) + \varepsilon;$$

so

$$\mu_f\{(a, b]\} < f(b) - f(a) + \varepsilon$$

for every $\varepsilon > 0$. Thus, $\mu_f\{(a, b]\} \leq f(b) - f(a)$. This, combined with the previous result, gives

$$\mu_f\{(a, b]\} = f(b) - f(a).$$

That is, if f is a distribution function, then $f(b) - f(a)$ gives the measure of the half-open interval $(a, b]$.

This shows us how to construct a distribution function to represent a given physical situation. Suppose we want to construct a measure to describe a given mass distribution on the real line. We choose a distribution function f so that $f(b) - f(a)$ is the mass of the half-open interval $(a, b]$. In particular, if the mass of the whole line is finite, we can accomplish this by letting $f(x)$ equal the total mass at and to the left of x. For the infinite case, we can set $f(0) = 0$. Then for $x > 0$, we set $f(x)$ equal to the mass of $(0, x]$; for $x < 0$, we set $f(x)$ equal to minus the mass of $(x, 0]$. With the distribution function f thus defined, we proceed to construct the measure μ_f. For the half-open intervals, μ_f will have values equal to the masses originally assumed for these sets; furthermore, μ_f will assign, in a quite natural manner, a mass to every measurable set.

If we want to describe a one-dimensional electrostatic field in which there are both positive and negative charges, we construct, as above, a measure μ_f from the positive charges and a measure μ_g from the absolute values of the negative charges. Then, if either the total positive charge or the total negative charge is finite, $\mu_f - \mu_g$ will be a completely additive set function which describes the field quite satisfactorily.

More generally, if f is any right continuous function of bounded variation defined on R_1, we can take its Jordan decomposition $f = f_1 - f_2$ (see Exercise 10.u) and construct measures μ_{f_1} and μ_{f_2}. The completely additive set function $\mu_{f_1} - \mu_{f_2}$ is usually called the Lebesgue-Stieltjes measure induced by f, though unless f is monotone, it is not, strictly speaking, a measure in the sense in which we are using the word.

A continuous mass distribution in R_1 is one for which the distribution function f is continuous. It is easily seen that in such a case μ_f vanishes on every one-point set. Since f is monotone, its only discontinuities are "jumps":

$$\lim_{x \to a^-} f(x) < \lim_{x \to a^+} f(x).$$

Such jumps represent what are called mass points. If E consists of the single mass point a, then

$$\mu_f(E) = \lim_{x \to a^+} f(x) - \lim_{x \to a^-} f(x) > 0.$$

EXERCISES

a. Let f be defined by

$$f(x) = \begin{cases} 0 & \text{for } x < 0, \\ 1 & \text{for } x \geq 0. \end{cases}$$

Show that $\mu_f\{(-1, 0)\} < f(0) - f(-1)$. [*Hint:* $(-1, 0) = \bigcup_{n=1}^{\infty} (-1, -1/n)$.]

b. Give an example of a right continuous, monotone distribution function f and a pair of points a and b such that

$$\mu_f\{(a, b)\} < f(b) - f(a) < \mu_f\{[a, b]\}.$$

c. Give an example of a right continuous, monotone distribution function f such that for one interval

$$\mu_f\{[a, b)\} < f(b) - f(a),$$

while for another interval

$$\mu_f\{[c, d)\} > f(d) - f(c).$$

d. Write out the details for the development of a Lebesgue-Stieltjes outer measure from a left continuous, monotone function f. In this case $f(b) - f(a)$ gives the measure of what set?

e. Show that no matter which type of one-sided continuity we assume for f, we always have

$$\mu_f\{(a, b)\} \leq f(b) - f(a) \leq \mu_f\{[a, b]\}.$$

f. Show by an example that if the one-sided continuity requirement of f is dropped, there may be a pair of points a and b such that neither of the half-open intervals determined by a and b has μ_f-measure $f(b) - f(a)$.

g. Let f be defined by $f(x) = x$. Show that μ_f^* is identically equal to Lebesgue outer measure.

h. Take a uniform mass distribution in R_1 (mass of an interval proportional to its length), and add a unit mass point at the origin. Describe this distribution by a distribution function f.

i. If f describes a mass distribution, what is the physical significance of an interval in which f is constant?

j. Take an electrostatic field in which there is a uniform positive charge on the interval $[0, 10]$. Add unit point charges at each of the integer points, those on the odd integers being positive and those on the even integers being negative. Define a function of bounded variation whose induced Lebesgue-Stieltjes measure will describe this field.

k. Describe a Lebesgue-Stieltjes measure on R_1 which will generate a discrete probability measure on the set of positive integers.

For the construction of Lebesgue-Stieltjes measures in Euclidean n-space, R_n, the basic ideas are the same as in the case of R_1. Only the details are more complicated. A point $x \in R_n$ is an n-tuple $(x_1, x_2, \ldots, x_n)$ of real numbers. An *open interval* in R_n is a set of the form

$$\{x \mid a_i < x_i < b_i; i = 1, 2, \ldots, n\}.$$

Let us designate such an interval by (a, b). In R_1, (a, b) is an open interval as we have previously used the term; in R_2 it is the interior of a rectangle; in R_3, the interior of a rectangular parallelepiped. Every such interval is an open set in R_n. We shall also be interested in *half-open intervals*:

$$(a, b] = \{x \mid a_i < x_i \leq b_i; i = 1, 2, \ldots, n\}.$$

Half-open intervals may be fitted together in such a way that, even though they are disjoint, there are no points between them. Thus, any half-open interval is the union of a finite number of disjoint half-open intervals with arbitrarily small diameters. We now follow a procedure analogous to that which we employed in the one-dimensional case. We seek a nonnegative function τ_0 defined on the half-open intervals such that if

$$I = \bigcup_{k=1}^{m} I_k$$

expresses the half-open interval I as the union of disjoint half-open intervals I_k, then

$$\tau_0(I) = \sum_{k=1}^{m} \tau_0(I_k).$$

We also want τ_0 to have the property that it is increased by an arbitrarily small amount by a sufficiently small outward movement of the closed faces of the interval. If τ_0 has these properties, we define τ on the open intervals by

$$\tau\{(a, b)\} = \tau_0\{(a, b]\}.$$

We then let $\mathscr{C}$ be the class consisting of ϕ together with the open intervals and apply Method I to construct an outer measure. With τ_0 having the properties listed, the argument to show that 13.8 applies is completely analogous to that given for the one-dimensional case. Thus, we get a regular, metric outer measure to which 13.5 and 13.7 apply.

The problem, then, is to construct functions τ_0 on the half-open intervals so that they exhibit the properties listed above. The simplest such function is the one that gives *Lebesgue n-dimensional measure*:

$$\tau_0\{(a, b]\} = \prod_{i=1}^{n} (b_i - a_i).$$

In this case, we associate with each interval its n-dimensional "volume." In one dimension this function τ_0 gives the length of the interval; in two, its area; in three, its volume.

Other suitable functions τ_0, and consequently other measures, may be constructed from real valued functions of n real variables. In order to see what we are doing, let us think momentarily of a mass distribution in which the total mass is finite, and let $f(a_1, a_2, \ldots, a_n)$ represent the mass of the semi-infinite interval

$$(-\infty, a] = \{x \mid x_i \leq a_i; i = 1, 2, \ldots, n\}.$$

Half-open intervals can be constructed by subtracting semi-infinite intervals one from the other; so we shall construct τ_0 by subtracting the corresponding values of f one from the other. The two-dimensional case illustrates all the points involved, and in that case we can draw a picture (see Fig. 3). For each of the points a, b, c, d

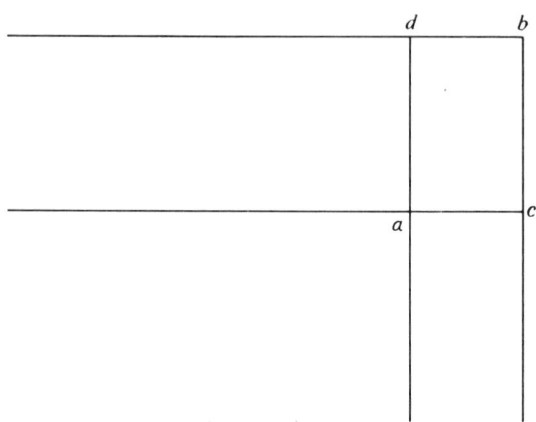

Figure 3

in Fig. 3, let the corresponding capital letter stand for the closed quadrant to the left of and below that point. Then,

$$(a, b] = B - D - (C - A).$$

If a has coordinates (a_1, a_2) and b has coordinates (b_1, b_2), then c has coordinates (b_1, a_2) and d has coordinates (a_1, b_2). So, we set

14.1 $\quad \tau_0\{(a, b])\} = f(b_1, b_2) - f(a_1, b_2) - [f(b_1, a_2) - f(a_1, a_2)].$

It is clear from Fig. 3 that we could describe $(a, b]$ equally well as $B - C - (D - A)$. Subtracting masses in this order gives the same value for τ_0 as does 14.1.

The general case may be treated as follows. For any interval $(a, b]$ we define a set of difference operators $\delta_1, \delta_2, \ldots, \delta_n$. Specifically, if g is a function of x_k (and perhaps others of the coordinate variables), we define

$$\delta_k(g) = g(\ldots, b_k, \ldots) - g(\ldots, a_k, \ldots);$$

that is, the operation δ_k consists of subtracting the values obtained by shifting x_k and leaving the other arguments alone. In this notation, τ_0 is very easily defined:

14.2 $\quad \tau_0\{(a, b]\} = \delta_1(\delta_2(\cdots(\delta_n(f)))\cdots).$

The operators δ_k may be applied in any order, for if g depends on x_j and x_k, the relation

$$\delta_j(\delta_k(g)) = \delta_k(\delta_j(g))$$

amounts to no more than changing the order of subtraction in 14.1. It then follows by induction that any permutation of the δ's in 14.2 leaves τ_0 unchanged. So, τ_0 depends on the coordinate system in R_n (a rotation, for example, changes the whole concept of interval), yet it is independent of the numbering of the coordinate axes. Thus, for example, in three dimensions a shift from a left-handed to a right-handed coordinate system with a corresponding shift in the geometric significance of the arguments of f leaves 14.2 unchanged.

The formula 14.2 will construct a function of intervals from any function of points, but we want a function of intervals that will lead to a metric outer measure by Method I. We have already mentioned the properties of τ_0 that will bring this about, but now we should like to describe the situation in terms of properties of f.

First, τ_0 must be nonnegative. Monotonicity of f will not guarantee this as it does in the one-dimensional case. For example, let f be defined by

$$f(x_1, x_2) = \begin{cases} x_1 + x_2 & \text{for } x_1 + x_2 < 0, \\ 0 & \text{for } x_1 + x_2 \geq 0. \end{cases}$$

Here, f is monotone in each variable separately, but for any interval with its diagonal on the line $x_1 + x_2 = 0$, τ_0 is negative. On the other hand, f need not be monotone. For example, let f be defined by

$$f(x_1, x_2) = \begin{cases} x_2 & \text{for } x_1 \geq 0, \\ 0 & \text{for } x_1 < 0. \end{cases}$$

Here, τ_0 is nonnegative for every interval; but for $x_2 > 0$, f is nondecreasing as a function of x_1, while for $x_2 < 0$, f is nonincreasing as a function of x_1. Thus, the best plan seems to be merely to require:

14.3. *The function f is such that 14.2 yields a nonnegative function τ_0.*

Second, τ_0 must be additive for every decomposition of a half-open interval into a finite number of disjoint half-open intervals. This follows automatically from 14.2 with no further restriction on f. If $I = I_1 \cup I_2$, the values of f at the corner points of the common face appear with one set of signs to form $\tau_0(I_1)$ and with exactly the opposite set to form $\tau_0(I_2)$. Thus, they cancel to give

$$\tau_0(I) = \tau_0(I_1) + \tau_0(I_2).$$

Repeated application of this result for two intervals will give the general result.

Third, a small outward movement of the closed faces of an interval must give a small increase in τ_0. An outward movement of the closed faces of the interval $(a, b]$ means an increase in the coordinates of b, so a right continuity condition seems called for. At first, we might think (since the whole face is carried out) that this continuity in each variable would have to be uniform in the other variables; but if we note that τ_0 is defined in terms of only a finite number of values of f, it appears that the following condition is sufficient to guarantee that τ satisfies 13.8:

14.4. *f is right continuous in each variable separately.*

To sum up, we refer to a finite, real valued function f of n real variables as an *n-dimensional distribution function* if it satisfies 14.3 and 14.4. From such a function we define τ_0 on the half-open intervals by 14.2. We then set

$$\tau\{(a, b)\} = \tau_0\{(a, b]\}.$$

Letting $\mathscr{C}$ be the class consisting of ϕ and the open intervals, we apply Method I to define μ_f^*, the *Lebesgue-Stieltjes outer measure induced by f*. $\mathscr{C}$ and τ satisfy 13.8, so μ_f^* is a regular, metric outer measure satisfying 13.5 and 13.7.

An important case of a Lebesgue-Stieltjes measure in n-space is that in which the distribution function f is the product of n functions, each depending on one of the coordinate variables; that is,

$$f(x_1, x_2, \ldots, x_n) = \prod_{i=1}^{n} f_i(x_i).$$

In such a case, μ_f is called a *product measure*. In particular, if

$$f(x_1, x_2, \ldots, x_n) = \prod_{i=1}^{n} x_i,$$

the resulting product measure is Lebesgue n-dimensional measure.

EXERCISES

l. Show that if f and g are n-dimensional distribution functions, then $f + g$ is a distribution function and $\mu_{f+g} = \mu_f + \mu_g$.

m. Let f be defined over R_2 by

$$f(x_1, x_2) = \begin{cases} 1 & \text{for } x_1 \geq 0 \text{ and } x_2 \geq 0, \\ 0 & \text{otherwise}. \end{cases}$$

Show that f is a distribution function describing a unit mass point at the origin.

n. Give distribution functions describing mass distributions in R_2 consisting of unit mass points: (1) at $(0, 0)$ and $(1, 1)$; (2) at $(0, 0)$ and $(1, 0)$; (3) at $(0, 0)$ and $(1, -1)$.

o. Give distribution functions describing uniform mass distributions with unit mass points added at the points given in Exercise *n*.

p. Let f be defined over R_2 by

$$f(x_1, x_2) = \begin{cases} x_1\sqrt{2} & \text{for } x_2 > x_1, \\ x_2\sqrt{2} & \text{for } x_2 \leq x_1. \end{cases}$$

Show that f is a distribution function describing a mass distribution in which all the mass is on the line $x_1 = x_2$ and the mass of a segment of that line is equal to its length.

q. Give a distribution function for a mass distribution in R_2 in which all the mass is uniformly distributed (1) on the line $x_1 = -x_2$, (2) on the line $x_1 = 0$. Note that one of these distribution functions is continuous while the other one is not.

r. In terms of values of f, write out 14.2 for three dimensions. Draw a picture and check the result.

s. In terms of values of f, write out 14.2 for four dimensions.

t. Write out the proof that if f satisfies 14.3 and 14.4, then 14.2 leads to a function τ which satisfies 13.8.

u. Show that if f satisfies 14.3 and 14.4, then
$$\mu_f\{(a, b]\} = \tau_0\{(a, b]\}.$$

v. Show that if
$$f(x_1, x_2, \ldots, x_n) = \prod_{i=1}^{n} f_i(x_i),$$
then
$$\tau_0\{(a, b]\} = \prod_{i=1}^{n} [f_i(b_i) - f_i(a_i)].$$

w. Show that if each f_i ($i = 1, 2, \ldots, n$) is nondecreasing and right continuous and if f is defined by
$$f(x_1, x_2, \ldots, x_n) = \prod_{i=1}^{n} f_i(x_i),$$
then f satisfies 14.3 and 14.4.

x. Give distribution functions describing mass distributions in R_3: (1) consisting of a single mass point, (2) consisting of a uniform distribution on a coordinate axis, (3) consisting of a uniform distribution in a coordinate plane. Repeat parts (2) and (3) with tilted lines and planes.

In developing the construction methods for Lebesgue-Stieltjes outer measures, we pointed out only that each new step introduced was sufficient to produce the end we had in mind. We conclude this section with a proof that our method of construction is completely general. First, we need a result of general interest concerning open sets and half-open intervals.

Theorem 14.5. *Every open set in R_n is the union of a countable class of disjoint half-open intervals.*

Proof. Let G be an open set in R_n. For each positive integer k, the hyperplanes

$$x_i = \frac{m}{2^k} \quad (m = \ldots, -1, 0, 1, 2, \ldots; i = 1, 2, \ldots, n) \tag{1}$$

partition R_n into a countable class of disjoint half-open intervals. Let $I_1^1, I_1^2, I_1^3, \ldots$ be the class of intervals generated by (1) for $k = 1$ which are contained in G. For each $k > 1$, let $I_k^1, I_k^2, I_k^3, \ldots$ be the class of intervals generated by (1) which are contained in G but not contained in any interval I_q^p with $q < k$. If $x \in G$, then x is an interior point of G; so there is a partition of R_n given by (1) such that the interval containing x is contained in G. Thus,

$$\bigcup_{k=1}^{\infty} \bigcup_{j} I_k^j \supset G.$$

Since $I_k^j \subset G$ for each j, k, we have

$$\bigcup_{k=1}^{\infty} \bigcup_{j} I_k^j \subset G.$$

This is clearly a countable class of intervals, and we have constructed them so that they are disjoint. ∎

Theorem 14.6. *If $\mathscr{C}$ is the class consisting of ϕ and the open intervals in R_n, if τ is any nonnegative function defined on $\mathscr{C}$ and vanishing on ϕ, and if the outer measure μ^* constructed from $\mathscr{C}$ and τ by Method I is a metric outer measure with $\mu^*\{(a, b]\} < \infty$ for every half-open interval $(a, b]$, then there is a distribution function f such that $\mu^* = \mu_f^*$.*

Proof. Let f be identically zero on the coordinate hyperplanes $x_k = 0$. An interval with one corner at the origin has each of its corners except one in some coordinate hyperplane. Thus, if τ_0 is given for all such intervals, 14.2 gives the value of f at the corner point off the coordinate hyperplanes. For every half-open interval I, we define

$$\tau_0(I) = \mu(I). \tag{1}$$

14.2 then defines f everywhere, and it is easily checked (using additivity of μ) that the reapplication of 14.2 gives values of τ_0 consistent with (1).

The condition 14.3 for f follows from the fact that μ is nonnegative. Right continuity of f follows from 10.3.1 applied to μ.

For Lebesgue-Stieltjes measures, $\mu_f = \tau_0$ on the half-open intervals; thus for half-open intervals I, $\mu_f(I) = \mu(I)$ because of (1). Using 14.5 and the complete additivity of both μ_f and μ, we have that $\mu_f(G) = \mu(G)$ for any open set G. The desired result now follows by 11.4. ∎

EXERCISES

y. Prove the following theorem, similar to 14.6 but showing the generality of Lebesgue-Stieltjes measure from a different viewpoint: If μ^* is a metric outer measure in R_n such that every set has a measurable cover of class $\mathscr{G}_\delta$ and such that $\mu^*\{(a, b]\} < \infty$ for every half-open interval $(a, b]$, then there exists a distribution function f such that $\mu^* = \mu_f^*$. [*Hint:* Carry out the steps in 14.6; take limits and measurable covers to show that the measures μ and μ_f are identical; then use 12.4.5 to show that the outer measures are identical.]

z. The outer measure of Exercise 13.d is not a Lebesgue-Stieltjes outer measure. Why?

*15 PROBABILITY

A completely general description of probability is given by saying that we take a space Ω (of any kind) and define an outer measure in it for which $\mu(\Omega) = 1$. We then define an *event* as a measurable set and the *probability of an event* E as $\mu(E)$.

Suppose we are given an experiment whose results are described by a single numerical measurement. Let Ω be the real number system, and let the event (point set) E have the physical interpretation, "The result was a number from the set E." Then, a measure in Ω defines probabilities for various physical events, namely, for all physical events represented by measurable sets. If we know anything about the experiment, we should like to have these probabilities coincide with our conception of the likelihood of occurrence of the various physical events; and if our ideas about these likelihoods are sufficiently well crystallized, we can construct a Lebesgue-Stieltjes measure which will bring this about. In particular, if for each $x \in \Omega$, we decide on the value of the probability that the result is less than or equal to x and assign $f(x)$ this value, then the measure μ_f will give probabilities consistent with our assumptions.

Suppose we consider n experiments. (One common case is n repetitions of the same experiment.) A result in this situation consists of n measurements, and it is natural to represent such physical results by points in R_n—each coordinate giving the result of a single experiment. An event (measurable set) $E \subset R_n$ has the physical interpretation, "The n measurements give coordinates for a point in E." A function f of n real variables such that $f(x_1, x_2, \ldots, x_n)$ gives the probability that for each i, the ith result is less than or equal to x_i is called a *joint distribution function*, and the Lebesgue-Stieltjes measure μ_f defines probabilities consistent with those assumed in setting up the function f. If μ_f is a product measure, we say the experiments are *independent*.

In many probability problems we want to consider an infinite sequence of experiments. Though it may seem so at first, such an infinite sequence is not necessarily a figment of the imagination. Suppose we consider the experiment of choosing a number at random between 0 and 1 and noting its nth decimal place. Then, a single random choice of a number determines the result for each of an infinite sequence of such experiments. More often, however, we are concerned with situations in which the actual performance of an infinite sequence of experiments is impossible; yet we still want to speak of the probability of a certain result if our experiments should be continued indefinitely. A practical sort of question would be this: If we know enough about the experiments and their interrelationships to assign a joint probability distribution to any space representing a finite set of results, can we construct a measure which will give probabilities for events involving infinite sequences of results?

Again we assume that the result of each experiment is a real number; then the space s (see Section 7) represents the set of all possible sequences of results. If $x \in s$, then x is a sequence of real numbers; conversely, if x is a sequence of real numbers, then $x \in s$. Let $i_1, i_2, \ldots, i_n$ be any finite set of positive integers, and let $a_{i_1}, a_{i_2}, \ldots, a_{i_n}$ and $b_{i_1}, b_{i_2}, \ldots, b_{i_n}$ be two sets of real numbers with $a_{i_k} < b_{i_k}$ ($k = 1, 2, \ldots, n$). A set $I \subset s$ of the form

$$I = \{x \mid a_{i_k} < x_{i_k} < b_{i_k}; k = 1, 2, \ldots, n\}$$

will be called an *open interval in s*. Any such interval is an open set in s. We define *half-open intervals* as sets of the form

$$\{x \mid a_{i_k} < x_{i_k} \leq b_{i_k}; k = 1, 2, \ldots, n\}.$$

Thus, an interval $I \subset s$ represents the physical event that the results of experiments $i_1, i_2, \ldots, i_n$ fall within certain specified bounds, with nothing at all said about the other results. This is precisely the kind of physical event for which we are assuming that the probabilities are known. Our procedure, then, will be to take this known probability for a half-open interval I and call it $\tau_0(I)$. We then transfer this function to the open intervals by setting $\tau(I^o) = \tau_0(I)$. Finally, from this function τ and the class $\mathscr{C}$ consisting of ϕ and the open intervals, we construct an outer measure μ^* on s by Method I.

This answers our question about probabilities for infinite sequences of results, but there are two points that we should check. First, we should show that $\mathscr{C}$ and τ satisfy 13.8 so that μ^* is a metric outer measure. Second, we should show that for half-open intervals I, $\mu^*(I) = \tau_0(I)$; that is, that the probability measure on s is consistent with the probabilities assumed in constructing it.

It follows from the nature of the distance function in s that given $\varepsilon > 0$, if we choose n_0 so that

$$\sum_{n=n_0}^{\infty} \frac{1}{2^n} < \frac{\varepsilon}{2}$$

and choose a_n and b_n so that

$$0 < b_n - a_n < \frac{\varepsilon}{2n_0} \quad (n = 1, 2, \ldots, n_0 - 1),$$

then

$$\{x \mid a_n < x_n \leq b_n; n = 1, 2, \ldots, n_0 - 1\}$$

is an interval of diameter less than ε. Let us call these ε-*intervals*.

To express an arbitrary half-open interval I as a union of disjoint ε-intervals, let us first take n_0 at least as large as the largest index used in defining I. Then we take each component that is restricted in the definition of I and partition the range of permissible values into sections of length less than $\varepsilon/2n_0$. For the unrestricted components with indices less than n_0, we partition the entire real number system into a countable class of such sections. The interval I is then the union of all the ε-intervals obtained by using one of these sections for each index less than n_0. To see this, we note that if $x \in I$, its first $n_0 - 1$ components lie each in one of the sections just constructed. Some ε-interval from the class in question is determined by exactly these sections, so x is in that ε-interval. Conversely, if x is in one of the ε-intervals just defined, its components for the indices determining I are properly restricted; thus $x \in I$. These ε-intervals are disjoint because if x belongs to one of them, then, given any other one of them, at least one component of x is in the wrong section. Finally, these ε-intervals form a countable class because there are a countable number of choices of sections for each index less than n_0, and n_0 is finite.

Let I be an arbitrary half-open interval; let n be any positive integer; and let

$$I = \bigcup_{k=1}^{\infty} I_k$$

express I as a union of disjoint half-open intervals, each of diameter less than $1/n$. If the intervals I_k are constructed as indicated above, they are all determined by indices from some one finite set; so τ_0 is defined for all of them from some fixed finite-dimensional probability measure. The I_k are Borel sets, so τ_0 is completely additive over them. Since τ_0 is the restriction of an everywhere finite measure function, it follows from 10.3.1 that given $\eta > 0$, we may expand each interval I_k to an interval J_k by increasing each upper bound on the components, and have

$$I \subset \bigcup_{k=1}^{\infty} J_k^o,$$

and also for each k, have $d(J_k) < 1/n$ and

$$\tau(J_k^o) = \tau_0(J_k) < \tau_0(I_k) + \frac{\eta}{2^k}.$$

Then

$$\sum_{k=1}^{\infty} \tau(J_k^o) < \sum_{k=1}^{\infty} \left[\tau_0(I_k) + \frac{\eta}{2^k}\right] = \tau_0(I) + \eta = \tau(I^o) + \eta,$$

so 13.8 is satisfied.

For half-open intervals I, the proof that $\mu(I) = \tau_0(I)$ is simpler in this case than it was for Lebesgue-Stieltjes measure, because we know here that τ_0 is the restriction of a measure function; so 10.7 applies. Let J be a half-open interval, and let I be a sequence of half-open intervals such that

$$J \subset \bigcup_{k=1}^{\infty} I_k^o;$$

then

$$\tau_0(J) \leq \sum_{k=1}^{\infty} \tau_0(I_k) = \sum_{k=1}^{\infty} \tau(I_k^o).$$

Since this is true for every such sequence I, we have

$$\tau_0(J) \leq \mu(J).$$

It follows from 10.3.1 that given $\varepsilon > 0$, there is a half-open interval K such that $K^o \supset J$ and

$$\tau(K^o) = \tau_0(K) < \tau_0(J) + \varepsilon.$$

Thus, $\mu(J) \leq \tau_0(J)$, and the equality is proved.

We have now accomplished the main purpose of this section, which was to describe a mathematical model in which probabilities can be assigned to events involving infinite sequences of experimental results. In conclusion, let us look at one interesting set in this space. If m, n, and k are three positive integers, then

$$\left\{x \mid |x_m - x_n| \geq \frac{1}{k}\right\}$$

is a closed set in s, because if x is a sequence of points in s with

$$\lim_j x_j = x_0,$$

then by 7.1,

$$\lim_j x_{j,m} = x_{0,m} \quad \text{and} \quad \lim_j x_{j,n} = x_{0,n};$$

so if $|x_{j,m} - x_{j,n}| \geq 1/k$ for each j, then $|x_{0,m} - x_{0,n}| \geq 1/k$. Thus, for each m, n, k,

$$\left\{ x \mid |x_m - x_n| < \frac{1}{k} \right\}$$

is open, and hence measurable. Now, the set of elements of s consisting of convergent sequences can be written as

15.1
$$\bigcap_{k=1}^{\infty} \bigcup_{n=1}^{\infty} \bigcap_{m=n}^{\infty} \left\{ x \mid |x_m - x_n| < \frac{1}{k} \right\}.$$

Hence, this set is measurable, and we can speak of the probability that the sequence of experimental results will be convergent.

The construction we have given here of a probability measure in a sequence space can be generalized to give a probability measure in more general function spaces. Probability in a space of functions defined on the real number system is called for in the study of continuous stochastic processes. The physical picture of such a process is usually that of a chance phenomenon depending on a time parameter. If $x(t)$ is the result at time t, then the function x represents the entire succession of results. We shall not pursue this subject any further except to say that there are additional difficulties here, not the least of which are measurability problems. For example, the set of functions x such that

$$\lim_{t \to \infty} x(t)$$

exists may be written in terms of unions and intersections of measurable sets as we described the corresponding set in the sequence space above. However, in the more general case there are noncountable unions and intersections involved, so measurability does not follow as it does for the set of convergent sequences.

EXERCISES

a. Show that the measure μ constructed in s is a true probability measure; that is, $\mu(s) = 1$. *Hint:*

$$s = \bigcup_{n=-\infty}^{\infty} \{x \mid n - 1 < x_1 \leq n\}.$$

b. Let μ_f be a product probability measure in R_2; let f_1 and f_2 be the factors of f; let

$$\lim_{x_1 \to -\infty} f_1(x_1) = \lim_{x_2 \to -\infty} f_2(x_2) = 0.$$

Show that
$$\lim_{x_1 \to \infty} f_1(x_1) = \lim_{x_2 \to \infty} f_2(x_2) = 1.$$

c. Show that if μ_f is a product probability measure in R_2 and if
$$E_1 = \{x \mid a_1 < x_1 \le b_1\} \quad \text{and} \quad E_2 = \{x \mid a_2 < x_2 \le b_2\},$$
then
$$\mu_f(E_1 \cap E_2) = \mu_f(E_1)\mu_f(E_2).$$

d. Generalize Exercise c. Let μ_f be a product probability measure in R_2 with $f(x_1, x_2) = f_1(x_1)f_2(x_2)$. Let A_1 be a μ_{f_1}-measurable set in R_1, and let A_2 be a μ_{f_2}-measurable set in R_1. Let $E_1 = \{x \mid x_1 \in A_1\}$ and $E_2 = \{x \mid x_2 \in A_2\}$. Show that
$$\mu_f(E_1 \cap E_2) = \mu_f(E_1)\mu_f(E_2).$$

e. A probability measure in s is called a *product measure* if each of the finite-dimensional measures used in its construction is a product measure. Let μ be a product measure in s, and let K_1 and K_2 be two disjoint, finite sets of indices. For $j = 1, 2$, let $A_j \subset s$ be a measurable set in the finite-dimensional space of components with indices in K_j, and for each $x \in s$ let x_{K_j} be the set of components of x with indices in K_j. Finally, let $E_j = \{x \mid x_{K_j} \in A_j\}$. Show that $\mu(E_1 \cap E_2) = \mu(E_1)\mu(E_2)$.

f. Let μ be a product probability measure in R_3, and let
$$E_1 = \{x \mid |x_1 - x_3| < \varepsilon\} \quad \text{and} \quad E_2 = \{x \mid |x_2 - x_3| < \varepsilon\},$$
where $\varepsilon > 0$. Show that $\mu(E_1 \cap E_2) = \mu(E_1)\mu(E_2)$. Note: This is not a special case of Exercise e because E_1 and E_2 are not defined from disjoint sets of components.

g. Show that if μ is a product probability measure in s and if c is the set of convergent sequences, then
$$\mu(c) = \lim_{k \to \infty} \lim_{n \to \infty} \prod_{m=n}^{\infty} \mu\left\{x \mid |x_m - x_n| < \frac{1}{k}\right\}.$$

h. Show that if μ is a product probability measure in s, if I is an interval in s, and if c is the set of convergent sequences, then $\mu(I \cap c) = \mu(I)\mu(c)$.

i. Prove the *zero-one law*: For a sequence of independent experiments (product measure in s), the probability of convergence of the sequence of results is either 0 or 1. *Hint:* If $\mu(c) > 0$, define
$$\mu_0(E) = \frac{\mu(E \cap c)}{\mu(c)}$$
for every measurable $E \subset s$. Note that $\mu_0(c) = 1$. By Exercise h, $\mu_0 = \mu$ on the intervals; use 14.5 and 10.8.1 to extend this to c.

*16 HAUSDORFF MEASURES

Let Ω be any separable metric space. Then, the class $\mathscr{G}$ of all open sets is a sequential covering class; furthermore, for each n, the class of all open sets with diameter less

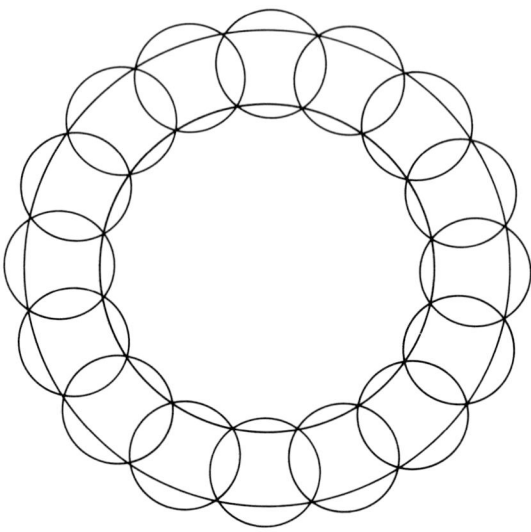

Figure 4

than or equal to $1/n$ is also a sequential covering class.† Let p be any nonnegative number, and for each nonvacuous $G \in \mathscr{G}$, let

$$\tau(G) = [d(G)]^p.$$

Let $\tau(\phi) = 0$. If we set $\mathscr{C} = \mathscr{G}$ and use this function τ, the conditions of 13.8 are not necessarily satisfied. For example, let $p = 1$ and let G be a ring in R_2 (see Fig. 4). The best that can be done to cover such a ring with open sets of small diameter is to use small circles such as those shown in Fig. 4, but the sum of the diameters of the circles is roughly π times that of the ring.

This is the first example we have encountered in which it is necessary to resort to Method II in order to construct a metric outer measure, but we do so now. For $\mathscr{C} = \mathscr{G}$ and τ defined as above, we call the metric outer measure constructed by Method II *Hausdorff p-dimensional outer measure*. We shall designate it by $\mu^{*(p)}$. In general, 13.7 does not apply to $\mu^{*(p)}$ (see Exercise e below); however, it follows from 13.4 that $\mu^{*(p)}$ is regular and, indeed, that every set has a measurable cover of class $\mathscr{G}_\delta$. Then from Exercise 12.b it follows that for sets on which $\mu^{*(p)}$ is finite, or for countable unions of such sets, 13.5.1 and 13.5.2 give necessary and sufficient conditions for measurability.

Hausdorff measure may be used to define dimensionality‡ of a set in any

---

† We make the requirement of separability in order to guarantee that these are sequential covering classes. In the nonseparable case we can carry out the construction given here for all sets with countable coverings and assign outer measure ∞ to all other sets.

‡ The definition of dimension given here differs from the usual topological one. For comments on the relation between these definitions of dimension, see Hurewicz and Wallman [15].

separable metric space. This is because a set can have finite, nonzero, p-dimensional measure for at most one value of p.

Theorem 16.1. *If $\mu^{*(p)}(E) < \infty$ and if $q > p$, then $\mu^{*(q)}(E) = 0$.*

Proof. Let n be any positive integer, and let G be a sequence of open sets covering E, with $d(G_k) \leq 1/n$ for each k. Then, for each k,

$$\frac{[d(G_k)]^q}{[d(G_k)]^p} = [d(G_k)]^{q-p} \leq \left(\frac{1}{n}\right)^{q-p};$$

so

$$\mu_n^{*(q)}(E) \leq \sum_{k=1}^{\infty} [d(G_k)]^q \leq \left(\frac{1}{n}\right)^{q-p} \sum_{k=1}^{\infty} [d(G_k)]^p.$$

Since this holds for every such covering of E,

$$\mu_n^{*(q)}(E) \leq \left(\frac{1}{n}\right)^{q-p} \mu_n^{*(p)}(E).$$

Letting $n \to \infty$ and remembering that $q - p > 0$ and that

$$\lim_n \mu_n^{*(p)}(E) = \mu^{*(p)}(E) < \infty,$$

we have the desired result. ∎

For a given set E, there may be no number p such that $\mu^{*(p)}(E)$ is finite and different from zero; but if we define the *dimension of E* by

$$\dim(E) = \sup\{p \mid \mu^{*(p)}(E) = \infty\},$$

then we have that $\mu^{*(q)}(E) = 0$ if $q > \dim(E)$, while $\mu^{*(q)}(E) = \infty$ if $q < \dim(E)$.

EXERCISES

a. Show that in R_1, $\mu^{*(1)}$ is Lebesgue outer measure.

b. Let $\Omega = R_2$, and let λ be Lebesgue 2-dimensional measure. Show that if J is a square,
$$\mu^{(2)}(J) \leq 2\lambda(J) \leq 2\mu^{(2)}(J).$$

c. Generalize the result in Exercise b to any open set.

d. Generalize the result in Exercise c to n dimensions, and show that the sets $E \subset R_n$ for which $\mu^{(n)}(E) = 0$ are precisely the sets of Lebesgue n-dimensional measure zero.

e. Show that if G is any open set in R_2, then $\mu^{(1)}(G) = \infty$; hence 13.7 does not apply to $\mu^{*(1)}$ in R_2.

f. Show that R_1 is one-dimensional, but $\mu^{(1)}(R_1) = \infty$.

g. Show that if for each set E_k in a sequence E,

$$\dim(E_k) = p,$$

then

$$\dim\left(\bigcup_{k=1}^{\infty} E_k\right) = p.$$

h. Show that if E consists of a single point, then $\dim(E) = 0$; hence every countable set has dimension zero.

*17 HAAR MEASURE

In this section we want to present the development of a measure in a space Ω which has the algebraic structure of a *group*. That is, there is defined in Ω an operation called multiplication such that for every pair of points x and y in Ω, the product xy is also a point in Ω; furthermore, the following postulates are satisfied:

G–I. Multiplication is associative; that is, for every x, y, $z \in \Omega$, $x(yz) = (xy)z$.

G–II. There is a unique identity element $e \in \Omega$ such that for every $x \in \Omega$, $xe = ex = x$.

G–III. To each $x \in \Omega$ there corresponds a unique inverse x^{-1} such that $x^{-1}x = xx^{-1} = e$.

The set of all real numbers except 0 forms a group with multiplication defined in the usual way. The set of all real numbers forms another group if "multiplication" is interpreted to mean addition. Similarly, any linear space forms a group with addition as the group operation. Group multiplication need not be commutative; that is, it is not necessarily true that $xy = yx$. One of the most familiar examples of a noncommutative group is the set of all nonsingular $n \times n$ matrices.

If Ω is a group, if $x \in \Omega$, and if $A \subset \Omega$, then we define

$$xA = \{xz \mid z \in A\}$$

and

$$Ax = \{zx \mid z \in A\}.$$

Similarly for $A \subset \Omega$ and $B \subset \Omega$, we define

$$AB = \{xy \mid x \in A; y \in B\}$$

and

$$A^{-1} = \{x^{-1} \mid x \in A\}.$$

The sets xA and Ax are called, respectively, the *left translation* and *right translation* of A by x. The terminology here is mixed, to say the least. The group operation is usually called multiplication, but the term translation is motivated by the example of the real numbers with addition as the group operation.

If Ω is simultaneously a group and a Hausdorff space such that the function f from $\Omega \times \Omega$ to Ω defined by

$$f(x, y) = x^{-1}y$$

is continuous, then Ω is a *topological group*. For the remainder of this section we assume that Ω is a locally compact topological group.

A measure μ in Ω is a *left Haar measure* if it satisfies the following conditions:

H–I.   Every compact set is measurable.

H–II. If C is compact, $\mu(C) < \infty$.

H–III. μ is not identically zero.

H–IV. μ is invariant under left translations; that is, $\mu(xE) = \mu(E)$.

Our principal aim here is to show that every locally compact topological group has a left Haar measure. We shall use the theory of Chapter 2 (indeed, we shall use Method I), but Section 13 will not be applicable because we are not assuming that the topology in Ω comes from a metric. The idea of metric outer measure is to give topological conditions for measurability. This is important, but in the theory of Haar measure it is embodied in condition H–I which we shall prove directly.

In a metric space neighborhoods have "sizes" because they have diameters. However, this is the mechanism for constructing Hausdorff measures. We want to emphasize that we start here with a space in which neighborhoods have no intrinsic sizes attached. Nevertheless, Haar measures are constructed from the topology by a very clever use of two concepts which are available here. These two concepts are translation (available because we have a group) and compactness (available in abundance because we have a locally compact topology).

Very roughly, translation and compactness yield a notion of relative size of sets as follows. Let C be compact and let N be a neighborhood of e. The class

$$\{xN \mid x \in C\}$$

is an open cover of C and hence is reducible to a finite cover. Thus, there is a minimum number of left-translates of N that will cover C; we denote this minimum number by

$$(C : N).$$

In a rough sense, then, C is $(C : N)$ "times as big as" N. This is very rough, however, because translates of N may not cover very efficiently. For example, how many circles of radius 1 does it take to cover a circle of radius 2? Some of this difficulty is compensated for by turning from a simple comparison of C and N to the idea of using N as a mechanism to compare C with a fixed reference set C_0. Specifically, let C_0 be a compact set with nonvacuous interior; we let this reference set stay fixed for the remainder of the discussion. Now, let C be any compact set and N any neighborhood of e; we define

$$\tau_N(C) = \frac{(C:N)}{(C_0:N)}.$$

The hope is that translates of N are just as inefficient in covering C_0 as they are in covering C so that $\tau_N(C)$ is a reasonable measure of the relative sizes of C and C_0. Needless to say, this is not accurate enough yet, but it is the way to get started.

Theorem 17.1. *Let C_0 and N be fixed; then the function τ_N has the following properties:*

17.1.1 $0 \leq \tau_N(C) < \infty$ *for every compact C.*

17.1.2 *If C and D are compact and $C \subset D$, then $\tau_N(C) \leq \tau_N(D)$.*

17.1.3 *If C and D are compact, then $\tau_N(C \cup D) \leq \tau_N(C) + \tau_N(D)$.*

17.1.4 *If C is compact and $x \in \Omega$, then $\tau_N(xC) = \tau_N(C)$.*

17.1.5 *If C and D are compact and $CN^{-1} \cap DN^{-1} = \phi$, then $\tau_N(C \cup D) = \tau_N(C) + \tau_N(D)$.*

Proof. 17.1.1 is obvious. 17.1.2 follows because a cover of D is a cover of C. In 17.1.3 a cover of C together with a cover of D constitutes a cover of $C \cup D$ and the result follows. In 17.1.4 if $\{y_iN\}$ covers C, then $\{xy_iN\}$ covers xC; and if $\{z_iN\}$ covers xC, then $\{x^{-1}z_iN\}$ covers C. In 17.1.5, if $y \in C \cap xN$, then $y = xn$ where $n \in N$; so $x = yn^{-1} \in CN^{-1}$. Similarly, if $D \cap xN \neq \phi$, then $x \in DN^{-1}$. Thus the hypothesis implies that no one translate of N can intersect both C and D and the result follows. ∎

Theorem 17.2. *Let C_0 and C be fixed; there exists $k_C < \infty$ such that $\tau_N(C) \leq k_C$ for all neighborhoods N of e.*

Proof. Let k_C be the minimum number of translates of C_0^0 that covers C. Then, $k_C(C_0:N)$ translates of N will cover C; that is, $(C, N) \leq k_C(C_0, N)$. ∎

Let $\mathcal{K}$ be the class of all compact subsets of Ω and for each $C \in \mathcal{K}$ let X_C be the closed interval $[0, k_C]$. Form the product space

$$P = \underset{C \in \mathcal{K}}{\times} X_C.$$

Recall that P is a set of real-valued functions on $\mathcal{K}$ and note that by 17.1.1, 17.2, and the definition of X_C, $\tau_N \in P$ for each neighborhood N of e. For each such N, we define $T_N \subset P$ by

$$T_N = \{\tau_M \mid M \subset N, M \text{ a neighborhood of } e\}.$$

If $N_1, N_2, \ldots, N_n$ is any finite set of neighborhoods of e, there is a neighborhood M of e such that

$$M \subset \bigcap_{i=1}^{n} N_i;$$

thus $M \subset N_i$ for each i, which is to say

$$\tau_M \in \bigcap_{i=1}^{n} T_{N_i}.$$

Thus the class

$$\{T_N \mid N \text{ a neighborhood of } e\}$$

has the finite intersection property, and P is compact by the Tychonoff theorem (8.4); so there exists

$$\tau_0 \in \bigcap_{N} \overline{T}_N.$$

Now, $\tau_0 \in P$, which is to say that τ_0 is a function on $\mathscr{K}$, the class of compact sets. As usual, we want a τ-function defined on open sets. In the construction of Lebesgue-Stieltjes measure, for example, it sufficed to transfer a function τ_0 on the half-open intervals to one on the corresponding open intervals. This worked because with the distribution function right-continuous, we "did not lose very much." In the present situation we have a promising-looking τ_0 on the compact sets and we want a τ-function on their interiors. However, we have no real picture of what $C - C^o$ looks like; so we resort to the following slightly more elaborate procedure. If C is compact, we define

$$\tau(C^o) = \sup \{\tau_0(D) \mid D \subset C^o, D \text{ compact}\}.$$

Clearly, $\tau(\phi) = 0$. If Ω is not compact, we extend τ by setting $\tau(\Omega) = \infty$. Now, let $\mathscr{C}$ be Ω together with the interiors of compact sets. Then $\mathscr{C}$ is a sequential covering class and τ is a nonnegative function on $\mathscr{C}$ that vanishes on ϕ. We construct μ^* from $\mathscr{C}$ and τ by Method I. In the usual way, μ^* determines a measure μ, and we shall show that μ is a left Haar measure.

Theorem 17.3. *The function τ_0 has the following properties:*

17.3.1 *If C and D are compact, then $\tau_0(C \cup D) \leq \tau_0(C) + \tau_0(D)$.*

17.3.2 *If C and D are compact and $C \cap D = \phi$, then $\tau_0(C \cup D) = \tau_0(C) + \tau_0(D)$.*

17.3.3 *If C and D are compact and $C \subset D$, then $\tau_0(C) \leq \tau_0(D)$.*

17.3.4 *If C is compact and $x \in \Omega$, then $\tau_0(xC) = \tau_0(C)$.*

17.3.5 $\tau_0(C_0) = 1$.

Proof. For each $C \in \mathscr{K}$, the function f_C on the product space P defined by $f_C(\sigma) = \sigma(C)$ is continuous (because convergence in P means convergence on every compact set and convergence of values of f_C means convergence at C). Thus

$$\Delta_1 = \{\sigma \mid \sigma(C \cup D) \leq \sigma(C) + \sigma(D)\}$$

is closed in P. By 17.1.3, $\tau_N \in \Delta_1$ for every N; so $T_N \subset \Delta_1$ and with Δ_1 closed we have $\tau_0 \in \overline{T}_N \subset \Delta_1$. This proves 17.3.1. If $C \cap D = \phi$, there exists (see final paragraph of Section 8) a neighborhood N_0 of e such that $CN_0^{-1} \cap DN_0^{-1} = \phi$ and so by 17.1.5,

$$T_{N_0} \subset \Delta_2 = \{\sigma \mid \sigma(C \cup D) = \sigma(C) + \sigma(D)\}.$$

Continuity of the functions f_C implies that Δ_2 is closed; so

$$\tau_0 \in \overline{T_{N_0}} \subset \Delta_2,$$

and 17.3.2 is established. Let

$$\Delta_3 = \{\sigma \mid \sigma(C) \leq \sigma(D)\},$$

$$\Delta_4 = \{\sigma \mid \sigma(xC) = \sigma(C)\},$$

$$\Delta_5 = \{\sigma \mid \sigma(C_0) = 1\}.$$

Each of these is closed in P; so by 17.1.2,

$$\tau_0 \in \overline{T}_N \subset \Delta_3.$$

By 17.1.4,

$$\tau_0 \in \overline{T}_N \subset \Delta_4;$$

and (since obviously $\tau_N(C_0) = 1$),

$$\tau_0 \in \overline{T}_N \subset \Delta_5. \quad \blacksquare$$

Corollary 17.3.6. *If C and D are compact, then*

$$\tau(C^0 \cup D^0) \leq \tau(C^0) + \tau(D^0).$$

Proof. Let $E \subset C^0 \cup D^0$ with E compact. The sets $E - C^0$ and $E - D^0$ are disjoint and compact; so there are disjoint open sets $G_1 \supset E - C^0$ and $G_2 \supset E - D^0$. A routine check shows that $E - G_1 \subset C^0$, $E - G_2 \subset D^0$, and $(E - G_1) \cup (E - G_2) = E$; so

$$\tau_0(E) \leq \tau_0(E - G_1) + \tau_0(E - G_2)$$
$$\leq \tau(C^0) + \tau(D^0).$$

Since E is an arbitrary compact subset of $C^0 \cup D^0$, the result follows. $\blacksquare$

Corollary 17.3.7. *If C is compact, then $\mu^*(C) \geq \tau_0(C)$.*

Proof. Let

$$C \subset \bigcup_{i=1}^{\infty} K_i^o;$$

then C is covered by a finite set of the K_i^o:

$$C \subset \bigcup_{j=1}^{n} K_{i_j}^o;$$

so, by 17.3.6,

$$\tau_0(C) \leq \tau\left(\bigcup_{j=1}^{n} K_{i_j}^o\right) \leq \sum_{j=1}^{n} \tau(K_{i_j}^o) \leq \sum_{i=1}^{\infty} \tau(K_i^o).$$

Therefore,

$$\mu^*(C) = \inf\left\{\sum_{i=1}^{\infty} \tau(K_i^o) \mid C \subset \bigcup_{i=1}^{\infty} K_i^o\right\} \geq \tau_0(C). \blacksquare$$

Theorem 17.4. *Every compact set is μ^*-measurable.*

Proof. First we prove that if C and K are both compact, then

17.4.1 $\mu^*(K^o) \geq \mu^*(K^o \cap C) + \mu^*(K^o - C).$

That is, we show that the condition for measurability of C is satisfied with interiors of compact sets as test sets. Let D be a compact subset of $K^o - C$ and let E be a compact subset of $K^o - D$. We note the following: (i) $K^o - C$ and $K^o - D$ are both interiors of compact sets; (ii) $D \cap E = \phi$; (iii) $D \cup E \subset K^o$. Thus, by 17.3.7 and 17.3.2,

$$\mu^*(K^o) \geq \mu^*(D \cup E) \geq \tau_0(D \cup E) = \tau_0(D) + \tau_0(E);$$

therefore

$$\mu^*(K^o) \geq \tau_0(E) + \sup\{\tau_0(D) \mid D \subset K^o - C\}$$
$$= \tau_0(E) + \tau(K^o - C)$$
$$\geq \tau_0(E) + \mu^*(K^o - C).$$

From this and the fact that $K^o - D \supset K^o \cap C$, it follows that

$$\mu^*(K^o) \geq \mu^*(K^o - C) + \sup\{\tau_0(E) \mid E \subset K^o - D\}$$
$$= \mu^*(K^o - C) + \tau(K^o - D)$$
$$\geq \mu^*(K^o - C) + \mu^*(K^o \cap C).$$

This establishes 17.4.1. Now, let A be any set with $\mu^*(A) < \infty$; then given $\varepsilon > 0$,

$$A \subset \bigcup_{i=1}^{\infty} K_i^o,$$

where each K_i is compact and, by 17.4.1,

$$\mu^*(A) + \varepsilon > \sum_{i=1}^{\infty} \tau(K_i^0) \geq \sum_{i=1}^{\infty} \mu^*(K_i^0)$$

$$\geq \sum_{i=1}^{\infty} [\mu^*(K_i^0 \cap C) + \mu^*(K_i^0 - C)]$$

$$\geq \mu^*\left[\left(\bigcup_{i=1}^{\infty} K_i^0\right) \cap C\right] + \mu^*\left[\left(\bigcup_{i=1}^{\infty} K_i^0\right) - C\right]$$

$$\geq \mu^*(A \cap C) + \mu^*(A - C).$$

Since ε is arbitrary, the result follows by 11.1. ∎

Theorem 17.5. μ *is a left Haar measure.*

Proof. 17.4 establishes H–I. Since $\tau_0 \in P$, we have $\tau_0(C) \leq k_C$ for each compact C. If D is compact, then D is covered by a finite number of open sets with compact closures; so $D \subset C^0$, where C is compact. Thus, $\mu^*(D) \leq \tau(C^0) \leq \tau_0(C) \leq k_C$, and H–II is established. H–III follows from 17.3.5 and 17.3.7. H–IV follows by routine computation from 17.3.4; that is, once τ_0 is invariant under translation, so is τ and then so is μ^*. ∎

EXERCISES

a. Let Ω be the set of all positive real numbers and let multiplication and distance be defined in the usual manner. Let μ be the Haar measure defined in the text with $[1, e]$ taken as the reference set C_0. (Note that 1 is the identity element here; e in this case is the Naperian logarithm base.) Show that $\mu[(a, b)] = \log(b/a)$.

b. Let Ω be the set of all real numbers with distance defined in the usual way, and let "multiplication" be defined as addition. Let μ be the Haar measures defined in the text with $[0, 1]$ taken as the reference set C_0. Show that μ is Lebesgue measure.

c. Let Ω be a subset of R_n; let multiplication be defined in such a way that Ω is a metric group. Let f be a Lebesgue-Stieltjes distribution function on Ω such that

$$f(yx) = a + f(x),$$

where a depends only on y. Show that the iterated difference τ_0 defined by 14.2 is invariant under left translations; hence the Lebesgue-Stieltjes measure μ_f is a left Haar measure.

d. Matrix multiplication for 2×2 matrices is defined as follows:

$$\begin{pmatrix} a_{11} & a_{12} \\ a_{21} & a_{22} \end{pmatrix} \begin{pmatrix} b_{11} & b_{12} \\ b_{21} & b_{22} \end{pmatrix} = \begin{pmatrix} a_{11}b_{11} + a_{12}b_{21} & a_{11}b_{12} + a_{12}b_{22} \\ a_{21}b_{11} + a_{22}b_{21} & a_{21}b_{12} + a_{22}b_{22} \end{pmatrix}.$$

Show that this multiplication is not commutative.

e. Let Ω be the set of all matrices of the form

$$\begin{pmatrix} x & y \\ 0 & x \end{pmatrix},$$

where $x > 0$ and y is any real number. There is an obvious mapping of Ω onto the right half plane. Let Ω be metrized by this mapping and the Euclidean metric in the half plane. Show that Ω is a locally compact metric group.

f. Let f be defined over the space Ω of Exercise e by

$$f\begin{pmatrix} x & y \\ 0 & x \end{pmatrix} = \frac{-y}{x}.$$

Show that the Lebesgue-Stieltjes measure μ_f is both a left and a right Haar measure.

g. Let Ω be the set of all matrices of the form

$$\begin{pmatrix} x & y \\ 0 & 1 \end{pmatrix}.$$

Show that with matrix multiplication (Exercise d) and with a metric defined as in Exercise e, Ω is a noncommutative, locally compact, metric group.

h. Let f be defined over the space Ω of Exercise g by

$$f\begin{pmatrix} x & y \\ 0 & 1 \end{pmatrix} = \frac{-y}{x}.$$

Show that the Lebesgue-Stieltjes measure μ_f is a left Haar measure which is not invariant under right translations.

i. Let g be defined over the space of Exercise g by

$$g\begin{pmatrix} x & y \\ 0 & 1 \end{pmatrix} = y \log x.$$

Show that the Lebesgue-Stieltjes measure μ_g is a right Haar measure which is not left invariant. Compare Exercise h.

j. Show that in any group $(xy)^{-1} = y^{-1}x^{-1}$; hence $(xA)^{-1} = A^{-1}x^{-1}$.

k. Show that if μ is a left Haar measure and if ν is defined by $\nu(E) = \mu(E^{-1})$, then ν is a right Haar measure.

l. Prove that a Haar measure is positive on every nonvacuous open set. *Hint:* If a finite number of translates of G covers the reference set C_0 and if $\mu(G) = 0$, then we have a contradiction.

18 NONMEASURABLE SETS

We have given a number of artificial examples of outer measures for which there are many obviously nonmeasurable sets (see Exercises 11.g through 11.j). However, it is not at all obvious that for something like Lebesgue measure there are any such sets. Many investigations would be simplified to a considerable extent if all sets were

Lebesgue measurable, but unfortunately this is not the case. However, a Lebesgue nonmeasurable set is such a weird thing that to date none has been constructed without the axiom of choice.

For the construction of such a set, we let $\Omega = [0, 1)$ and let μ be Lebesgue measure. We introduce the notion of *translation of a set, modulo* 1. Let $E \subset [0, 1)$, and let $a \in [0, 1)$; we say that

$$H = E + a \, [\text{mod. } 1]$$

provided

$$H = \{y \mid y = x + a; x \in E; x + a < 1\}$$
$$\cup \{y \mid y = x + a - 1; x \in E; x + a \geq 1\}.$$

Theorem 18.1. *If $E \subset [0, 1)$, if $a \in [0, 1)$, if $H = E + a \, [\text{mod. } 1]$, and if μ^* is Lebesgue outer measure, then $\mu^*(H) = \mu^*(E)$.*

Proof. The set H consists of two parts, each of which is a translation of a part of E. Lebesgue outer measure is obviously invariant under translations; so we have $H = H_1 \cup H_2$ and $E = E_1 \cup E_2$, where

$$\mu^*(H_1) = \mu^*(E_1) \quad \text{and} \quad \mu^*(H_2) = \mu^*(E_2).$$

However, H_1 is the intersection of H and a measurable set, and a similar statement holds for H_2, E_1, and E_2. Therefore, by 11.2.4,

$$\mu^*(H) = \mu^*(H_1) + \mu^*(H_2) = \mu^*(E_1) + \mu^*(E_2) = \mu^*(E). \quad \blacksquare$$

Corollary 18.1.1. *If $H = E + a \, [\text{mod. } 1]$, then H is measurable if and only if E is measurable.*

Proof. This follows at once from 18.1 and the fact that measurability is defined in terms of values of the outer measure function. $\blacksquare$

To proceed with the construction of a nonmeasurable set, let R be the set of rational numbers in $[0, 1)$. For each $x \in [0, 1)$ let

$$E_x = R + x \, [\text{mod. } 1].$$

Suppose $y \in E_{x_1} \cap E_{x_2}$ and let $z \in E_{x_1}$; then, since

$$z - x_2 = (z - x_1) + (x_1 - y) + (y - x_2)$$

and since each of these latter three numbers is rational, it follows that $z - x_2$ is rational. That is, $z \in E_{x_2}$. Thus, two sets E_{x_1} and E_{x_2} are either identical or disjoint. Let $\mathscr{E}$ be the class of disjoint sets consisting of all sets of the form E_x. By the axiom of choice there is a set Q_0 consisting of one point from each of the sets of $\mathscr{E}$. Let r be the sequence of all rational numbers in $(0, 1)$, and for each n, let

$$Q_n = Q_0 + r_n \, [\text{mod. } 1].$$

If $y \in Q_{n_1} \cap Q_{n_2}$, then $y - r_{n_1}$ and $y - r_{n_2}$ either belong to Q_0 or differ by 1 from points in Q_0, so we have two points in Q_0 whose difference is rational. However, this is impossible because these two points would belong to the same set E_x, and Q_0 contains only one point from each set in $\mathscr{E}$. Thus, the sets Q_n are disjoint.

Since $Q_n \subset [0, 1)$ for each n,

$$\bigcup_{n=0}^{\infty} Q_n \subset [0, 1).$$

However, if $x \in [0, 1)$, then for some rational number r_n, $x - r_n$ is (or differs by 1 from) the point of E_x that belongs to Q_0; thus $x \in Q_n$. So, we have

$$\bigcup_{n=0}^{\infty} Q_n = [0, 1).$$

By 18.1.1 the sets Q_n are either all measurable or all nonmeasurable. If they are all measurable, we have

18.2 $$\sum_{n=0}^{\infty} \mu^*(Q_n) = \mu\{[0, 1)\} = 1;$$

however, by 18.1 the sets Q_n all have the same outer measure a. If $a = 0$, 18.2 says $0 = 1$; if $a > 0$, 18.2 says $\infty = 1$. Therefore, the sets Q_n must be nonmeasurable.

EXERCISES

a. Let K be a sequence of Cantor sets such that for each n, $\mu(K_n) \geq 1 - (1/n)$ (see Exercise 13.n). Let Q_0 be the nonmeasurable set constructed above. Show that for some n, $K_n \cap Q_0$ is nonmeasurable.

b. Show that if in Exercise 12.q μ^* is Lebesgue outer measure, then the outer measure v^* constructed there is invariant under translations, modulo 1.

c. Show that the sets Q_n are nonmeasurable with respect to the outer measure v^* of Exercise 12.q. Note: v^* is thus an example of a metric outer measure which is not regular; see Exercises 13.b and 12.r.

d. Let Q be the sequence of nonmeasurable sets constructed above. Let $\mathscr{C}$ be the class of all sets of the form $Q_n \cap I$ where I is either an interval or the vacuous set. Let λ^* be Lebesgue outer measure, and define τ on $\mathscr{C}$ by setting $\tau(Q_n \cap I) = \lambda^*(Q_n \cap I)$. Let μ^* be constructed from $\mathscr{C}$ and τ by Method I. Show that for every $E \subset [0, 1)$, $\mu^*(E) \geq \lambda^*(E)$.

e. Show that the function τ of Exercise d satisfies the conditions in 13.8; therefore, the outer measure μ^* of Exercise d is a metric outer measure.

f. Show that the sets of $\mathscr{C}$ in Exercise d are μ^*-measurable; therefore μ^* is regular.

g. Show that for the regular, metric outer measure μ^* of Exercise d, not every measurable set is the union of a Borel set and a set of measure zero. Compare Exercise 13.h.

REFERENCES FOR FURTHER STUDY

On Lebesgue measure in R_n:
  Kestelman [17]
On Lebesgue-Stieltjes measures in R_n:
  McShane and Botts [20]
  von Neumann [23]
  Saks [27]
On the general concept of probability:
  Halmos [10], [11]
  Kolmogoroff [18]

On probability in function spaces:
  Doob [6]
  Hewitt and Stromberg [13]
  Kolmogoroff [18]
On Hausdorff measures:
  Hurewicz and Wallman [15]
On Haar measure:
  Berberian [3]
  Halmos [11]
On nonmeasurable sets:
  Hahn and Rosenthal [9]

CHAPTER 4

MEASURABLE FUNCTIONS

An integral, $\int f d\mu$, might be characterized as an operation on a point function f generated by a measure function μ. For a given outer measure μ^* and corresponding measure μ, the only functions susceptible to this treatment are those which are called measurable functions. It is the purpose of this chapter to define such functions and study some of their properties. Measurability of a point function f is determined by the class $\mathcal{M}$ of measurable sets, so, strictly speaking, we should refer to μ^*-measurable functions. However, where there is no chance of ambiguity, we shall usually delete the reference to μ^*.

Despite the similarity in terminology, the reader should not confuse measure functions with measurable functions. The former are set functions; the latter are point functions.

19 DEFINITIONS AND BASIC PROPERTIES

If Ω is a metric space, then (see 5.3) a continuous function f is characterized by the condition that for every open set G in the range space, $f^{-1}(G)$ is open. By means of similar characterizations of inverse images, we can define two other important classes of functions.

Let Ω be a metric space, and let f be a finite, real valued function whose domain is a Borel set in Ω. We say that f is a *Baire function* if for every open set G in the real number system, $f^{-1}(G)$ is a Borel set.

Let Ω be any space in which an outer measure is defined, and let f be a finite, real valued function whose domain is a measurable set in Ω. We say that f is a *measurable function* if for every open set G in the real number system, $f^{-1}(G)$ is a measurable set.

To extend these last two concepts to the case of a function which assumes infinite values, we merely add the requirement that $f^{-1}(\infty)$ and $f^{-1}(-\infty)$ be Borel sets or measurable sets, as the case may be.

The notions of continuous function and Baire function depend on topological properties of inverse images, while that of measurable function depends on a measure theoretic property of inverse images. Therefore, in general, there may be no particular relationship between the class of measurable functions and that of Baire functions (or that of continuous functions). However, we do well to think of these concepts in the order listed (continuous, Baire, measurable), because in the important case of a metric space with a metric outer measure, this is the order of increasing generality. This fact may be stated more precisely as follows.

Theorem 19.1. *Let Ω be a metric space. Every continuous, real valued function with domain Ω is a Baire function. If, in addition, μ^* is a metric outer measure, then every Baire function on Ω is measurable. Finally, if every continuous, real valued function on Ω is measurable, then μ^* is a metric outer measure.*

Proof. As usual, let $\mathscr{G}$, $\mathscr{B}$, and $\mathscr{M}$ be, respectively, the classes of open, Borel, and measurable sets in Ω. We always have $\mathscr{G} \subset \mathscr{B}$; so if f is continuous and U is open in the real number system, then

$$f^{-1}(U) \in \mathscr{G} \subset \mathscr{B},$$

and f is a Baire function. If μ^* is a metric outer measure, then $\mathscr{B} \subset \mathscr{M}$; and the same argument shows that Baire functions must be measurable. Finally, if every continuous function is measurable, let G be an arbitrary open set in Ω, and define f by the formula

$$f(x) = \rho(x, -G).$$

Then, f is continuous, hence measurable, and

$$f^{-1}[(0, \infty)] = G$$

must be a measurable set. That is, every open set in Ω is measurable. If $\rho(A, B) > 0$, there is an open set G such that $G \supset A$ and $-G \supset B$. Measurability of G yields the result that

$$\mu^*(A \cup B) = \mu^*[(A \cup B) \cap G] + \mu^*[(A \cup B) - G]$$
$$= \mu^*(A) + \mu^*(B),$$

and this is postulate M–IV; so μ^* is a metric outer measure. ∎

A more important role played by the Baire functions in the theory of measurable functions is indicated by the following results.

Theorem 19.2. *If f is a measurable function on Ω and if B is a Borel set in the real number system, then $f^{-1}(B)$ is a measurable set in Ω.*

Proof. Let $\mathscr{E}$ be the class of all sets in the real number system whose inverse images under f are measurable. If $A \in \mathscr{E}$, then

$$f^{-1}(-A) = -f^{-1}(A)$$

is a measurable set; so $-A \in \mathscr{E}$. If E is a sequence of sets from $\mathscr{E}$, then

$$f^{-1}\left(\bigcup_{n=1}^{\infty} E_n\right) = \bigcup_{n=1}^{\infty} f^{-1}(E_n)$$

is a measurable set; so

$$\bigcup_{n=1}^{\infty} E_n \in \mathscr{E}.$$

Thus, $\mathscr{E}$ is a completely additive class. We are given that $\mathscr{E}$ contains all open sets; therefore it contains all Borel sets. ∎

Corollary 19.2.1. *If f is a measurable function on any space Ω and if g is a Baire function on the real number system, then the composite function $g \circ f$ is a measurable function on Ω.*

Proof. Let U be any open set in the real number system; then, since

$$(g \circ f)^{-1}(U) = f^{-1}[g^{-1}(U)]$$

and since $g^{-1}(U)$ is a Borel set, it follows from 19.2 that $(g \circ f)^{-1}(U)$ is measurable. ∎

The notion of a measurable function is quite analogous to that of a continuous function, and we chose the definition used above with the idea of emphasizing that analogy. However, the following characterization of a measurable function is extremely useful and is frequently given as a definition.

Theorem 19.3. *In order that an extended real valued function f defined on a measurable set D be measurable, it is necessary and sufficient that for every finite real number a, $f^{-1}\{[-\infty, a]\}$ be a measurable set.*

Proof. To prove necessity, we note that

$$f^{-1}\{[-\infty, a]\} = D - [f^{-1}\{(a, \infty)\} \cup f^{-1}(\infty)].$$

If f is measurable, each set on the right is measurable.

Conversely, if $f^{-1}\{[-\infty, a]\}$ is measurable for each a, then each of the sets

$$f^{-1}(-\infty) = \bigcap_{n=1}^{\infty} f^{-1}\{[-\infty, -n]\},$$

$$f^{-1}(\infty) = D - \bigcup_{n=1}^{\infty} f^{-1}\{[-\infty, n]\}$$

is measurable. For any half-open interval $(a, b]$,

$$f^{-1}\{(a, b]\} = f^{-1}\{[-\infty, b]\} - f^{-1}\{[-\infty, a]\}$$

is measurable. Finally, if G is any open set in the real number system,

$$G = \bigcup_{i=1}^{\infty} (a_i, b_i]$$

by 14.5; so

$$f^{-1}(G) = \bigcup_{i=1}^{\infty} f^{-1}\{(a_i, b_i]\}$$

is a measurable set. ∎

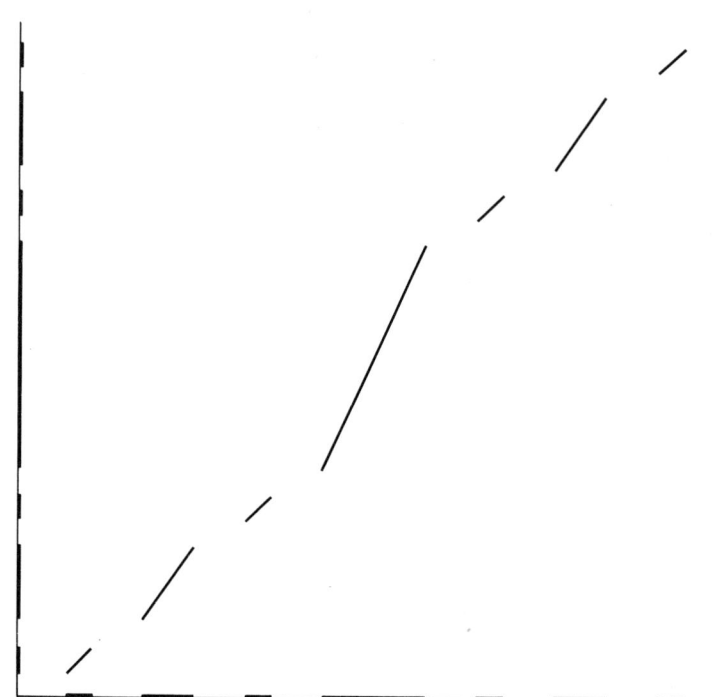

Figure 5

We conclude this section with an example. Let $\Omega = [0, 1]$ and let μ^* be Lebesgue outer measure. Following Exercise 18.a, let K be a nowhere dense perfect set containing a nonmeasurable set Q. Let K_0 be a nowhere dense perfect set of measure zero (see Exercise 13.l). The intervals of $-K$ may be put into an obvious one-to-one correspondence with those of $-K_0$ so as to preserve the natural ordering of the points. Let g be the function on $-K$ which maps each interval of $-K$ linearly onto the corresponding interval of $-K_0$ (see Fig. 5). Since g is monotone, its one-sided limits exist at every point of $\overline{(-K)} = [0, 1]$. If, at any point, these one-sided limits were different, there would be a gap of positive length in the range of g; however, this is impossible because K_0 is nowhere dense. So, the limit of g exists at every point of $[0, 1]$. Let f be the extension of g obtained by setting

$$f(x) = \lim_{y \to x} g(y).$$

The function f is continuous, and therefore measurable. It maps Q onto some set $Q_0 \subset K_0$. Since $\mu(K_0) = 0$, Q_0 is Lebesgue measurable; however, f is one-to-one, so $f^{-1}(Q_0) = Q$. Thus, we have a measurable (indeed, continuous) function whose inverse mapping carries a measurable set into a nonmeasurable one. It now follows from 19.2 that Q_0 is not a Borel set, even though it is Lebesgue measurable. Finally, we might observe that

$$C_{Q_0}[f(x)] = C_Q(x),$$

so we have a measurable function of a continuous function producing a nonmeasurable function. Thus, the roles of Baire functions and measurable functions cannot be reversed in 19.2.1.

EXERCISES

a. Let Ω be the set of positive integers, and let an outer measure μ^* be defined by 10.1. Show that every real valued function on Ω is measurable.

b. Let Ω be any space and μ^* be an outer measure for which the only measurable sets are ϕ and Ω. What functions are measurable in this case?

c. Describe the measurable functions on the space of Exercise 11.i.

d. Prove that for a real valued function f defined on a Borel set to be a Baire function, it is necessary and sufficient that, for every finite real number a, $f^{-1}\{[-\infty, a]\}$ be a Borel set.

e. Prove that if f is a Baire function and B is a Borel set in the real number system, then $f^{-1}(B)$ is a Borel set.

f. Show that in order that the characteristic function C_E be measurable, it is necessary and sufficient that E be a measurable set. What is a similar condition for C_E to be a Baire function?

g. Let Q and Q_0 be, respectively, the nonmeasurable and measurable but non-Borel sets of the example at the end of this section. Let h be the inverse of the function in that example. For $x \in Q_0$, let $f(x) = h(x)$; and for $x \in [0, 1] - Q_0$, let $f(x) = 1 + h(x)$. For $x - 1 \in Q_0$, let $f(x) = f(x - 1) + 1$; and for $x - 1 \in [0, 1] - Q_0$, let $f(x) = f(x - 1) - 1$. Show that f is a Lebesgue measurable function.

h. Show that there is a Borel set B such that under the measurable function f of Exercise g the direct image of B is nonmeasurable.

i. Let g be the inverse of the function f of Exercise g. Show that g is single valued but nonmeasurable on $[0, 2]$.

j. Show that every Borel set in $[0, 2]$ has a measurable direct image under the nonmeasurable function g of Exercise i.

k. Sierpinski has given an example† of a Lebesgue nonmeasurable set in R_2 which has exactly one point in common with each line parallel to either of the coordinate axes. Use this to show that a function of two variables may be a Baire function in each variable separately, yet be nonmeasurable on R_2.

---

† For a description of this example, see Hahn and Rosenthal [9], pp. 241–242.

20 OPERATIONS ON MEASURABLE FUNCTIONS

Let f be an extended real valued function on Ω, and let a be a real number. We define the functions af, $a + f$, $|f|^a$, f^+, and f^-, respectively, for each $x \in \Omega$ by the formulas $af(x)$, $a + f(x)$, $|f(x)|^a$, $\max[f(x), 0]$, and $\max[-f(x), 0]$. Each of these functions is defined everywhere f is with the exceptions that if $a = 0$, af and $|f|^a$ are not defined at points x for which $|f(x)| = \infty$; and if $a < 0$, $|f|^a$ is not defined at points x for which $f(x) = 0$. These exceptions are not serious because if f is measurable, $\{x \mid |f(x)| < \infty\}$ and $\{x \mid f(x) \neq 0\}$ are measurable; so af and $|f|^a$ always have measurable domains.

Theorem 20.1. *If f is a measurable function and a is a real number, then each of the functions af, $a + f$, $|f|^a$, f^+, and f^- is measurable.*

Proof. Each of the formulas ay, $a + y$, $|y|^a$, $\max[y, 0]$, and $\max[-y, 0]$ defines a continuous function of y, so the result follows from 19.2.1. ∎

Let f and g be two extended real valued functions on Ω. We define the functions $f + g$ and fg, respectively, for each x by the formulas $f(x) + g(x)$ and $f(x)g(x)$. The function $f + g$ is undefined on $\{x \mid f(x) = -g(x) = \pm\infty\}$, and fg is undefined on $\{x \mid f(x) = 0; g(x) = \pm\infty\} \cup \{x \mid f(x) = \pm\infty; g(x) = 0\}$. However, for f and g measurable, each of these is a measurable set, so $f + g$ and fg each have a measurable domain.

Theorem 20.2. *If f and g are measurable functions, then $f + g$ is a measurable function.*

Proof. We prove this by means of the following lemma:

Lemma 20.2.1. *If f and g are measurable functions, then $\{x \mid f(x) \leq g(x)\}$ is a measurable set.*

This follows from the fact that

$$\{x \mid f(x) > g(x)\} = \bigcup_{m=-\infty}^{\infty} \bigcup_{n=1}^{\infty} \left[\left\{ x \mid f(x) > \frac{m}{n} \right\} \cap \left\{ x \mid g(x) < \frac{m}{n} \right\} \right].$$

Each set on the right is the inverse map of an open set under a measurable function, so $\{x \mid f(x) > g(x)\}$ is a measurable set; hence, so is its complement. This proves the lemma.

The proof of 20.2 may now be completed as follows:

$$\{x \mid f(x) + g(x) \leq a\} = \{x \mid f(x) \leq a - g(x); |g(x)| < \infty\}$$
$$\cup \{x \mid f(x) < \infty; g(x) = -\infty\}.$$

Since $a - g$ is measurable by 20.1, the first set on the right is measurable by 20.2.1. The second is obviously measurable, and the result follows by 19.3. ∎

Corollary 20.2.2. *In order that f be measurable, it is necessary and sufficient that both f^+ and f^- be measurable.*

Proof. Necessity of this condition is included in 20.1. Sufficiency follows from 20.2 and the fact that $f = f^+ - f^-$. ∎

Corollary 20.2.3. *If f and g are measurable functions, then fg is measurable.*

Proof. This follows from 20.1, 20.2, and the identity

$$fg = \tfrac{1}{4}(|f + g|^2 - |f - g|^2). \quad \blacksquare$$

EXERCISES

a. Show that if f and g are measurable functions, then $\{x \mid f(x) = g(x)\}$ is a measurable set.

b. Show that for any real valued function f, $f = f^+ - f^-$, $|f| = f^+ + f^-$, $f^+ = (|f| + f)/2$, and $f^- = (|f| - f)/2$.

c. Give an example in which $|f|$ is measurable but f is not.

Let f be a sequence of functions on some fixed set E. We define the functions $\sup_n f_n$, $\inf_n f_n$, $\overline{\lim}_n f_n$, and $\underline{\lim}_n f_n$ by performing the indicated operations on the sequence $f(x)$ of numbers for each $x \in E$. Each of these functions is defined over the whole of E.

Theorem 20.3. *If f is a sequence of measurable functions on some common domain, then $\sup_n f_n$ and $\inf_n f_n$ are measurable functions.*

Proof. This follows by 19.3 from the fact that

$$\{x \mid \sup_n f_n(x) \leq a\} = \bigcap_{n=1}^{\infty} \{x \mid f_n(x) \leq a\}.$$

This gives measurability for $\sup_n f_n$; the result for $\inf_n f_n$ follows by 20.1 from the fact that

$$\inf_n f_n = -\sup_n (-f_n). \quad \blacksquare$$

Corollary 20.3.1. *If f is a sequence of measurable functions on some common domain, then $\overline{\lim}_n f_n$ and $\underline{\lim}_n f_n$ are measurable functions.*

Proof. This follows from 20.3 and the fact that

$$\overline{\lim_n} f_n = \inf_k \sup_{n \geq k} f_n,$$

$$\underline{\lim_n} f_n = \sup_k \inf_{n \geq k} f_n. \quad \blacksquare$$

From 20.3.1 and Exercise a, it follows that $\{x \mid \underline{\lim}_n f_n(x) = \overline{\lim}_n f_n(x)\}$ is a measurable set whenever f is a sequence of measurable functions; and it also follows from 20.3.1 that the function $\lim_n f_n$, defined in an obvious way on this set, is a measurable function. However, in this connection we want to introduce certain restrictions which lead to a more useful concept.

If P is a proposition concerning the points x of a set E, we say that P is true *almost everywhere* in E (abbreviated† a. e. in E) if there is a subset E_0 of E such that $\mu(E_0) = 0$ and $P(x)$ is true for each $x \in E - E_0$. Sometimes it is more convenient to say "$P(x)$ is true for *almost all* $x \in E$." By definition, this will mean the same thing as "P is true a. e. in E." Clearly, the meaning of the statement "P holds a. e." depends on the measure μ, but as usual, when there is only one measure function in the discussion, we shall omit the reference to μ.

Let f be a sequence of functions, each defined on some fixed set E and each having finite values a. e. in E. Let f_0 be a function whose values are finite a. e. in E. We say that f *converges* a. e. to f_0, or

$$\lim_n f_n = f_0 \ [\text{a. e.}],$$

provided the relation

$$\lim_n f_n(x) = f_0(x)$$

holds for almost all $x \in E$.

Convergence a. e. can be defined for the case in which some, or all, of the functions concerned have infinite values on a set of positive measure. We have restricted our definition to the a. e. finite case for two reasons. In this case we can give an ε, n_ε description of a. e. convergence in one statement as follows: Given a sequence f of functions on a common domain E,

$$\lim_n f_n = f_0 \ [\text{a. e.}]$$

if and only if there is a set $E_0 \subset E$ such that $\mu(E_0) = 0$ and such that for every $x \in E - E_0$ and every $\varepsilon > 0$, there is a positive integer $n_{\varepsilon,x}$ such that for all $n > n_{\varepsilon,x}$,

$$|f_n(x) - f_0(x)| < \varepsilon.$$

A more important reason for the a. e. finiteness restriction is that in this case a Cauchy theorem holds. We shall return to this consideration in Chapter 7. For the present we merely want the fundamental result on the a. e. limit of a sequence of measurable functions.

Corollary 20.3.2. *If f is a sequence of measurable functions on some common domain and if $\lim_n f_n = f_0$ [a. e.], then f_0 is a measurable function.*

Proof. The values of f_0 for x in the exceptional set of measure zero are immaterial, since every subset of a set of measure zero is measurable. For other values of x, $f_0(x) = \overline{\lim}_n f_n(x)$, and the result follows from 20.3.1. ∎

---

† The phrase *almost everywhere* was first introduced in French (*presque partout*); therefore in the older works on Lebesgue integration the French abbreviation p. p. is frequently used. The modern practice is to use the English abbreviation when writing in English.

The reader will recall that the class of continuous functions on Ω is closed under the operation of taking uniform limits, but not under that of taking pointwise limits. Corollary 20.3.2 tells us that the class of measurable functions on Ω is closed under an even more general operation—that of taking a. e. limits.

EXERCISES

d. Let f be a sequence of functions on Ω, and let G be an open set in the real number system. Show that

$$\{x \mid \lim_n f_n(x) \in G\} = \bigcup_{m=1}^{\infty} \bigcup_{k=1}^{\infty} \bigcap_{n=k}^{\infty} \left\{ x \mid \rho[f_n(x), -G] > \frac{1}{m} \right\}.$$

e. The class of Baire functions on Ω is frequently defined as the smallest class which contains the continuous functions and is closed under the operation of taking pointwise limits. Show that this definition is equivalent to that given in Section 19.

f. Prove 19.2.1 from the definition of Baire functions given in Exercise e.

g. Show that the class of Baire functions is not closed under the operation of taking a. e. limits.

h. Certain Baire functions may be classified as follows: The continuous functions are of *Baire class zero*, and the functions of *Baire class n* are those which are pointwise limits of sequences of functions of class $n - 1$. Show from Exercise d that if f is a Baire function of finite class and G is an open set in the real number system, then $f^{-1}(G)$ is a Borel set of finite order.

i. Deduce from 6.4 and 6.5 that on a complete metric space any function of Baire class one is continuous on a residual set.

j. Show that despite the result in Exercise i, a function of Baire class one may be discontinuous on a dense set. *Hint:* Let x be a sequence whose terms form a dense set in R_1. Let

$$f_{nk}(t) = \begin{cases} 0 & \text{for } t < x_n, \\ (kt - x_n) & \text{for } x_n \leq t \leq x_n + \frac{1}{k}, \\ 1 & \text{for } t > x_n + \frac{1}{k}. \end{cases}$$

Let

$$g_k(t) = \sum_{n=1}^{k} \frac{1}{2^n} f_{nk}(t).$$

The sequence g is monotone in k and bounded, hence convergent for each t, but the limit function is discontinuous at each point x_n.

k. Show that if f is a measurable function and if $g = f$ a. e., then g is measurable.

l. In defining convergence a. e., we specified that for each n, f_n be finite a. e. Show that this implies that for almost all x, $f_n(x)$ is finite for every n.

21 APPROXIMATION THEOREMS

The reader will recall that the characteristic function C_E of a set E is defined by

$$C_E(x) = \begin{cases} 1 & \text{for } x \in E, \\ 0 & \text{for } x \in -E. \end{cases}$$

Let $E_1, E_2, \ldots, E_n$ be any finite class of disjoint, measurable sets; and let $a_1, a_2, \ldots, a_n$ be a set of finite real numbers. A function f defined by

$$f(x) = \sum_{k=1}^{n} a_k C_{E_k}(x)$$

is called a *simple function*. Informally, a simple function is one which assumes a finite number of values and assumes each of these values on a measurable set. Clearly, a simple function is bounded and measurable.

Theorem 21.1. *If f_0 is a nonnegative, measurable function on a set E, then there exists a nondecreasing sequence f of nonnegative, simple functions such that for every $x \in E$,*

$$\lim_n f_n(x) = f_0(x).$$

Proof. For each integer n and each $x \in E$, let

$$f_n(x) = \begin{cases} \dfrac{i-1}{2^n} & \text{for } \dfrac{i-1}{2^n} \leq f_0(x) < \dfrac{i}{2^n}; \quad i = 1, 2, \ldots, n2^n; \\ n & \text{for } f_0(x) \geq n. \end{cases}$$

It is easily seen that each function f_n is simple and nonnegative and that the sequence f is nondecreasing. If $f_0(x) < \infty$, then for $n > f_0(x)$,

$$0 \leq f_0(x) - f_n(x) < \frac{1}{2^n}.$$

If $f_0(x) = \infty$, then for each n, $f_n(x) = n$. So, in either case

$$\lim_n f_n(x) = f_0(x). \quad \blacksquare$$

In Chapter 5 our development of the integral is based on monotone sequences of simple functions, and in this connection, 21.1 is vital. It tells us that every nonnegative, measurable function is the pointwise limit of such a sequence. The following embellishments to 21.1 will also be of use to us in discussing the definition of an integral.

Theorem 21.2. *If, in 21.1, $\{x \mid f_0(x) \neq 0\}$ is the union of a countable class of measurable sets of finite measure, then f may be constructed so as to have the additional property that for each n, $\{x \mid f_n(x) \neq 0\}$ has finite measure. Conversely, if for each simple function f_n, $\{x \mid f_n(x) \neq 0\}$ has finite measure, and if* for each x, $\lim_n f_n(x) = f_0(x)$, then $\{x \mid f_0(x) \neq 0\}$ is the union of a countable class of measurable sets of finite measure.

Proof. To prove the first statement, let

$$\{x \mid f_0(x) \neq 0\} = \bigcup_{k=1}^{\infty} E_k,$$

and let

$$A_n = \bigcup_{k=1}^{n} E_k$$

for each n. Then the sequence A is an expanding sequence of sets, so the characteristic functions C_{A_n} form a nondecreasing sequence of functions. If f is the sequence constructed in 21.1, then the sequence g defined by

$$g_n = f_n C_{A_n}$$

has all the required properties.

To prove the second statement, we note that if $f_n(x) = 0$ for every n, then $f_0(x) = 0$; so

$$\{x \mid f_0(x) \neq 0\} \subset \bigcup_{n=1}^{\infty} \{x \mid f_n(x) \neq 0\}.$$

Setting

$$E_n = \{x \mid f_0(x) \neq 0\} \cap \{x \mid f_n(x) \neq 0\},$$

we have

$$\{x \mid f_0(x) \neq 0\} = \bigcup_{n=1}^{\infty} E_n;$$

the sets E_n are measurable, and each has finite measure. ∎

EXERCISES

a. Show that if, in 21.1, the requirement that f_0 be nonnegative is removed, then f_0 is still the pointwise limit of a sequence of simple functions, but this sequence may no longer be monotone.

b. Show that if, in Exercise a, f_0 is unbounded both above and below, then the approximating sequence of simple functions cannot be monotone.

c. An *elementary function* is one which assumes a countable number of values and assumes each value on a measurable set. Show that if f_0 is everywhere finite, measurable, and nonnegative, then f_0 is the uniform limit of a monotone sequence of elementary functions.

d. Show that if f_0 is finite but unbounded, then it cannot be the uniform limit of a sequence of simple functions.

To illustrate the distinction between pointwise and uniform convergence, a standard example is the sequence f defined on $[0, 1]$ by

$$f_n(x) = x^n,$$

which converges pointwise, but not uniformly. We might say that the non-uniformity here is localized in the neighborhood of the point 1 in the sense that for $\varepsilon > 0$ we have uniform convergence on $[0, 1 - \varepsilon]$. This suggests the possibility that in any case of pointwise convergence, we can obtain uniform convergence by removing a rather small piece of the domain. This conjecture is essentially correct; the result is stated precisely as follows.

Theorem 21.3 (Egoroff). *If $\mu(\Omega) < \infty$ and if f is a sequence of measurable functions on Ω converging a. e. to f_0, then given $\varepsilon > 0$, there exists a measurable set E_0 such that $\mu(-E_0) < \varepsilon$ and such that f converges to f_0 uniformly on E_0.*

Proof. For each pair (k, n) of positive integers, let us define the set

$$E_{kn} = \bigcap_{m=n}^{\infty} \left\{ x \mid |f_m(x) - f_0(x)| < \frac{1}{k} \right\}.$$

Clearly, each of these sets is measurable. If

$$H = \{x \mid \lim_m f_m(x) = f_0(x)\},$$

then for each k,

$$\bigcup_{n=1}^{\infty} E_{kn} \supset H.$$

For each k, E_k is an expanding sequence of measurable sets; so we have from 10.3 that for each k,

$$\lim_n \mu(E_{kn}) = \mu\left(\bigcup_{n=1}^{\infty} E_{kn}\right) \geq \mu(H) = \mu(\Omega),$$

whence

$$\lim_n \mu(-E_{kn}) = 0. \tag{1}$$

Thus, given $\varepsilon > 0$, we have that for each k there is a positive integer n_k such that

$$\mu(-E_{kn_k}) < \frac{\varepsilon}{2^k}.$$

Let

$$E_0 = \bigcap_{k=1}^{\infty} E_{kn_k};$$

then E_0 is measurable, and

$$\mu(-E_0) = \mu\left[\bigcup_{k=1}^{\infty} (-E_{kn_k})\right] \leq \sum_{k=1}^{\infty} \mu(-E_{kn_k}) < \sum_{k=1}^{\infty} \frac{\varepsilon}{2^k} = \varepsilon.$$

Furthermore, given any integer k, it follows from the definition of the sets E_{kn} that for all $m \geq n_k$,

$$|f_m(x) - f_0(x)| < \frac{1}{k} \tag{2}$$

for every $x \in E_{kn_k}$. Since $E_0 \subset E_{kn_k}$ for each k, the condition $m \geq n_k$ yields (2) for every $x \in E_0$, and this is the definition of uniform convergence on E_0. ∎

There are two restrictive hypotheses in Egoroff's theorem; namely, that $\mu(\Omega) < \infty$ and that the functions f_n be measurable. The exercises that follow indicate some of the effects of removing either or both of these hypotheses. Briefly, the situation is this. If either the finiteness or the measurability condition is removed, the result as stated in 21.3 may fail (see Exercises e and i). However, if the conclusion is stated differently, we can still prove the theorem in certain more general cases.

21.3.1 Alternate statement of conclusion to Egoroff's theorem. Given $K < \mu(\Omega)$, there exists a set E_0 such that $\mu^*(E_0) > K$ and such that f converges to f_0 uniformly on E_0.

The reader should note that if E_0 is measurable and $\mu(\Omega) < \infty$, then 21.3.1 is equivalent to the conclusion in 21.3. However, if E_0 is nonmeasurable or if $\mu(\Omega) = \infty$, then 21.3.1 indicates a weaker result. This weakening of the conclusion is one way of extending the theorem. A different (and probably more useful) approach to the case of infinite measure is given in 37.1.

EXERCISES

e. Let Q be the sequence of nonmeasurable sets of Section 18, and let f_n be the characteristic function of the set $\bigcup_{i=n}^{\infty} Q_i$. Show that the result in 21.3 fails for this sequence of nonmeasurable functions.

f. Impose the outer measure of Exercise 11.h on the set of positive integers, and let f_n be the characteristic function of the set $\{k \mid k \geq n\}$. Show that even 21.3.1 fails in this case.

g. Generalize Exercise f to the case of any outer measure for which the limit theorem 12.1.1 fails.

h. Show that if Ω is a countable space and if μ^* is an outer measure for which $\lim_n \mu^*(E_n) = \mu^*(\lim_n E_n)$ for every expanding sequence E, then 21.3.1 holds for every sequence f such that $\lim_n f_n = f_0$ [a. e.]. Thus, in particular, this result applies to every countable space with a regular outer measure.

i. Use Lebesgue measure in R_1, and let f_n be the characteristic function of the set $[n, \infty]$. Show that the result in 21.3 fails for this sequence of measurable functions on a space of infinite measure.

j. Give an example to show that 21.3 cannot be strengthened to say that $\mu(-E_0) = 0$.

k. Noting the conclusion to 21.3 reapply it on the set $-E_0$. Continue this process to obtain the result that there is a set $H_0 = \bigcup_{i=1}^{\infty} H_i$ such that $\mu(-H_0) = 0$ and such that for each i, f converges to f_0 uniformly on H_i (compare Exercise j).

l. Show that if the functions f_n are measurable, then the result in Exercise k holds provided only that Ω is the union of a countable class of measurable sets of finite measure.

m. Show that the restriction on Ω in Exercise l is essential. *Hint:* Impose the outer measure of Exercise 11.f on the unit interval. Consider the sequence of continuous functions in Exercise 20.j. Show that this sequence cannot converge uniformly on any interval; thus the sets H_i (which may be considered closed) are nowhere dense in $[0, 1]$. So if the result in Exercise k held, $[0, 1]$ would be of Cat. I in itself.

n. Show that if Ω is as in Exercise l, then 21.3.1 holds for sequences f of measurable functions for which $\lim_n f_n = f_0$ [a. e.].

Egoroff's theorem localizes the nonuniformity of convergence for sequences of measurable functions. In the following theorem, we take a single measurable function and localize its discontinuities.

Theorem 21.4 (Lusin). *Let μ^* be an outer measure to which 13.7 applies, and let $\mu(\Omega)$ be finite. If f is an a. e. finite, measurable function on Ω, then given $\varepsilon > 0$, there is a closed set F such that $\mu(-F) < \varepsilon$ and such that the restriction of f to F is continuous.*

Proof. First, let us consider the case in which f is a simple function:

$$f = \sum_{k=1}^{n} a_k C_{E_k}.$$

Let $\varepsilon > 0$ be given; by 13.7.2 there is, for each $k (k = 1, 2, \ldots, n)$, a closed set $F_k \subset E_k$ such that

$$\mu(E_k - F_k) < \frac{\varepsilon}{n+1}.$$

We add one more closed set

$$F_{n+1} \subset -\bigcup_{k=1}^{n} E_k$$

such that

$$\mu\left[\left(-\bigcup_{k=1}^{n} E_k\right) - F_{n+1}\right] < \frac{\varepsilon}{n+1}.$$

Then

$$F = \bigcup_{k=1}^{n+1} F_k$$

is closed, and $\mu(-F) < \varepsilon$. Clearly, the restriction of f to F is continuous; the sets F_k are disjoint and closed, and f is constant on each.

For the general case, we note that $f = f^+ - f^-$, so we may restrict ourselves to the case of nonnegative f. By 21.1,

$$f = \lim_n f_n \text{ [a. e.]},$$

where the functions f_n are simple functions. Using the result already established for simple functions, given $\varepsilon > 0$, we have for each n, a closed set F_n such that

$$\mu(-F_n) < \frac{\varepsilon}{2^{n+1}}$$

and such that the restriction of f_n to F_n is continuous. Now, we let

$$F_0 = \bigcap_{n=1}^{\infty} F_n;$$

then F_0 is closed, and

$$\mu(-F_0) = \mu\left[\bigcup_{n=1}^{\infty}(-F_n)\right] \leq \sum_{n=1}^{\infty} \mu(-F_n) < \sum_{n=1}^{\infty} \frac{\varepsilon}{2^{n+1}} = \frac{\varepsilon}{2}.$$

By Egoroff's theorem (using F_0 for the space) we see that there is a measurable set $E \subset F_0$ such that $\mu(F_0 - E) < \varepsilon/4$ [hence $\mu(-E) < 3\varepsilon/4$] and such that $\lim_n f_n = f$ uniformly on E. Since the restriction of f_n to E is continuous for each n, the restriction of f to E is continuous. To complete the proof, we take a closed set $F \subset E$ such that $\mu(E - F) < \varepsilon/4$. Then, $\mu(-F) < \varepsilon$, and the restriction of f to F is continuous. ∎

EXERCISES

o. Show that if Ω is the space R_n, the requirement $\mu(\Omega) < \infty$ may be dropped from Lusin's theorem. [*Hint:* Partition R_n into disjoint half-open intervals, and show that if there is a closed set contained in each of these intervals, the union of this particular sequence of closed sets is closed.]

p. Given an outer measure μ^* for which 13.7 holds and $\mu(\Omega) < \infty$, an a. e. finite measurable function f, and $\varepsilon > 0$, show that there is a continuous function g such that $\mu\{x \mid f(x) \neq g(x)\} < \varepsilon$.

q. Let μ^* satisfy 13.7, and let $\mu(\Omega) < \infty$. Show that if f is an a. e. finite, measurable function, then there is a set $H \in \mathscr{F}_\sigma$ such that $\mu(-H) = 0$ and such that the restriction of f to H is a function of Baire class one.

r. Point out the distinction between the result in Exercise q and the result that f be equal a. e. to a function of Baire class one.

s. Give an example to show that the result in Exercise q cannot be extended to that suggested in Exercise r. *Hint:* Construct $E \subset [0, 1]$ (as a union of Cantor sets, for example) so that $\mu(E) = \mu(-E) = \frac{1}{2}$ and so that both E and $-E$ are dense in $[0, 1]$. Consider C_E in the light of Exercise 20.i.

t. Show that, given the hypotheses of Exercise *q*, it follows that *f* is equal a. e. to a function of Baire class two. [*Hint:* Show that if *F* is closed and the restriction of *f* to *F* is continuous, then the function *g*, equal to *f* on *F* and 0 on $-F$, is of Baire class one.]

*22 RANDOM VARIABLES

We have already pointed out (Section 15) that the general mathematical model for probability consists of a space Ω and a measure μ for which $\mu(\Omega) = 1$. In probability theory a measurable function is called a *random variable*. Roughly, the picture is this. The points of Ω represent the various physical results with which we are concerned, so a measurable function over Ω associates a number with each of these physical results. In other words, the values of a random variable *f* are numerical measurements associated with the physical results. Thus we might say that a random variable is a real valued function whose values are determined by chance.

For example, the statement, "A point is chosen at random in the interval $[0, 1]$," may be explained rigorously as follows. Let μ be Lebesgue measure on $[0, 1]$ and for each measurable set $E[0, 1]$ define the probability that the chosen point will be in E to be $\mu(E)$. Many random variables could be defined over these points. For instance, the first entry in the decimal expansion of the chosen point is given by the value of a random variable *f* which is a step function. Specifically, $f(1) = 0$, and $f(x) = k/10$ for $k/10 \leq x < (k + 1)/10$ ($k = 0, 1, 2, \ldots, 9$).

To turn to a more important example, the statistical process known as *sampling* consists of taking *n* numerical measurements whose values are governed by chance. As we pointed out in Section 15, a situation of this kind is naturally represented by the space R_n. A point $x \in R_n$ represents an aggregate of *n* sample values, the coordinates of *x* being given by the individual samples. An important set of random variables over this space is the set of coordinate variables; that is, the set of functions f_i defined by

$$f_i(x) = x_i.$$

The value of f_i at any point is the *i*th sample measurement associated with the point.

The statistician is interested in many random variables over this sample space. One of the most important is the *sample mean*. This is the function *f* defined by

$$f(x) = \frac{1}{n} \sum_{i=1}^{n} x_i.$$

The value of this random variable at any point *x* is the arithmetic mean of the sample measurements associated with *x*.

We do not intend to go into statistical theory here. We only want to point out that in the modern theory of probability the "variates" of the statistician are measurable functions over a measure space Ω.

One question of measure theoretic interest in connection with random variables is the following. We have said informally that a random variable is a function whose values are governed by chance. Does this mean that we can assign to every measurable set $E \subset R_1$ a probability that the value of the random variable will lie in E? We can answer this in the affirmative as follows.

Let μ be a probability measure in Ω, and let f be a random variable over Ω. For each $t \in R_1$, we define

$$p(t) = \mu[f^{-1}\{(-\infty, t]\}];$$

p is monotone because μ is, and p is right continuous because of 10.3.1. The function p is called a *distribution function for the random variable f*, and the Lebesgue-Stieltjes measure μ_p (see Section 14) gives a suitable probability for any measurable set of values of f.

To generalize this idea, let $f_1, f_2, \ldots, f_n$ be any set of n random variables over Ω, and let μ be a probability measure in Ω. For each point $t \in R_n$, let

$$p(t) = \mu\{x \mid f_i(x) \leq t_i; i = 1, 2, \ldots, n\}.$$

The function p is called a *joint distribution function for the random variables $f_1, f_2, \ldots, f_n$*. It follows from the additivity and nonnegativity of μ that p satisfies 14.3, and it follows from 10.3.1 that p satisfies 14.4. So, it may be used to generate a Lebesgue-Stieltjes measure μ_p in R_n. If the measure μ_p is a product measure, $f_1, f_2, \ldots, f_n$ are called *independent random variables*.

The gist of the above discussion is that any set of n random variables over a space Ω of any sort may be represented as the coordinate variables in R_n with a Lebesgue-Stieltjes probability measure. Finally, we might note that if f is a sequence of random variables over Ω, the representation of each finite set of them as coordinate variables in a space of n-tuples and the construction given in Section 15 will give a representation of f as the sequence of coordinate variables in s, with an appropriate probability measure in s.

EXERCISES

a. Let p be a distribution function for the random variable f. Show that a distribution function q for $|f|$ is given by

$$q(x) = \begin{cases} p(x) - \lim_{y \to x^+} p(-y) & \text{for } x > 0, \\ 0 & \text{for } x \leq 0; \end{cases}$$

one for f^2 is given by

$$q(x) = \begin{cases} p(\sqrt{x}) - \lim_{y \to x^+} p(-\sqrt{y}) & \text{for } x > 0, \\ 0 & \text{for } x \leq 0; \end{cases}$$

one for f^3 is given by

$$q(x) = p(\sqrt[3]{x}).$$

b. If two honest dice are thrown, the possible results are naturally represented by the 36 points in the xy plane given by $x = 1, 2, \ldots, 6$; $y = 1, 2, \ldots, 6$. The measure is described by giving each of these points measure $1/36$. The random variable f defined by $f(x, y) = x + y$ represents the total thrown. Describe the measure μ_p in R_1 determined by f.

c. Let Ω be the unit square, let μ be Lebesgue measure, and let f be the random variable defined by $f(x, y) = x + y$. Show that a distribution function for f is given by

$$p(t) = \begin{cases} t^2/2 & \text{for } 0 \le t \le 1, \\ 1 - (1-t)^2/2 & \text{for } 1 \le t \le 2, \\ 0 & \text{otherwise.} \end{cases}$$

d. Let μ be a probability measure in Ω; let f be a random variable on Ω; and let μ_p be the measure in R_1 determined by μ and f. Show that if E is any Borel set in R_1, then

$$\mu_p(E) = \mu[f^{-1}(E)].$$

e. With μ, f, and μ_p as in Exercise d, show that if $\mu_p(E) = 0$, then $\mu[f^{-1}(E)] = 0$. [*Hint:* Use $\mathscr{G}_\delta$ covering sets.]

f. Show that if μ, f, and μ_p are related as in Exercise d and if E is any μ_p^*-measurable set in R_1, then $f^{-1}(E)$ is μ^*-measurable, and

$$\mu_p(E) = \mu[f^{-1}(E)].$$

g. Contrast the measurability situation in Exercise f with that in the example given at the end of Section 19. Explain the seeming discrepancy.

h. State and prove a generalization of Exercise f to the case of n random variables and a measure μ_p in R_n.

i. Let $f_1, f_2, \ldots, f_n$ be independent random variables; let $g_1, g_2, \ldots, g_n$ be Baire functions on R_1. Show that the random variables $g_i \circ f_i$ are independent.

REFERENCES FOR FURTHER STUDY

On measurable functions:
Hahn and Rosenthal [9]
Halmos [11]
Hewitt and Stromberg [13]
McShane and Botts [20]

On random variables:
Halmos [10]

CHAPTER 5

INTEGRATION

One of the simplest interpretations of a definite integral is that of the area under a curve. In one sense the Lebesgue integral, or more generally any integral generated by a measure function, is a generalization of this notion of area. Indeed, the definition of an integral presented in this chapter is rather obviously just such a generalization. We define the integral of a simple function as the sum of the "areas" under its constant sections; then we employ 21.1 and take limits to get the integral of a more general function. Incidentally, the monotonicity of the sequence f in 21.1 is an invaluable aid in this procedure, because it simplifies the question of the existence of the limits in which we are interested.

However, there is another point of view, one suggested by the notion of an indefinite integral. In the usual notation of elementary calculus,

$$F(x) = \int_a^x f(t)\, dt$$

defines a function F of a real variable, called the indefinite integral of f. By using measure theory we shall define integrals over more general sets than intervals, so it is more apropos to think of F as a distribution function and designate as the indefinite integral of f the measure function (function of sets) induced by F.

Specifically, our notation for an integral will be $\int_E f$, read "the integral over E of f." In the language of function theory, to assign a value of $\int_E f$ for every measurable set E and every integrable function f is to define a function $\int$ over the set of all ordered pairs (E, f), where E is a measurable set and f is an integrable function. Then, according to the functional notation we are using here, $\int f$ is a function over the class of measurable sets. It is this function that we should naturally call the indefinite integral of f. In Sections 26 and 27 we turn from the picture of area under a curve and characterize the integral in terms of properties of the function $\int f$.

The fact that $\int$ is a function of two arguments suggests still another point of view. That is, what are the properties of the function $\int_E$ over the set of all integrable functions? Can the integral be characterized by the properties of this function? Results in this direction must be postponed until Chapter 8, but they are discussed there (Section 46).

The values of the function $\int$, as well as the determination of its domain, depend on the measure function μ. As in previous chapters, we shall speak of measurable sets, measurable functions, and integrals with the tacit assumption that some outer measure is given. Where two different measure functions appear in one discussion, we shall modify our notation slightly and write $\int_E f \, d\mu_1$, $\int_E f \, d\mu_2$. So far as we are concerned, the "d" in this notation signifies nothing except a concession to established conventions.

23 THE INTEGRAL OF A SIMPLE FUNCTION

Let f be a simple function, and let E be a measurable set. We say that f is *integrable on E* if

$$\mu(E \cap \{x \mid f(x) \neq 0\}) < \infty.$$

Clearly, this is equivalent to the condition that if

23.1 $$f = \sum_{k=1}^{n} a_k C_{E_k},$$

where $a_k \neq 0$ ($k = 1, 2, \ldots, n$), then

$$\mu\left[\bigcup_{k=1}^{n} (E \cap E_k)\right] < \infty.$$

If a simple function f is integrable on Ω, we shall say merely that f is *integrable*. Obviously, if a simple function is integrable, it is integrable on every measurable set. If the simple function f described by 23.1 is integrable on a measurable set E, we define

23.2 $$\int_E f = \sum_{k=1}^{n} a_k \mu(E \cap E_k).$$

If, in 23.1, $a_j \neq a_k$ for $j \neq k$, then any other description of f as a linear combination of characteristic functions amounts to a partitioning of the sets E_k. From this comment and the additivity of μ, it follows at once that $\int_E f$ is independent of the description of f in terms of characteristic functions.

Theorem 23.3. *Let $E_1, E_2, \ldots, E_n$ be disjoint measurable sets with*

$$E_0 = \bigcup_{k=1}^{n} E_k,$$

and let f be a simple function which is integrable on each set E_k; then f is integrable on E_0, and

$$\int_{E_0} f = \sum_{k=1}^{n} \int_{E_k} f.$$

Proof. Let $A = \{x \mid f(x) \neq 0\}$; then

$$\mu(E_0 \cap A) = \sum_{k=1}^{n} \mu(E_k \cap A) < \infty,$$

so f is integrable on E_0. Let
$$f = \sum_{i=1}^{m} a_i C_{A_i};$$
then
$$\sum_{k=1}^{n} \int_{E_k} f = \sum_{k=1}^{n} \sum_{i=1}^{m} a_i \mu(E_k \cap A_i) = \sum_{i=1}^{m} a_i \sum_{k=1}^{n} \mu(E_k \cap A_i)$$
$$= \sum_{i=1}^{m} a_i \mu(E_0 \cap A_i) = \int_{E_0} f. \blacksquare$$

Theorem 23.4. *If f and g are simple functions, each integrable on E, and if a and b are real numbers, then $af + bg$ is integrable on E and*
$$\int_E (af + bg) = a \int_E f + b \int_E g.$$

Proof. Integrability of $af + bg$ follows from the fact that
$$\{x \mid af(x) + bg(x) \neq 0\} \subset \{x \mid f(x) \neq 0\} \cup \{x \mid g(x) \neq 0\}.$$
To prove the equation, let
$$f = \sum_{k=1}^{n} a_k C_{A_k}, \quad g = \sum_{i=1}^{m} b_i C_{B_i}.$$
Then for each i, k, we have
$$\int_{A_k \cap B_i \cap E} (af + bg) = (aa_k + bb_i)\mu(A_k \cap B_i \cap E)$$
$$= a \int_{A_k \cap B_i \cap E} f + b \int_{A_k \cap B_i \cap E} g.$$
By introducing zero coefficients a_k and b_i, if necessary, we may assume that
$$\bigcup_{k=1}^{n} \bigcup_{i=1}^{m} (A_k \cap B_i) \supset E;$$
so the result follows from 23.3. $\blacksquare$

Theorem 23.5. *If f and g are simple functions, if f is integrable on E, and if for each $x \in E$, $f(x) \geq |g(x)|$, then g is integrable on E, and*
$$\int_E f \geq \int_E g.$$

Proof. Integrability of g follows from the fact that
$$\{x \mid g(x) \neq 0\} \subset \{x \mid f(x) \neq 0\}.$$
If $h = f - g$, then h is a nonnegative simple function. It is obvious from 23.2 that $\int_E h \geq 0$; so by 23.4,
$$\int_E f - \int_E g = \int_E h \geq 0. \blacksquare$$

EXERCISES

a. A *step function* on R_1 is a function which is constant on each of a finite class of intervals. Show that the integral of such a function with respect to Lebesgue measure gives the area under the curve.

b. Let Ω be a finite point set and let $\mu(E)$ be the number of points in E. Show that all functions on Ω are simple functions and that the theory of integration reduces to the theory of finite sums.

c. Let μ_1 and μ_2 be two measures with the same domain, and let $v = \mu_1 + \mu_2$. Show that if f is a simple function and is both μ_1- and μ_2-integrable on E, then f is v-integrable on E, and

$$\int_E f\, dv = \int_E f\, d\mu_1 + \int_E f\, d\mu_2.$$

The following results on integrals of simple functions are required for our general discussion of integration in the next section.

Theorem 23.6. *Let f be a nondecreasing sequence of nonnegative simple functions, each integrable on E; and let f_0 be a nonnegative simple function such that for each $x \in E$,*

$$\lim_n f_n(x) \geq f_0(x).$$

If f_0 is nonintegrable on E, then

$$\lim_n \int_E f_n = \infty;$$

if f_0 is integrable on E, then

$$\lim_n \int_E f_n \geq \int_E f_0.$$

Proof. Existence of $\lim_n \int_E f_n$ follows from 23.5 and the monotonicity of f.

If f_0 is nonintegrable on E, there is a set A_0 such that $f_0(x) = a > 0$ for $x \in A_0$ and such that $\mu(A_0 \cap E) = \infty$. Let

$$A_n = \left\{ x \mid f_n(x) \geq \frac{a}{2} \right\};$$

then A is an expanding sequence of measurable sets with $\lim_n A_n \supset A_0$, so we have

$$\lim_n \int_E f_n \geq \lim_n \left(\frac{a}{2}\right) \mu(A_n \cap E) \geq \left(\frac{a}{2}\right) \mu(A_0 \cap E) = \infty.$$

If f_0 is integrable on E and if $E_0 = \{x \mid f_0(x) > 0\}$, then $\mu(E \cap E_0) < \infty$. Furthermore, there is a number M such that $f_0(x) \leq M$ for all $x \in E$. If $\mu(E \cap E_0) = 0$, the theorem is trivial, so we assume that $\mu(E \cap E_0) > 0$. Then, given $\varepsilon > 0$, we let

$$B_n = \left\{ x \mid f_n(x) \geq f_0(x) - \frac{\varepsilon}{2\mu(E \cap E_0)} \right\}.$$

Then, B is an expanding sequence of measurable sets, and
$$\lim_n B_n \supset E \cap E_0;$$
therefore by 10.8.1,
$$\lim_n \mu(E \cap E_0 - B_n) = 0.$$
Thus, there is an integer n_0 such that for all $n > n_0$,
$$\mu(E \cap E_0 - B_n) < \frac{\varepsilon}{2M};$$
so for all $n > n_0$,
$$\int_E f_n \geq \int_{E \cap E_0 \cap B_n} f_n \geq \int_{E \cap E_0 \cap B_n} f_0 - \frac{\varepsilon \mu(E \cap E_0 \cap B_n)}{2\mu(E \cap E_0)}$$
$$\geq \int_{E \cap E_0 \cap B_n} f_0 - \frac{\varepsilon}{2} = \int_{E \cap E_0} f_0 - \int_{E \cap E_0 - B_n} f_0 - \frac{\varepsilon}{2}$$
$$\geq \int_{E \cap E_0} f_0 - \frac{M\varepsilon}{2M} - \frac{\varepsilon}{2} = \int_E f_0 - \varepsilon.$$

Therefore, $\lim_n \int_E f_n \geq \int_E f_0 - \varepsilon$ for every $\varepsilon > 0$, and the result follows at once. ∎

Corollary 23.6.1. *Let f be a nondecreasing sequence of nonnegative simple functions, each integrable on E; let f_0 be a simple function such that for each $x \in E$,*
$$\lim_n f_n(x) = f_0(x).$$
Then, f_0 is integrable on E if and only if
$$\lim_n \int_E f_n < \infty,$$
and if this is the case,
$$\lim_n \int_E f_n = \int_E f_0.$$

Proof. If f_0 is nonintegrable, it follows from 23.6 that
$$\lim_n \int_E f_n = \infty.$$
If f_0 is integrable, then by 23.5 and the monotonicity of f,
$$\int_E f_n \leq \int_E f_0$$
for every n; so
$$\lim_n \int_E f_n \leq \int_E f_0 < \infty.$$
The reverse inequality comes from 23.6. ∎

Corollary 23.6.2. *If f and g are two nondecreasing sequences of nonnegative, integrable, simple functions such that for each* $x \in E$,

$$\lim_n f_n(x) = \lim_n g_n(x),$$

then

$$\lim_n \int_E f_n = \lim_n \int_E g_n.$$

Proof. For any fixed positive integer k,

$$\lim_n f_n(x) \geq g_k(x)$$

for each $x \in E$; therefore by 23.6,

$$\lim_n \int_E f_n \geq \int_E g_k$$

for each k. Thus

$$\lim_n \int_E f_n \geq \lim_k \int_E g_k.$$

By interchanging the roles of f and g in this argument, we get the reverse inequality. ∎

EXERCISES

d. Give a separate proof of 23.6.1, not using 23.6.

e. Show that the hypothesis that f is monotone may be removed in 23.6. [*Hint:* Replace the sets B_n by $\bigcap_{k=n}^{\infty} B_k$.]

f. Show that the hypothesis that each f_n is nonnegative may be removed in 23.6. [*Hint:* Replace the functions f_n by $f_n - f_1$.]

g. Show that the modifications of 23.6 suggested in Exercises e and f cannot be made simultaneously. [*Hint:* Let $f_n = -nC_{[0, 1/n]}$.]

24 INTEGRABLE FUNCTIONS

Let f_0 be any nonnegative, measurable function, and let E be any measurable set. We say that f_0 is *integrable on* E if there exists a nondecreasing sequence f of nonnegative simple functions, each integrable on E, such that for each $x \in E$,

$$\lim_n f_n(x) = f_0(x)$$

and such that†

$$\lim_n \int_E f_n < \infty.$$

---

† This condition is a standard one in the definition of integrability. Many authors distinguish between the statements "f is integrable on E" and "$\int_E f$ is defined," thereby allowing the possibility that $\int_E f = \infty$. We shall regard these statements as equivalent and so require that $\int_E f$ always be finite.

If this is the case, we define

24.1
$$\int_E f_0 = \lim_n \int_E f_n.$$

This definition calls for several comments. First, it follows from 23.5 that for any monotone sequence f of integrable simple functions, $\lim_n \int_E f_n$ exists in the extended real number system. Next, it follows from the second part of 21.2 that f_0 cannot be integrable unless $\{x \mid f_0(x) > 0\}$ is the union of a countable class of sets of finite measure. Furthermore, because of 21.1 and the first part of 21.2, f_0 is the pointwise limit of a monotone sequence of nonnegative, integrable, simple functions whenever $\{x \mid f_0(x) > 0\}$ has this property; though certainly, not every such f_0 is integrable. From 23.6.1 it follows that for simple functions the definition 24.1 is consistent with 23.2. Finally, 23.6.2 tells us that the definition 24.1 is independent of the choice of the sequence f.

If f is any measurable function, we say that f is *integrable on E* if and only if f^+ and f^- are both integrable on E. If this is the case, we define

24.2
$$\int_E f = \int_E f^+ - \int_E f^-.$$

Theorem 24.3. *If f_0 is integrable on E, it is integrable on every measurable subset of E; furthermore, if f_0 is nonnegative and integrable on E, then $\int f$ is a nondecreasing set function on the measurable subsets of E.*

Proof. For nonnegative simple functions, both results are obvious. Thus, if f is a nondecreasing sequence of nonnegative, integrable, simple functions and if $A \subset E$, then

$$\lim_n \int_A f_n \leq \lim_n \int_E f_n,$$

and both results are extended to arbitrary, nonnegative, integrable functions. The proof of the theorem is completed by noting that in 24.2 integrability for any function is defined in terms of integrability of nonnegative functions. ∎

In the light of 24.3, we frequently speak of an *integrable function*, meaning integrable on Ω. Such a function is integrable on every measurable set.

EXERCISES

a. Show that if $\mu(\Omega) < \infty$, then every bounded measurable function is integrable.

b. Generalize Exercise 23.c to the case of an arbitrary integrable function f.

Let Ω be the real number system, and let μ be Lebesgue measure; then the integral defined in this section is called the *Lebesgue integral*. The integral of elementary calculus is called the *Riemann integral*. It may be defined as follows.

122 INTEGRATION

Let f be a bounded function on a closed interval $[a, b]$. Define a partition of $[a, b]$ by a finite set K of points $x_1, x_2, \ldots, x_n$ such that

$$a = x_0 < x_1 < x_2 < \cdots < x_n = b.$$

For each $x_i \in K$, let

$$M_i = \sup \{f(x) \mid x_{i-1} \leq x \leq x_i\},$$
$$m_i = \inf \{f(x) \mid x_{i-1} \leq x \leq x_i\}.$$

The *upper and lower Darboux sums* are given by

$$\bar{S} = \sum_{x_i \in K} M_i(x_i - x_{i-1}),$$

$$\underline{S} = \sum_{x_i \in K} m_i(x_i - x_{i-1}).$$

If, for every sequence of partitions of $[a, b]$ for which the maximum length of the subdivisions tends to zero, the sums $\bar{S}$ and $\underline{S}$ tend to the same limit, then we say that f is *Riemann integrable* on $[a, b]$ and we define

$$\int_a^b f(x)\, dx$$

as this common limit of $\bar{S}$ and $\underline{S}$.

In order to fit the general theory of integration we are developing here to the "ordinary" integrals in everyday use, we shall show that the Lebesgue integral is a generalization of the Riemann integral. That is, if f is Riemann integrable on $[a, b]$, then it is Lebesgue integrable on $[a, b]$, and

$$\int_a^b f(x)\, dx = \int_{[a,b]} f.$$

The standard theorem on Riemann integrability is as follows:

Theorem 24.4 (Lebesgue). *A necessary and sufficient condition that a bounded function f be Riemann integrable on $[a, b]$ is that it be continuous a. e. in $[a, b]$.*

Proof. First, we prove the condition necessary. For each positive integer k, let

$$E_k = \left\{ x \mid \frac{1}{k} > \lim_{\delta \to 0} \sup_{\substack{|y-x|<\delta \\ |z-x|<\delta}} |f(y) - f(z)| \right\};$$

then if E_0 is the set on which f is discontinuous,

$$E_0 = \bigcup_{k=1}^{\infty} E_k.$$

We now assume that $\mu^*(E_0) > 0$; then for some k_0,

$$\mu^*(E_{k_0}) = c > 0.$$

Let K be an arbitrary finite set defining a partition of $[a, b]$; the open intervals of this partition cover all of E_{k_0} except a finite set, so they cover a subset of outer measure c. Therefore, there is a set $K' \subset K$ of partition points such that for $x_i \in K'$,
$$(x_{i-1}, x_i) \cap E_{k_0} \neq \phi$$
and such that
$$\sum_{x_i \in K'} (x_i - x_{i-1}) \geq c.$$
It follows from the definition of E_{k_0} that for $x_i \in K'$,
$$M_i - m_i > \frac{1}{k_0};$$
so if $\bar{S}$ and $\underline{S}$ are the Darboux sums for the partition K,
$$\bar{S} - \underline{S} = \sum_{x_i \in K} (M_i - m_i)(x_i - x_{i-1})$$
$$\geq \sum_{x_i \in K'} (M_i - m_i)(x_i - x_{i-1}) > \frac{c}{k_0}.$$

Since $c/k_0 > 0$ is fixed and since this holds for every partition K of $[a, b]$, $\bar{S}$ and $\underline{S}$ cannot tend to the same limit; so f is not Riemann integrable, and necessity of the condition is proved.

To prove sufficiency, let $\varepsilon > 0$ be given, and let
$$M = \sup \{|f(x)| \mid x \in [a, b]\}.$$
We cover the set E_0 of discontinuities of f by an open set G with
$$\mu(G) < \frac{\varepsilon}{16M}. \tag{1}$$
Then, $F = [a, b] - G$ is closed, hence compact, and f is continuous on F; so by 5.6 there is a $\delta > 0$ such that if $x \in F$ and $|x - y| < \delta$, then
$$|f(x) - f(y)| < \frac{\varepsilon}{8(b - a)}. \tag{2}$$
Let K_1 and K_2 be two finite sets of points, each defining a partition of $[a, b]$ for which the maximum length of a subinterval is less than δ, and let $K = K_1 \cup K_2$. We now form a Riemann sum,
$$S_1 = \sum_{x_i^1 \in K_1} f(y_i^1)(x_i^1 - x_{i-1}^1),$$
where for each i, $y_i^1 \in [x_{i-1}^1, x_i^1]$. We form a similar sum,
$$S = \sum_{x_j \in K} f(z_j)(x_j - x_{j-1}),$$

with the added restriction that $z_j \in F \cap [x_{j-1}, x_j]$ unless
$$F \cap [x_{j-1}, x_j] = \phi. \tag{3}$$
Since each subinterval determined by K is a subset of one determined by K_1, we can (by repeating the value of y when necessary) write S_1 as a sum over K:
$$S_1 = \sum_{x_j \in K} f(y_j)(x_j - x_{j-1}),$$
where $y_j = y_i^1$ for $[x_{j-1}, x_j] \subset [x_{i-1}^1, x_i^1]$. Then,
$$|S - S_1| \le \sum_{x_j \in K} |f(z_j) - f(y_j)|(x_j - x_{j-1}).$$
Now, we let K' be the set of points $x_j \in K$ for which (3) holds; then it follows from (1) that
$$\sum_{x_j \in K'} |f(z_j) - f(y_j)|(x_j - x_{j-1}) \le 2M\mu(G) < \frac{\varepsilon}{8}.$$
Recalling that $|z_j - y_j| < \delta$ and that $z_j \in F$ for $x_j \in K - K'$, we have from (2) that
$$\sum_{x_j \in K - K'} |f(z_j) - f(y_j)|(x_j - x_{j-1}) < \frac{(b-a)\varepsilon}{8(b-a)} = \frac{\varepsilon}{8}.$$
Thus
$$|S - S_1| < \frac{\varepsilon}{4}.$$

It follows in a similar manner that $|S - S_2| < \varepsilon/4$ where S_2 is any Riemann sum for the partition defined by K_2; therefore
$$|S_1 - S_2| \le |S_1 - S| + |S - S_2| < \frac{\varepsilon}{2}.$$

Since each point y_i^1 was chosen arbitrarily in $[x_{i-1}^1, x_i^1]$, we can clearly make this choice so that $|\bar{S}_1 - S_1| < \varepsilon/4$ where $\bar{S}_1$ is the upper Darboux sum for the partition defined by K_1. Similarly, we can get $|\bar{S}_2 - S_2| < \varepsilon/4$. Thus, we have
$$|\bar{S}_1 - \bar{S}_2| \le |\bar{S}_1 - S_1| + |S_1 - S_2| + |S_2 - \bar{S}_2| < \varepsilon, \tag{4}$$
provided only that all subintervals in the partitions defining $\bar{S}_1$ and $\bar{S}_2$ are less than δ in length. So, (4) gives us a Cauchy condition for any significant sequence of upper sums, and it follows that every such sequence tends to a limit. Regarding $\bar{S}_1$ and $\bar{S}_2$ as sample terms from different sequences of upper sums, it follows from (4) that all such sequences tend to the same limit. Finally, letting $K_1 = K_2$ and choosing points y_i^1 so that $|S_1 - \underline{S}_1| < \varepsilon/4$, we have (with S_2 as above),
$$|\bar{S}_1 - \underline{S}_1| \le |\bar{S}_1 - S_2| + |S_2 - S_1| + |S_1 - \underline{S}_1| < \varepsilon,$$
so any sequence of lower sums tends to the same limit as the corresponding sequence of upper sums; therefore f is Riemann integrable. ∎

Theorem 24.5. *If a bounded function f_0 is Riemann integrable on $[a, b]$, then it is Lebesgue integrable on $[a, b]$, and*

$$\int_a^b f_0(x)\, dx = \int_{[a,b]} f_0.$$

Proof. Let E_1 and E_2 be, respectively, the sets where f_0 is continuous and discontinuous; then by 24.4, $\mu(E_2) = 0$. For $i = 1, 2$, let f_{0i} be the restriction of f_0 to E_i, and let G be an open set of real numbers; then

$$f_0^{-1}(G) = f_{01}^{-1}(G) \cup f_{02}^{-1}(G).$$

Now, f_{01} is continuous; therefore $f_{01}^{-1}(G)$ is the intersection of an open set with the measurable set E_1. Since $f_{02}^{-1}(G)$ has measure zero, it is also measurable. Thus, f_0 is a measurable function. If f is a nondecreasing sequence of integrable simple functions converging pointwise to f_0^+, then for each n,

$$\int_{[a,b]} f_n \leq (b - a) \sup_{a \leq x \leq b} f_0^+(x),$$

and the same can be said for f_0^-. Thus, f_0 is Lebesgue integrable.

To see that the integrals are equal, we note first that if f_0 is bounded, then

$$f_0 - \inf_{a \leq x \leq b} f_0(x)$$

is nonnegative; so it suffices to consider the nonnegative case. Let $\underline{S}$ be a sequence of lower Darboux sums converging to $\int_a^b f_0(x)\, dx$. Each sum $\underline{S}_n$ is the integral of a nonnegative simple function $s_n \leq f_0$. Let g be a nondecreasing sequence of nonnegative simple functions converging pointwise of f_0; then the sequence f defined by

$$f_n = \max\left[g_n, \max_{i \leq n} s_i\right]$$

is a monotone sequence of nonnegative simple functions converging pointwise to f_0, and for each n,

$$\int_{[a,b]} f_0 = \lim_n \int_{[a,b]} f_n \geq \int_{[a,b]} f_n \geq \int_{[a,b]} s_n = \underline{S}_n.$$

Therefore

$$\int_{[a,b]} f_0 \geq \lim_n \underline{S}_n = \int_a^b f_0(x)\, dx.$$

To prove the reverse inequality, we note that any upper Darboux sum $\bar{S}$ is the integral of a simple function $h \geq f_0$. Thus, if f is any nondecreasing sequence of nonnegative simple functions converging pointwise to f_0, then for each n,

$$\int_{[a,b]} f_n \leq \int_{[a,b]} h = \bar{S},$$

whence
$$\int_{[a,b]} f_0 = \lim_n \int_{[a,b]} f_n \le \bar{S}.$$

Since this is true for every upper sum $\bar{S}$, it follows that
$$\int_{[a,b]} f_0 \le \int_a^b f_0(x)\, dx. \quad \blacksquare$$

Thus, for bounded functions on a bounded set the Lebesgue theory of integration contains the Riemann theory. For comments on the relation of the Lebesgue integral to the Cauchy-Riemann integral (the improper integral of elementary calculus) see Section 26.

Two definitions of an integral are *equivalent* if they lead to the same class of integrable functions and yield the same values of the function $\int$. There are in the literature quite a number of definitions equivalent to that given here. A few of these are summarized in the exercises that follow. For a proper appreciation of the subject of integration the reader should not only work out some of these exercises but also check the references listed to see how the properties of the integral are developed from these various definitions.

EXERCISES

c. Lebesgue [19][†] defines the integral of a bounded, nonnegative, measurable function on a space of finite measure as follows:
$$\int_E f = \lim_{n \to \infty} \sum_{k=0}^{nM} \frac{k}{n} \mu \left(E \cap \left\{ x \,\Big|\, \frac{k}{n} \le f(x) < \frac{k+1}{n} \right\} \right),$$
where M is an upper bound for f. Prove that this definition yields the same values for the integral as the one given in the text.

d. De la Vallée Poussin [29][‡] extends the above definition to unbounded, nonnegative, measurable functions by considering *truncated functions*:
$$f_n(x) = \begin{cases} f(x) & \text{for } f(x) < n, \\ n & \text{for } f(x) \ge n. \end{cases}$$
He then defines $\int_E f$ as $\lim_n \int_E f_n$. For functions with both positive and negative values, he uses 24.2. Show that this definition is equivalent to that given in the text.

e. Hobson [13] uses a modification of de la Vallée Poussin's approach that circumvents 24.2. For any measurable function, he employs the double truncates:
$$f_{mn}(x) = \begin{cases} n & \text{for } f(x) \ge n, \\ f(x) & \text{for } -m < f(x) < n, \\ -m & \text{for } f(x) \le -m. \end{cases}$$

---

† See also Hobson [14], Titchmarsh [28], and de la Vallée Poussin [29].
‡ See also Titchmarsh [28].

He then uses a double limit to define

$$\int_E f = \lim_{\substack{m \to \infty \\ n \to \infty}} \int_E f_{mn}.$$

Show that this definition is equivalent to that given in the text.

f. Saks [27] defines the integral of a nonnegative, measurable function as follows:

$$\int_E f = \sup \sum_{k=1}^{n} \mu(E_k) \inf_{x \in E_k} f(x),$$

where the supremum is taken over the collection of all finite classes of disjoint measurable sets such that

$$E = \bigcup_{k=1}^{n} E_k.$$

For functions with both positive and negative values he uses 24.2. Show that this definition is equivalent to the one given in the text.

g. Carathéodory [4]† defines the Lebesgue integral of a nonnegative, measurable function over R_n as the Lebesgue measure in R_{n+1} of the *ordinate set*: that is,

$$\int_E f = \mu\{(x, y) \mid x \in E; 0 \le y \le f(x)\}.$$

Prove that this definition is equivalent to the one given in the text.

25 ELEMENTARY PROPERTIES OF THE INTEGRAL

In 23.3 through 23.5 we proved that the integrals of simple functions have certain properties quite naturally associated with an integral. It is the purpose of this section to extend these and other properties to integrals in general.

Theorem 25.1. *Let* $E_1, E_2, \ldots, E_n$ *be disjoint measurable sets with*

$$E_0 = \bigcup_{k=1}^{n} E_k,$$

and let f be integrable on each set E_k; *then f is integrable on* E_0, *and*

$$\int_{E_0} f = \sum_{k=1}^{n} \int_{E_k} f.$$

Proof. By 23.3, this holds for simple functions; by passing to the limit, we see that it holds for nonnegative functions. The general case then follows by 24.2. ∎

Theorem 25.2. *If f and g are each integrable on E and if a and b are real numbers, then af + bg is integrable on E, and*

$$\int_E (af + bg) = a \int_E f + b \int_E g.$$

---

† See also Hahn and Rosenthal [9], Kestelman [17], von Neumann [23], and Saks [27].

Proof. For $a \geq 0$, $(af)^+ = af^+$ and $(af)^- = af^-$; for $a < 0$, $(af)^+ = af^-$ and $(af)^- = af^+$. From these facts and 23.4, a passage to the limit shows that $\int_E af = a \int_E f$ in general. Thus, we need consider only the case $a = b = 1$.

We divide Ω into three sets:

$$A = \{x \mid f(x)g(x) \geq 0\},$$
$$B = \{x \mid f(x) \geq 0, g(x) < 0\},$$
$$C = \{x \mid f(x) < 0, g(x) \geq 0\}.$$

On A, $(f + g)^+ = f^+ + g^+$ and $(f + g)^- = f^- + g^-$; so it follows at once from 23.4 that

$$\int_{E \cap A} (f + g) = \int_{E \cap A} f + \int_{E \cap A} g.$$

We divide B into two sets:

$$B_1 = B \cap \{x \mid f(x) + g(x) \geq 0\},$$
$$B_2 = B \cap \{x \mid f(x) + g(x) < 0\}.$$

Each of the following relations now follows from 23.4 because each involves the addition theorem for nonnegative functions only:

$$\int_{E \cap B_1} f = \int_{E \cap B_1} (f + g) + \int_{E \cap B_1} (-g)$$
$$= \int_{E \cap B_1} (f + g) - \int_{E \cap B_1} g;$$
$$-\int_{E \cap B_2} g = \int_{E \cap B_2} (-g) = \int_{E \cap B_2} (-f - g) + \int_{E \cap B_2} f$$
$$= -\int_{E \cap B_2} (f + g) + \int_{E \cap B_2} f.$$

The set C may be treated in a similar manner, and the final result now follows from 25.1. ∎

Theorem 25.3. *If f is any function and if $\mu(E) = 0$, then*

$$\int_E f = 0.$$

Proof. For simple functions this is obvious; so it follows for the general case by 24.1 and 24.2. ∎

Theorem 25.4. *If f is integrable on E and if $f \geq 0$ a. e. on E, then $\int f$ is a nonnegative, nondecreasing function on the measurable subsets of E.*

Proof. By 25.3 and 25.1, $\int f$ is finitely additive, so nonnegativity and monotonicity are equivalent. Let $A = \{x \mid f(x) < 0\}$; it follows from 24.3 that on the measurable subsets of $E - A$, $\int f$ is nondecreasing, hence nonnegative. Thus, by 25.3, on the measurable subsets of E, $\int f$ is nonnegative, hence nondecreasing. ∎

Corollary 25.4.1. *If f and g are each integrable on E and if $f \leq g$ a. e. on E, then*

$$\int_E f \leq \int_E g.$$

Proof. By 25.4, $\int_E (g - f) \geq 0$, and the result follows from 25.2. ∎

Theorem 25.5. *If $0 \leq f_0 \leq g$ a. e. on E, if f_0 is measurable, and if g is integrable on E, then f_0 is integrable on E.*

Proof. Clearly, $\{x \mid f_0(x) > 0\} \subset \{x \mid g(x) > 0\}$; and since g is integrable, this latter set is the union of a countable class of sets of finite measure. Therefore, by 21.1 and 21.2, there is a nondecreasing sequence f of nonnegative, integrable, simple functions converging pointwise to f_0. By 25.4.1,

$$\int_E f_n \leq \int_E g < \infty$$

for each n; thus

$$\lim_n \int_E f_n < \infty;$$

so f_0 is integrable on E. ∎

Corollary 25.5.1. *A measurable function f is integrable on E if and only if $|f|$ is integrable on E.*

Proof. Suppose f is integrable on E; then by definition f^+ and f^- are both integrable on E. Since $|f| = f^+ + f^-$, it follows from 25.2 that $|f|$ is integrable on E. Conversely, suppose $|f|$ is integrable on E. We have from 20.1 that f^+ is measurable if f is; so, since $0 \leq f^+ = |f| - f^- \leq |f|$, it follows from 25.5 that f^+ is integrable on E. A similar argument applies to f^-, so f is integrable on E. ∎

Corollary 25.5.2. *If f is measurable, if g is integrable on E, and if $|f| \leq g$ a. e. on E, then f is integrable on E.*

Proof. By 25.5, $|f|$ is integrable on E; so by 25.5.1, f is also. ∎

Corollary 25.5.3. *If $f = g$ a. e. on E and if g is integrable on E, then f is integrable on E and*

$$\int_E f = \int_E g.$$

Proof. We have a. e. in E that $0 \leq f^+ \leq g^+$ and $0 \leq f^- \leq g^-$; so integrability of f follows from 25.5. The equality of the integrals now follows from 25.4.1. ∎

Theorem 25.6. *If f is integrable on E, then*
$$\left| \int_E f \right| \leq \int_E |f|.$$

Proof. By 25.4, $\int_E f^+$ and $\int_E f^-$ are both nonnegative; so
$$\left| \int_E f \right| = \left| \int_E f^+ - \int_E f^- \right| \leq \max\left[\int_E f^+, \int_E f^- \right]$$
$$\leq \int_E f^+ + \int_E f^- = \int_E |f|. \blacksquare$$

Theorem 25.7. *If $f \geq 0$ a. e. in E and if $\int_E f = 0$, then $f = 0$ a. e. in E.*

Proof. Let $A_0 = \{x \mid x \in E; f(x) > 0\}$; for each positive integer n, let $A_n = \{x \mid x \in E; f(x) > 1/n\}$. Then, A is an expanding sequence of measurable sets, and $\lim_n A_n = A_0$, so $\lim_n \mu(A_n) = \mu(A_0)$. Thus, if $\mu(A_0) > 0$, there is an n such that $\mu(A_n) > 0$, and we have by 25.4 and 25.4.1 that
$$\int_E f \geq \int_{A_n} f \geq \frac{\mu(A_n)}{n} > 0.$$
Therefore, $\mu(A_0) = 0$. $\blacksquare$

Corollary 25.7.1. *If f and g are both integrable on E and if*
$$\int_A f = \int_A g$$
for every measurable $A \subset E$, then $f = g$ a. e. in E.

Proof. Letting $E_1 = \{x \mid f(x) \geq g(x)\}$, we have by hypothesis that
$$\int_{E_1} (f - g) = 0;$$
so by 25.7, $f = g$ a. e. in E_1. Similarly, $f = g$ a. e. in $\{x \mid f(x) \leq g(x)\}$, and the desired result follows. $\blacksquare$

So far in this chapter our definitions and theorems have been concerned with integrals with respect to measure functions. Unless otherwise noted, all subsequent discussions will be concerned with integrals of this type, but we want to pause here to describe briefly the notion of an integral with respect to an arbitrary completely additive set function. The phrase "Lebesgue-Stieltjes integral" is usually thought of as referring to such an integral. That is, given a right continuous function of bounded variation on R_1, we take its Jordan decomposition (Exercise 10.u) and generate measure functions μ_1 and μ_2; then we consider integrals with respect to the completely additive set function $\sigma = \mu_1 - \mu_2$.

More generally, let σ be any completely additive set function in any space Ω, and let $\sigma = \mu_1 - \mu_2$ be its Jordan decomposition (see 10.6). If f is both μ_1- and

μ_2-integrable, then we say that f is σ-integrable, and we define for each set E in the domain of σ,

$$\int_E f \, d\sigma = \int_E f \, d\mu_1 - \int_E f \, d\mu_2.$$

Some of the properties of integrals carry over to this more general case, while others do not. Roughly speaking, conclusions involving integrability and equalities may be generalized, while those involving inequalities may not. Specific results are indicated in Exercises a and b below. In connection with these exercises the reader should note that for a general completely additive set function σ the phrase "a. e." must be interpreted "except on a set E for which $V(\sigma, E) = 0$."

EXERCISES

a. Show that 25.1, 25.2, 25.3, and 25.5 and all its corollaries hold for integrals with respect to an arbitrary completely additive set function.

b. Show by specific examples that 25.4, 25.4.1, 25.6, and 25.7 may fail for integrals with respect to an arbitrary completely additive set function. Note: Even though 25.7 fails, 25.7.1 still holds; see Exercise 27.n.

c. Let Ω be the set of positive integers and let $\mu(E)$ be the number of points in E. Show that if f is a nonnegative function on Ω, then f is integrable if and only if the series $\sum f(n)$ is convergent.

d. Show that for the space of Exercise c the theory of integration reduces to the theory of absolutely convergent series.

e. Use Exercise d to show that if a series is absolutely convergent, then every rearrangement converges to the same sum.

f. From Exercise d show that a series is absolutely convergent if and only if every subseries is convergent.

g. A series is called *unconditionally convergent* if every rearrangement converges or if every subseries converges. Show that these definitions are equivalent.

h. Show that for series of real numbers absolute and unconditional convergence are equivalent.

i. Show that if f is bounded and measurable and if g is integrable, then fg is integrable.

j. Prove the *first mean value theorem for integrals*: If $m \leq f(x) \leq M$ for all $x \in E$ and if f is measurable and g is integrable, then there is a number a such that $m \leq a \leq M$ and

$$\int_E f|g| = a \int_E |g|.$$

k. Show by examples that the result in Exercise j may not hold if the absolute value signs are removed from g or if the integrals are taken with respect to an arbitrary completely additive set function.

l. Show that if f is continuous on $[a, b] \subset R_1$ and if μ is Lebesgue measure, then for each $x \in (a, b)$,

$$\lim_{y \to x} \frac{1}{y - x} \left[\int_{[a,y]} f - \int_{[a,x]} f \right] = f(x).$$

That is, for continuous integrands, the derivative of the indefinite integral is the integrand. [*Hint:* Use Exercise *j.*]

26 ADDITIVITY OF THE INTEGRAL

If f is integrable, then the set function $\int f$ has for its domain the class of all measurable sets. The main purpose of this section is to show that $\int f$ is a completely additive set function on this completely additive class of sets. In order to obtain this result we need another theorem that is of quite some interest in itself.

Theorem 26.1 (Lebesgue Monotone Convergence Theorem). *Let f be a nondecreasing sequence of nonnegative functions, each integrable on E, and let f_0 be a function such that*

$$\lim_n f_n = f_0 \ [a.\ e.].$$

Then, f_0 is integrable on E if and only if $\lim_n \int_E f_n < \infty$, and if this is the case, then

$$\lim_n \int_E f_n = \int_E f_0.$$

Proof. Since the sequence f is monotone, we have for each n, $f_n \leq f_0$ a. e. Thus, if f_0 is integrable, we have from 25.4.1 that

$$\int_E f_n \leq \int_E f_0$$

for each n; so

$$\lim_n \int_E f_n \leq \int_E f_0. \tag{1}$$

For each n, let g_n be a nondecreasing sequence of nonnegative, integrable, simple functions converging pointwise to f_n. For each n and each k, let h_{nk} be defined by

$$h_{nk}(x) = \max_{i \leq n} g_{ik}(x).$$

The double sequence h is obviously nondecreasing as a function of n. Since each sequence g_i is nondecreasing, it follows that h is also nondecreasing as a function of k. Furthermore, each function h_{nk} is a nonnegative, integrable, simple function. Since f is a nondecreasing sequence, we have for each n and each $k \geq n$,

$$g_{nk} \leq h_{nk} \leq h_{kk} \leq \max_{i \leq k} f_i = f_k. \tag{2}$$

So letting $k \to \infty$, we have a. e. that
$$f_n \leq \lim_k h_{kk} \leq f_0,$$
and letting $n \to \infty$, we have
$$\lim_k h_{kk} = f_0 \text{ [a. e.]}. \tag{3}$$

We let 25.3 take care of the exceptional set of measure zero, and since the h_{kk} are simple functions, we have from (3) and the definition 24.1 that if $\lim_k \int_E h_{kk} < \infty$, then f_0 is integrable on E and
$$\int_E f_0 = \lim_k \int_E h_{kk}.$$
However, it follows from (2) and 25.4.1 that
$$\int_E h_{kk} \leq \int_E f_k$$
for each k; so if $\lim_k \int_E f_k < \infty$, then f_0 is integrable on E and
$$\int_E f_0 \leq \lim_k \int_E f_k.$$
This, together with (1), completes the proof of the theorem. ∎

EXERCISES

a. Show that if f is a sequence of nonnegative functions, each integrable on E, if for almost all $x \in E$
$$\sum_{n=1}^{\infty} f_n(x) = f_0(x),$$
then f_0 is integrable on E if and only if $\sum_{n=1}^{\infty} \int_E f_n$ is convergent, and if this is the case, then
$$\int_E f_0 = \sum_{n=1}^{\infty} \int_E f_n.$$

b. For an arbitrary measurable function f on a space of finite measure, Fréchet [7] defines the sums
$$s_n(E) = \sum_{k=-\infty}^{\infty} \frac{k}{2^n} \mu\left(x \mid x \in E; \frac{k}{2^n} \leq f(x) < \frac{k+1}{2^n}\right).$$
He calls f *summable* on E if this series is absolutely convergent for every n. Show that if the series is absolutely convergent for some n, then it is absolutely convergent for every n.

c. Show that for $\mu(E) < \infty$, f is integrable on E if and only if it is summable on E, and that if this is the case, then
$$\int_E f = \lim_n s_n(E).$$

d. Show by an example that for $\mu(E) = \infty$, the result in Exercise b does not necessarily hold.

e. Show that even if $\mu(E) = \infty$, a necessary and sufficient condition for f to be integrable on E is that it be summable on E and that $\lim_n s_n(E)$ be finite. Show also that if this is the case, then

$$\int_E f = \lim_n s_n(E).$$

f. Royden [25] defines an integral as follows. Let B be the set of bounded, measurable functions that are different from zero on sets of finite measure. For $g \in B$, let

$$\int_E g = \inf \left\{ \int_E h \mid h \text{ simple}; h \geq g \right\}.$$

For any nonnegative, measurable f, let

$$\int_E f = \sup \left\{ \int_E g \mid g \in B; g \leq f \right\}.$$

He then uses 24.2. Show that this definition is equivalent to that given in the text.

We now turn to the main purpose of this section.

Theorem 26.2. *Let E be a sequence of disjoint measurable sets, and let*

$$E_0 = \bigcup_{n=1}^{\infty} E_n;$$

let f_0 be a nonnegative measurable function, integrable on each set E_n. Then, f_0 is integrable on E_0 if and only if

$$\sum_{n=1}^{\infty} \int_{E_n} f_0 < \infty,$$

and if this is the case,

$$\int_{E_0} f_0 = \sum_{n=1}^{\infty} \int_{E_n} f_0.$$

Proof. For each integer k, let

$$A_k = \bigcup_{n=1}^{k} E_n,$$

and let the sequence f be defined by

$$f_k(x) = \begin{cases} f_0(x) & \text{for } x \in A_k, \\ 0 & \text{otherwise.} \end{cases}$$

The sequence f then satisfies the conditions of 26.1, and the result follows immediately when we note that by 25.1,

$$\int_{E_0} f_k = \int_{A_k} f_0 = \sum_{n=1}^{k} \int_{E_n} f_0. \quad \blacksquare$$

Corollary 26.2.1. *If f is integrable, then $\int f$ is an everywhere finite, completely additive set function on the class of all measurable sets.*

Proof. This follows from the application of 26.2 to f^+ and f^-. ∎

The *Cauchy-Riemann integral* is the generalization of the Riemann integral to unbounded sets and unbounded functions. For an infinite interval, for instance, we define

$$\int_0^\infty f(x)\,dx = \lim_{t\to\infty} \int_0^t f(x)\,dx.$$

This definition permits *conditional convergence*, that is, integrability of f without that of $|f|$. So, for example,

$$\int_0^\infty \frac{\sin x}{x}\,dx$$

is defined in the Cauchy-Riemann sense, but not in the Lebesgue sense because

$$\int_0^\infty \left|\frac{\sin x}{x}\right|\,dx = \infty;$$

by 25.5.1 this means that the function f such that $f(x) = (\sin x)/x$ is not integrable on $[0, \infty)$.

The property of *absolute integrability*, described by 25.5.1, is a natural consequence of our procedure of defining integrability in terms of f^+ and f^-, but it rules out certain cases covered by the Cauchy-Riemann theory. Unfortunately, however, any modification of the definition to admit conditional convergence of integrals will rule out the property of complete additivity, described by 26.2.1. That is, if f^+ and f^- are measurable but not integrable, it is always possible to find a sequence E of disjoint measurable sets such that

$$\sum_{n=1}^k \int_{E_n} f$$

oscillates as $k \to \infty$.

We might list the following as three desirable properties for a theory of integration:

1) Integrability for every bounded measurable function on a set of finite measure.

2) Complete additivity of the integral.

3) Admission of conditionally convergent integrals.

The Lebesgue theory has only the first two of these properties, while the Cauchy-Riemann theory has only the third. It is possible to combine properties (1) and (3), that is, integrate functions too badly discontinuous to have Riemann integrals and yet admit conditional convergence. This is accomplished by integrals of Perron and Denjoy.† However, properties (2) and (3) are incompatible.

EXERCISES

g. Show that in 26.2 the condition that E be a sequence of disjoint sets may be replaced by the condition that $\mu(E_m \cap E_n) = 0$ for $m \neq n$.

h. Show that for Lebesgue measure in R_1,

$$\int_{[a,b]} f + \int_{[b,c]} f = \int_{[a,c]} f$$

whenever f is integrable.

i. Let $g = \sum_{n=-\infty}^{\infty} n C_{[n-1,n]}$, and let μ_g be the Lebesgue-Stieltjes measure induced by g. Show that in general the formula in Exercise h does not hold for integrals with respect to μ_g.

j. Show that for $\Omega = R_1$, a necessary and sufficient condition that

$$\int_{[a,b]} f \, d\mu_g + \int_{[b,c]} f \, d\mu_g = \int_{[a,c]} f \, d\mu_g$$

is that g be continuous at b.

k. Show that for nonnegative measurable functions f the Cauchy definition of an improper integral,

$$\int_{[0,\infty)} f = \lim_{t \to \infty} \int_{[0,t]} f,$$

is equivalent to the Lebesgue definition.

l. Hahn and Rosenthal [9] define $\int f$ in a space of finite measure as the (uniquely determined) completely additive set function σ such that if E is any measurable set and if $a \leq f(x) \leq b$ for all $x \in E$, then $a\mu(E) \leq \sigma(E) \leq b\mu(E)$. Show that this definition is equivalent to that given in Section 24.

27 ABSOLUTE CONTINUITY

If μ is a measure function and σ is any set function whose domain is a subclass of the class of measurable sets, then we say that σ is *absolutely continuous* with respect to μ, provided that given $\varepsilon > 0$, there exists a $\delta < 0$ such that for every set E in the domain of σ with $\mu(E) < \delta$ we have $|\sigma(E)| < \varepsilon$. Obviously, if σ is absolutely continuous with respect to μ, then $\sigma(E) = 0$ whenever $\mu(E) = 0$. The interesting thing is that under certain conditions this remark can go the other way.

Theorem 27.1. *If σ is an everywhere finite, completely additive function on the class of measurable sets and if $\sigma(E) = 0$ whenever $\mu(E) = 0$, then σ is absolutely continuous with respect to μ.*

---

† See Saks [27], Chapters VI–VIII.

Proof. It suffices to prove this for the case in which σ is nonnegative, because $V(\sigma, \ )$ is a nonnegative, completely additive set function; and if $\mu(E) = 0$ while $V(\sigma, E) > 0$, then there is a set $E_0 \subset E$ for which $|\sigma(E_0)| > 0$, but $\mu(E_0) = 0$. So, if σ satisfies the hypotheses, so does $V(\sigma, \ )$. Furthermore, for every E, $|\sigma(E)| \leq V(\sigma, E)$; so if $V(\sigma, \ )$ is absolutely continuous, so is σ.

Suppose, then, that σ is completely additive, nonnegative, everywhere finite, and not absolutely continuous. There exist a number $\varepsilon > 0$ and a sequence E of measurable sets such that for each n,

$$\mu(E_n) < \frac{1}{2^n}$$

and

$$\sigma(E_n) > \varepsilon.$$

Let

$$E_0 = \varlimsup_n E_n = \bigcap_{k=1}^{\infty} \bigcup_{n=k}^{\infty} E_n.$$

Then

$$\mu(E_0) \leq \sum_{n=k}^{\infty} \mu(E_n) \leq \sum_{n=k}^{\infty} \frac{1}{2^n} = \frac{1}{2^{k-1}}$$

for every k, so $\mu(E_0) = 0$. However, by 10.8.1,

$$\sigma(E_0) \geq \varlimsup_n \sigma(E_n) \geq \varepsilon > 0,$$

so the hypotheses are contradicted. ∎

Corollary 27.1.1. *If f is integrable, then the set function $\int f$ is absolutely continuous.*

Proof. This follows from 26.2.1 and 27.1. ∎

EXERCISES

a. Give an example in which 27.1 fails because σ is not everywhere finite.

b. For the measure μ_g of Exercise 26.i and Lebesgue measure, show that neither is absolutely continuous with respect to the other.

c. Prove that 27.1 holds provided only that σ is finite on every set of finite measure.

d. Prove that if f is a measurable function which is integrable on every set of finite measure, then $\int f$ is absolutely continuous though not necessarily completely additive.

Though absolute continuity is most naturally defined for set functions, in the special case of Lebesgue measure in R_1 a slightly different point of view is usually taken, and an *absolutely continuous function of a real variable* is defined as follows. A point function g on R_1 is absolutely continuous provided that, given $\varepsilon > 0$, there exists $\delta > 0$ such that for every finite set $(a_1, b_1], (a_2, b_2], \ldots, (a_n, b_n]$ of disjoint half-open intervals with

$$\sum_{i=1}^{n} (b_i - a_i) < \delta,$$

we have

$$\sum_{i=1}^{n} |g(b_i) - g(a_i)| < \varepsilon.$$

To see the connection between this and our original definition of absolute continuity for set functions, let us consider the case in which g is a distribution function inducing the Lebesgue-Stieltjes measure μ_g. If μ_g is absolutely continuous with respect to Lebesgue measure λ, then $\mu_g(E)$ is small for all E for which $\lambda(E)$ is small. Therefore, this certainly holds when E is a finite union of intervals, so g is absolutely continuous. Conversely, suppose μ_g is not absolutely continuous with respect to λ; then there exists $\varepsilon > 0$ such that for every $\delta > 0$ there is a set E with $\lambda(E) < \delta/2$ and $\mu_g(E) > 2\varepsilon$. By 13.7.1 there is an open set $G \supset E$ such that $\lambda(G) < \delta$. By 14.5,

$$G = \bigcup_{i=1}^{\infty} I_i,$$

where I is a sequence of disjoint half-open intervals. Since

$$\sum_{i=1}^{\infty} \mu_g(I_i) = \mu_g(G) \geq \mu_g(E) > 2\varepsilon,$$

there is an n such that

$$\sum_{i=1}^{n} \mu_g(I_i) > \varepsilon$$

while

$$\sum_{i=1}^{n} \lambda(I_i) \leq \lambda(G) < \delta;$$

so g is not absolutely continuous. Finally, if g is any function of bounded variation, the above argument may be applied to the monotone functions of its Jordan decomposition (Exercise 10.u) to show that a function of bounded variation is absolutely continuous if and only if its induced Lebesgue-Stieltjes measure is absolutely continuous with respect to Lebesgue measure.

If g is an absolutely continuous function of a real variable, then $|g(b) - g(a)|$ is small for every sufficiently small single interval $(a, b]$; that is, g is uniformly continuous. Thus, absolute continuity is at least formally stronger than uniform continuity; the following example shows that it is definitely stronger.

Let K be a nowhere dense perfect set of measure zero. On $[0, 1] - K$ we construct a monotone function g similar to that in Section 19; this time, however, we make g constant over each interval of $[0, 1] - K$. Specifically, we number the intervals E_n^k of $[0, 1] - K$ as in Section 4 and set

$$g(x) = \frac{2k - 1}{2^n}$$

for $x \in E_n^k$ (Fig. 6). The function g is monotone, and its range is dense in $[0, 1]$; therefore, for each $x \in [0, 1]$, we may set

$$f(x) = \lim_{t \to x} g(t)$$

and define a continuous function f on $[0, 1]$. This function f is called the Cantor function.

Figure 6

Since f is continuous on the closed interval $[0, 1]$, it is uniformly continuous; however, it is not absolutely continuous. Since $\mu(K) = 0$, given $\varepsilon > 0$, there is a sequence I of open intervals covering K such that

$$\sum_{k=1}^{\infty} \mu(I_k) < \varepsilon.$$

Since K is bounded and closed, it is compact, so there is a finite set of these intervals which still covers K. By amalgamating the overlapping ones, if any, we may assume that the intervals of this finite set are disjoint. Let us designate this finite set of intervals by $(a_1, b_1), (a_2, b_2), \ldots, (a_n, b_n)$ with $a_i > b_{i-1}$ for $i = 1, 2, \ldots, n$. Then, since these intervals cover K, a_i and b_{i-1} lie in the same interval of $[0, 1] - K$; so

$$f(a_i) = f(b_{i-1}),$$

and (setting $a_1 = 0$, $b_n = 1$), we have

$$\sum_{i=1}^{n} |f(b_i) - f(a_i)| = f(1) - f(0) = 1,$$

while

$$\sum_{i=1}^{n} (b_i - a_i) < \varepsilon.$$

Thus f is not absolutely continuous.

EXERCISES

e. Show that if f is Lebesgue integrable on R_1, then the function g defined by

$$g(x) = \int_{(-\infty, x]} f$$

is an absolutely continuous function of a real variable.

f. Show that if f is an absolutely continuous function of a real variable, then f is of bounded variation on every bounded interval.

g. Let f be defined for $x \in (0, 1]$ by

$$f(x) = x \sin \frac{1}{x};$$

show that f is uniformly continuous but not of bounded variation.

h. Discuss the relations among continuity, absolute continuity, and bounded variation for functions on a bounded closed interval in R_1.

i. Generalize the considerations in Exercise g. Let

$$f(x) = x^p \sin \frac{1}{x^q}$$

for $0 < x \leq 1$. Show that if $0 < q < p$, then f is absolutely continuous, but that if $0 < p \leq q$, then f is not even of bounded variation.

j. Let functions f and g be defined on $(0, 1]$ by

$$f(x) = x^2 \left| \sin \frac{1}{x} \right|,$$

$$g(x) = x^{\frac{1}{2}}.$$

Show that f and g are each absolutely continuous; the composite function $f \circ (g)$ is absolutely continuous, but $g \circ (f)$ is not.

k. Prove that if f and g are each absolutely continuous and g is monotone, then $f \circ (g)$ is absolutely continuous.

l. Let f be defined over R_2 by
$$f(x, y) = \begin{cases} x + y & \text{for } x + y > 0, \\ 0 & \text{for } x + y \leq 0. \end{cases}$$
Show that f is absolutely continuous in each variable separately, but as a Lebesgue-Stieltjes distribution function it induces a measure μ_f which is not absolutely continuous with respect to Lebesgue measure in R_2.

We have seen in 26.2.1 and 27.1.1 that if f is integrable, then the indefinite integral $\int f$ is an everywhere finite, completely additive, absolutely continuous function on the measurable sets. It turns out that these properties give a complete characterization of the integral as a function on the measurable sets. In order to prove this, we need the following result.

Theorem 27.2 (Hahn Decomposition Theorem). *If σ is an everywhere finite, completely additive set function on a completely additive class $\mathscr{A}$, then there is a set $E_0 \in \mathscr{A}$ such that $\sigma(A) \geq 0$ whenever $A \in \mathscr{A}$ and $A \subset E_0$, and such that $\sigma(A) \leq 0$ whenever $A \in \mathscr{A}$ and $A \subset -E_0$.*

Proof. Clearly, it is sufficient to show that
$$\overline{V}(\sigma, -E_0) = \underline{V}(\sigma, E_0) = 0.$$
For each positive integer n, let $E_n \in \mathscr{A}$ be a set such that
$$\sigma(E_n) > \overline{V}(\sigma, \Omega) - \frac{1}{2^n};$$
then for $A \in \mathscr{A}$ and $A \subset -E_n$, we have
$$\sigma(A) \leq \overline{V}(\sigma, \Omega) - \sigma(E_n) < \frac{1}{2^n},$$
so
$$\overline{V}(\sigma, -E_n) \leq \frac{1}{2^n}.$$
By 10.6, we have
$$\underline{V}(\sigma, E_n) = \sigma(E_n) - \overline{V}(\sigma, E_n) \geq \sigma(E_n) - \overline{V}(\sigma, \Omega) \geq -\frac{1}{2^n}.$$
Now, we let
$$E_0 = \overline{\lim_n} E_n;$$
then
$$-E_0 = \underline{\lim_n} (-E_n).$$
Therefore, from 10.4 and 10.8, we have
$$0 \leq \overline{V}(\sigma, -E_0) \leq \underline{\lim_n} \overline{V}(\sigma, -E_n) \leq \underline{\lim_n} \frac{1}{2^n} = 0,$$

and the first half of the theorem is proved. Now, $-\underline{V}(\sigma, \ )$ is completely additive and nonnegative; so for each positive integer k,

$$0 \le -\underline{V}(\sigma, E_0) \le -V\left(\sigma, \bigcup_{n=k}^{\infty} E_n\right) \le -\sum_{n=k}^{\infty} \underline{V}(\sigma, E_n) \le \sum_{n=k}^{\infty} \frac{1}{2^n} = \frac{1}{2^{k-1}}.$$

Therefore, $\underline{V}(\sigma, E_0) = 0$. ∎

We shall refer to the ordered pair $(E_0, -E_0)$ as a *Hahn decomposition of Ω induced by σ*. Such decompositions are not unique (see Exercise m below). We shall call the set E_0 a *positive set* for σ and call $-E_0$ a *negative set* for σ.

Corollary 27.2.1. *If $(E_0, -E_0)$ is a Hahn decomposition of Ω induced by σ, then*

$$\overline{V}(\sigma, \Omega) = \sigma(E_0) \quad \text{and} \quad \underline{V}(\sigma, \Omega) = \sigma(-E_0).$$

Proof. Since σ is additive and nonnegative, hence monotone, on subsets of E_0, we have for any A in the domain of σ

$$\overline{V}(\sigma, \Omega) \ge \sigma(E_0) \ge \sigma(A \cap E_0) \ge \sigma(A \cap E_0) + \sigma(A - E_0) = \sigma(A).$$

Thus the first result follows; proof of the other is similar. ∎

Corollary 27.2.2. *If f is integrable, then*

$$V\left(\int f, \Omega\right) = \int_{\Omega} |f|.$$

Proof. Let $E_0 = \{x \mid f(x) \ge 0\}$; then E_0 is a positive set for $\int f$ and the result follows by 27.2.1. ∎

EXERCISES

m. In the proof of 27.2 we construct a sequence E of sets and show that if $E_0 = \overline{\lim}_n E_n$, then $(E_0, -E_0)$ is a Hahn decomposition. Show that, for the same sequence E, setting $E_0 = \underline{\lim}_n E_n$ also describes a Hahn decomposition.

n. Show that 25.7.1 holds for integrals with respect to an arbitrary completely additive set function σ. [*Hint:* Consider a Hahn decomposition of Ω induced by σ.]

Theorem 27.3 (Radon-Nikodym). *If Ω is the union of a countable class of sets of finite measure, and if σ is an everywhere finite, completely additive, absolutely continuous function on the class of measurable sets, then there is an integrable function f_0 such that $\sigma(E) = \int_E f_0$ for every measurable set E.*

Proof. Because of 26.2 and 10.6 it suffices to consider only the case $\sigma \ge 0$ and $\mu(\Omega) < \infty$. We use the absolute continuity of σ only to note that if f_0 is defined on E_0 so that $\sigma(E \cap E_0) = \int_{E \cap E_0} f_0$ for each measurable set E and if $\mu(-E_0) = 0$, then the proof is completed by setting $f_0 = 0$ on $-E_0$.

The set E_0 on which f_0 will be defined is constructed as follows. Let A_n^k be a negative set for the function $\sigma - (k/n)\mu$, and set

$$E_0 = \bigcap_{n=1}^{\infty} \bigcup_{k=1}^{\infty} A_n^k. \tag{1}$$

For each fixed n,

$$\infty > \sigma\left[\bigcap_{k=1}^{\infty} (-A_n^k)\right] \geq \left(\frac{j}{n}\right)\mu\left[\bigcap_{k=1}^{\infty} (-A_n^k)\right] \tag{2}$$

for every j because $-A_n^j$ is a positive set for $\sigma - (j/n)\mu$ and $\bigcap_{k=1}^{\infty}(-A_n^k) \subset -A_n^j$. Therefore

$$\mu(-E_0) = \mu\left[\bigcup_{n=1}^{\infty} \bigcap_{k=1}^{\infty} (-A_n^k)\right] \leq \sum_{n=1}^{\infty} \mu\left[\bigcap_{k=1}^{\infty} (-A_n^k)\right] = 0,$$

because if (2) holds for every j, then each term in this last sum is zero.

We must now construct f_0 on E_0, and to simplify notation we may assume from here on that $E_0 = \Omega$. For each n, we use 2.2 to replace the sequence A_n^0 by a sequence E_n^0 of disjoint sets; then we set

$$g_n(x) = \frac{k-1}{n}$$

for $x \in E_n^k$, and by (1) this defines each g_n over $E_0 = \Omega$. Next, we obtain a monotone sequence f by setting

$$f_n(x) = \max_{i \leq n} g_i(x)$$

for each x. For any measurable set E we have $E = \bigcup_{i=1}^n B_i$ where $f_n(x) = g_i(x)$ for $x \in B_i$; so by 26.2,

$$\int_E f_n = \sum_{i=1}^n \int_{B_i} g_i = \sum_{i=1}^n \sum_{k=1}^{\infty} \int_{B_i \cap E_i^k} g_i = \sum_{i=1}^n \sum_{k=1}^{\infty} \frac{k-1}{i} \mu(B_i \cap E_i^k)$$

$$\leq \sum_{i=1}^n \sum_{k=1}^{\infty} \sigma(B_i \cap E_i^k) = \sigma(E).$$

The inequality $(k-1)\mu(B_i \cap E_i^k)/i \leq \sigma(B_i \cap E_i^k)$ comes from the fact that $E_i^k \subset -A_i^{k-1}$. Using the fact that $E_i^k \subset A_i^k$, we have in a similar way that

$$\int_E f_n \geq \int_E g_n = \sum_{k=1}^{\infty} \frac{(k-1)\mu(E \cap E_n^k)}{n}$$

$$\geq \sum_{k=1}^{\infty} \left[\sigma(E \cap E_n^k) - \frac{\mu(E \cap E_n^k)}{n}\right] = \sigma(E) - \frac{\mu(E)}{n}.$$

Therefore

$$\lim_n \int_E f_n = \sigma(E);$$

so if f_0 is defined as the limit of the monotone sequence f, the result follows by 26.1. ∎

Since $\int_E f_0$ is specified for each measurable E, it follows from 25.7.1 that for a given σ, the f_0 of 27.3 is essentially unique. That is, any two such functions must be equal a. e.

A natural follow-up to the Radon-Nikodym theorem would be to show that in some sense the function f_0 of 27.3 is a derivative of σ. The general proof of such a result calls for the development of a fairly elaborate theory. This discussion is postponed until Chapter 6. An important special case is treated by the very simple argument suggested in Exercise 25.*l*.

In the most direct applications of integral calculus to the physical sciences, a physical quantity is given by a definite integral because it appears as the limit of a particular sequence of finite sums. The integral as we have defined it (from sequences of simple functions) is certainly capable of this interpretation. However, the physical scientist wants to evaluate his definite integrals, and he does this by using a table of integrals. These tables come from derivative formulas, so the use of an integral formula is very definitely an application of the theory developed in Chapter 6.

If we confine ourselves to sufficiently simple cases, we can employ Exercise 25.*l* in place of Chapter 6 to justify the evaluation of Lebesgue integrals by means of the usual formulas. Exercises *q* to *t* below indicate how 27.3 may be used to reduce certain Lebesgue-Stieltjes integrals to forms that are easily evaluated.

EXERCISES

o. Show that the measure of Exercise 11.*i* is absolutely continuous with respect to that of Exercise 11.*f*, but that the Radon-Nikodym theorem fails. Why?

p. Show that if g is a bounded absolutely continuous function of a real variable, then there is a Lebesgue integrable function f such that for each $x \in R_1$,

$$g(x) = \int_{(-\infty, x]} f.$$

q. Show that if μ_1 is absolutely continuous with respect to μ_2, then there is a function g such that if f is μ_1-integrable,

$$\int_E f \, d\mu_1 = \int_E fg \, d\mu_2$$

for every u_1-measurable set E. *Hint*: If f is a characteristic function, this reduces to 27.3. Use this to define g; then obtain the general result by considering linear combinations and monotone sequences.

r. Let g be an absolutely continuous function of a real variable having a continuous derivative g'. Let μ_g be the Lebesgue-Stieltjes measure induced by g, and let μ be Lebesgue measure. Show that for any μ_g-integrable f and any Lebesgue measurable set E,

$$\int_E f \, d\mu_g = \int_E fg' \, d\mu.$$

s. Let g be any Lebesgue-Stieltjes distribution function on R_1, and let f be μ_g-integrable. Show that if E consists of the single point a, then

$$\int_E f \, d\mu_g = f(a)[g(a) - \lim_{x \to a^-} g(x)].$$

t. Let f and g be defined on $[0, 2\pi]$ by the following equations:

$$f(x) = \sin x \sum_{n=1}^{4} nC_{(n\pi/2 - \pi/2, n\pi/2]}(x),$$

$$g(x) = x^2 \sum_{n=1}^{4} nC_{(n\pi/2 - \pi/2, n\pi/2]}(x).$$

Compute $\int f \, d\mu_g$ and $\int g \, d\mu_f$ over each of the four intervals $(0, 2\pi)$, $(0, 2\pi]$, $[0, 2\pi)$, and $[0, 2\pi]$.

28 DOMINATED CONVERGENCE

A naturally important question in integration theory is that of "taking a limit under the integral sign." Roughly speaking, this means finding conditions under which $\lim \int f_n = \int \lim f_n$. This description is far too rough; first let us define the problem more precisely. Note, for example, that

$$\int_{-\pi}^{\pi} \sin n x \, dx = 0$$

for every n. Now, do we want to say that in some sense

$$\lim_n \sin n x = 0$$

and that we can take this limit under the integral sign? What we are trying to point out here is that

$$\lim_n \int_\Omega f_n = \int_\Omega f_0 \tag{1}$$

can happen quite by accident, so to speak. A more satisfactory theory involves conditions under which

$$\lim_n \int_\Omega |f_n - f_0| = 0. \tag{2}$$

Since

$$\left| \int_\Omega f_n - \int_\Omega f_0 \right| \le \int_\Omega |f_n - f_0|,$$

it is clear that (2) implies (1).

Theorems in which (2) is the conclusion form an important part of integration theory. In Section 40 we shall return to this question and give two really refined theorems in which (2) is the conclusion. We present here as a part of our discussion of basic properties of the integral a classical theorem of Lebesgue's giving sufficient conditions for (2).

Theorem 28.1 (Lebesgue Dominated Convergence Theorem). *Let f be a sequence of measurable functions such that*

$$\lim_n f_n = f_0 \text{ [a. e.]}$$

and such that there is an integrable function g with

$$|f_n| \leq g \text{ [a. e.]}$$

for every n. Then

$$\lim_n \int_\Omega |f_n - f_0| = 0.$$

Proof. First note that $|f_0| \leq g$ [a. e.]; so $|f_n - f_0| \leq 2g$ [a. e.] for every n. Now, $\int 2g$ is finite and completely additive, hence continuous at ϕ; and $\{x \mid g(x) \neq 0\}$ is a countable union of sets of finite measure. Thus there is an expanding sequence A of sets of finite measure with $\int_{-A_k} 2g \to 0$. Let $\varepsilon > 0$ be given. We set $E_1 = -A_k$ so that

$$\int_{E_1} |f_n - f_0| \leq \int_{E_1} 2g < \frac{\varepsilon}{3} \quad (1)$$

for all n and note that $\mu(-E_1) < \infty$. Since $\int 2g$ is absolutely continuous, there is a $\delta > 0$ such that $\mu(E_2) < \delta$ implies

$$\int_{E_2} |f_n - f_0| \leq \int_{E_2} 2g < \frac{\varepsilon}{3} \quad (2)$$

for all n. Applying Egoroff's theorem (21.3) to f on $-E_1$, we can get $E_2 \subset -E_1$ so that $\mu(E_2) < \delta$, thus guaranteeing (2), and also such that $f_n \to f_0$ uniformly on $E_3 = -(E_1 \cup E_2)$. Thus there exists n_0 such that $n > n_0$ implies

$$|f_n(x) - f_0(x)| < \frac{\varepsilon}{3\mu(-E_1)} \quad (3)$$

for every $x \in E_3$; hence for $n > n_0$

$$\int_{E_3} |f_n - f_0| < \frac{\varepsilon \mu(E_3)}{3\mu(-E_1)} \leq \frac{\varepsilon}{3}. \quad (4)$$

Since $\Omega = E_1 \cup E_2 \cup E_3$, the result follows from (1), (2), and (4). ∎

We want to use basically this same proof twice more (once in this section and again in Section 40). To this end we outline it as follows.

Theorem 28.2. *The following conditions are sufficient that*

$$\lim_n \int_\Omega |f_n - f_0| = 0.$$

Let $\varepsilon > 0$ be given.

28.2.1 *There exists E_1 with $\mu(-E_1) < \infty$ and*

$$\int_{E_1} |f_n - f_0| < \frac{\varepsilon}{3}$$

for every n.

28.2.2 *There exists $\delta > 0$ such that if $\mu(E_2) < \delta$, then*

$$\int_{E_2} |f_n - f_0| < \frac{\varepsilon}{3}$$

for every n.

28.2.3 *There exists n_0 such that for each $n > n_0$ we have*

$$-E_1 = E_2^n \cup E_3^n$$

where $\mu(E_2^n) < \delta$ and

$$|f_n(x) - f_0(x)| < \frac{\varepsilon}{3\mu(-E_1)}$$

for all $x \in E_3^n$.

Proof. Take $n > n_0$; then $\Omega = E_1 \cup E_2^n \cup E_3^n$ and the integral of $|f_n - f_0|$ over each of these three sets is less than $\varepsilon/3$. ∎

In the proof of 28.1 we did not use the full power of 28.2.3. As noted in 28.2.3, it is permissible that the decomposition of $-E_1$ (into an E_2 of small measure and an E_3 where $f_n - f_0$ is small) depend on n. Using Egoroff's theorem to get the decomposition, we have E_2 and E_3 set before we ever introduce n_0. The following modification of 28.1 is designed specifically to take advantage of this feature of 28.2.3.

Theorem 28.3. *Let f be a sequence of measurable functions such that for every $\eta > 0$,*

$$\lim_n \mu(\{x \mid |f_n(x) - f_0(x)| \geq \eta\}) = 0$$

and such that there is an integrable g with

$$|f_n| \leq g \, [\text{a. e.}]$$

for every n. Then,

$$\lim_n \int_\Omega |f_n - f_0| = 0.$$

Proof. The first hypothesis here is called convergence in measure. This concept is studied at greater length in Section 38 where it will be shown that we still have $|f_n - f_0| \leq 2g$ [a. e.]. Assuming this, we proceed as in the proof of 28.1. Complete additivity and absolute continuity of $\int 2g$ establish 28.2.1 and 28.2.2. Now, let $\eta = \varepsilon/[3\mu(-E_1)]$ and let n_0 be such that $n > n_0$ implies

$$\mu(\{x \mid |f_n(x) - f_0(x)| \geq \eta\}) < \delta.$$

If we set $E_2^n = \{x \mid |f_n(x) - f_0(x)| \geq \eta\} - E_1$ and $E_3^n = -(E_1 \cup E_2^n)$, we have 28.2.3 satisfied. ∎

By letting E_2 and E_3 depend on n we are able to replace the hypothesis of a. e. convergence in 28.1 with a formally weaker hypothesis in 28.3. We shall show in Section 39 that the second hypothesis in 28.3 is strictly weaker than a. e. convergence. The idea might occur, "Why not let E_1 depend on n?" At first this seems to have intriguing possibilities. If 28.2.1 reads merely that

$$\int_{E_1^n} |f_n - f_0| < \frac{\varepsilon}{3},$$

we can establish it from the continuity at ϕ of $\int |f_n - f_0|$ and avoid use of the dominant function. Similarly, if δ in 28.2.2 becomes δ_n, then it can come from absolute continuity of $\int |f_n - f_0|$ and the dominant function is no longer needed at all. The trouble with these modifications is that they would radically modify the logic of 28.2.3. As 28.2.3 now appears, E_1 and δ set the criteria determining n_0; so they must be given in advance. While it might be possible to frame a much more restrictive version of 28.2.3 in which E_1 and δ vary with n, such a procedure is not practical because we shall show in Section 40 that while the dominant function can be dispensed with, 28.2.1 and 28.2.2 themselves are both necessary for the desired conclusion.

EXERCISES

a. We say that a series of functions converges a. e. and write

$$\sum_{n=1}^{\infty} f_n = f_0 \text{ [a. e.]} \tag{1}$$

meaning that

$$\lim_k \sum_{n=1}^{k} f_n = f_0 \text{ [a. e.]}.$$

Prove that if (1) holds and if there is an integrable g such that

$$\left| \sum_{n=1}^{k} f_n \right| \leq g \text{ [a. e.]}$$

for every k, then

$$\int_\Omega f_0 = \sum_{n=1}^{\infty} \int_\Omega f_n.$$

b. Let $f_n = nC_{[0,1/n]}$. Show that $\lim_n f_n = 0$ [a. e.] on $[0, 1]$ but $\int_{[0,1]} f_n = 1$ for every n. The first hypothesis of 28.2 alone is not sufficient.

c. Let

$$f_n(x) = \begin{cases} 0 & \text{for } 0 < x < 1/(n+1), \\ x^{-3/2} & \text{for } 1/(n+1) \leq x < 1/n, \\ 0 & \text{for } 1/n \leq x \leq 1. \end{cases}$$

Show that there is no integrable g such that $f_n \leq g$ [a. e.] for every n but that $\lim_n \int_{[0,1]} f_n = 0$. The second hypothesis of 28.2 is not necessary. (More of this in Section 40.)

d. Prove that if $\mu(\Omega) < \infty$ and $\lim_n f_n = f_0$ [unif.], then

$$\lim_n \int_\Omega |f_n - f_0| = 0.$$

e. Let $f_n = (1/n) C_{[0,n]}$. Show that $\lim_n f_n = 0$ [unif.] on $[0, \infty)$ but $\int_{[0,\infty)} f_n = 1$ for every n.

f. Let $f_n(x) = (1/x) C_{[1,n]}(x)$. Show that each f_n is integrable, that f converges uniformly, but that the limit function is not integrable.

29 FUBINI'S THEOREM

In elementary calculus a so-called multiple integral (integral over a space of more than one dimension) is presented in terms of an iterated integration process in which each single integration is over a one-dimensional space. In the general presentation of measure and integration given here, the dimension (if any) of the underlying space is completely immaterial; so there is no distinction to be made between "single integrals" and "multiple integrals" so far as their definitions and fundamental properties are concerned. However, it is useful to have an iterated integral representation for an integral over a multidimensional space, and it is with this sort of problem that Fubini's theorem is concerned.

First, let us formulate the problem in terms of abstract spaces with measures defined in them. Let X and Y be two spaces; we form a space which might be called "two-dimensional" by taking the product space $X \times Y$ (see Section 1). Any set of the form $A \times B$ where $A \subset X$ and $B \subset Y$ we shall call a *generalized rectangle*, and we shall call A and B its sides. This terminology is motivated by the fact that if $X = Y = R_1$, then $X \times Y = R_2$; if A and B are intervals in R_1, then $A \times B$ is a genuine rectangle.

Let α^* be an outer measure in X generating a class $\mathscr{A}$ of measurable sets and a measure α. Similarly, let β^* generate a class $\mathscr{B}$ and a measure β in Y. Let us assume throughout this section that each of the spaces X and Y is a countable union of measurable sets of finite measure. Then, the class $\mathscr{C}$ of generalized rectangles with measurable sides of finite measure is a sequential covering class. For each $A \times B \in \mathscr{C}$, let us define

$$\tau(A \times B) = \alpha(A)\beta(B).$$

Now, we let μ^* be the outer measure in $X \times Y$ constructed from $\mathscr{C}$ and τ by Method I.

Our procedure will be as follows. We shall give another characterization of τ as an iterated integral and show that it is equivalent to the definition given above. From this integral representation we shall obtain an extension τ_0 of τ to the class

$\mathscr{C}_\sigma$. Using τ_0, we shall show that every set in $\mathscr{C}$ is μ^*-measurable and that $\tau = \mu$ on $\mathscr{C}$. Thus, μ will have an iterated integral representation on $\mathscr{C}$. Using simultaneously the complete additivity of μ and that of iterated integrals, we shall extend the integral representation of μ to $\mathscr{C}_{\sigma\delta}$. Then we shall employ 12.3 to show that all values of μ are given by iterated integrals.

This representation of μ by iterated integrals constitutes the nontrivial part of the proof of Fubini's theorem. Another way of arriving at this representation is to define μ straightaway as an iterated integral with no recourse to an outer measure at all. The problem then is to show that this definition is valid and unique. On $\mathscr{C}$ this is trivial, and the remaining problem is to extend the existence and uniqueness proof to the minimal completely additive class containing $\mathscr{C}$. This extension problem is essentially the same as our problem of extending the representation, but in our case the matter is simplified somewhat because we have to develop the extension only as far as $\mathscr{C}_{\sigma\delta}$.

Let us turn now to the notation for an *iterated integral*. Suppose f is a function over $X \times Y$ such that for almost all $x \in X$, $f(x,\ )$ is integrable on Y. Then, the formula

$$g(x) = \int_Y f(x,\ ) \, d\beta$$

defines a function g a. e. on X. Whenever g is integrable, we define

$$\int_X \int_Y f \, d\beta \, d\alpha = \int_X g \, d\alpha.$$

Similarly,

$$\int_Y \int_X f \, d\alpha \, d\beta$$

is defined as the integral over Y of the function defined for almost all $y \in Y$ by

$$\int_X f(\ , y) \, d\alpha.$$

Theorem 29.1. *If $E \in \mathscr{C}$, then*

$$\tau(E) = \int_X \int_Y C_E \, d\beta \, d\alpha = \int_Y \int_X C_E \, d\alpha \, d\beta.$$

Proof. Let $E = A \times B$. Then, for each $(x, y) \in X \times Y$,

$$C_E(x, y) = C_A(x) C_B(y);$$

so for each x, $C_E(x,\ )$ is integrable, and

$$\int_Y C_E(x,\ ) \, d\beta = C_A(x) \int_Y C_B \, d\beta = \beta(B) C_A(x).$$

Since C_A is integrable, the iterated integral exists, and

$$\int_X \int_Y C_E \, d\beta \, d\alpha = \beta(B) \int_X C_A \, d\alpha = \beta(B)\alpha(A) = \tau(E).$$

An analogous proof applies to the other iterated integral. ∎

Theorem 29.2. *Let f be a nondecreasing sequence of nonnegative functions on $X \times Y$ such that for each n,*

$$\int_X \int_Y f_n \, d\beta \, d\alpha = \int_Y \int_X f_n \, d\alpha \, d\beta;$$

let $f_0(x, y) = \lim_n f_n(x, y)$ for each $(x, y) \in X \times Y$. Then,

$$\int_X \int_Y f_0 \, d\beta \, d\alpha = \int_Y \int_X f_0 \, d\alpha \, d\beta = \lim_n \int_X \int_Y f_n \, d\beta \, d\alpha = \lim_n \int_Y \int_X f_n \, d\alpha \, d\beta$$

in the sense that if any one of these four quantities exists (finite), so do the others, and the equality holds.

Proof. By assumption, for each n, $f_n(x, \ )$ is measurable for almost all $x \in X$. The exceptional set may vary with n, but by taking a countable union of sets of measure zero, we may say that for almost all $x \in X$, $f_n(x, \ )$ is measurable for every n. Therefore, by 20.3.2, $f_0(x, \ )$ is measurable for almost all $x \in X$; thus by 26.1 it is integrable on Y for each $x \in X$ for which

$$g(x) = \lim_n \int_Y f_n(x, \ ) \, d\beta < \infty.$$

Now, g is measurable by 20.3.2, and if

$$\lim_n \int_X \int_Y f_n \, d\beta \, d\alpha < \infty, \tag{1}$$

then g is integrable because of 26.1, hence finite a. e. Thus, given (1), we have $f_0(x, \ )$ integrable for almost all x, and its integrals form an integrable function; that is,

$$\int_X \int_Y f_0 \, d\beta \, d\alpha$$

exists. Two applications of 26.1 now yield the result that

$$\int_X \int_Y f_0 \, d\beta \, d\alpha = \int_X \lim_n \int_Y f_n \, d\beta \, d\alpha = \lim_n \int_X \int_Y f_n \, d\beta \, d\alpha.$$

A similar argument applies to the integrals taken in the other order. Since the two iterated integrals of f_n are equal for each n, their limits are necessarily equal; so the result holds if either limit is finite. Finally, if either iterated integral of f_0 exists, it follows from 25.4.1 that the corresponding limit is finite; hence so is the other limit, and the result follows from what has already been proved. ∎

152   INTEGRATION

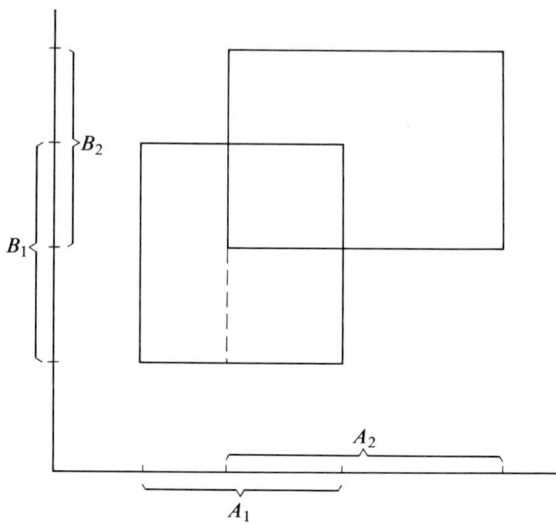

Figure 7

We now define $\mathscr{D}$ as the class of all sets in $X \times Y$ which are finite unions of disjoint sets from $\mathscr{C}$.

Theorem 29.3. *$\mathscr{D}$ is a finitely additive class, and every set in $\mathscr{C}_\sigma$ is the union of a sequence of disjoint sets from $\mathscr{D}$.*

Proof. First, we observe that

29.3.1 $\quad (A_1 \times B_1) \cap (A_2 \times B_2) = (A_1 \cap A_2) \times (B_1 \cap B_2),$

29.3.2 $\quad (A_1 \times B_1) - (A_2 \times B_2) = [(A_1 \cap A_2) \times (B_1 - B_2)]$
$$\cup [(A_1 - A_2) \times B_1].$$

These relations are easily checked from the definition of a product set, but it is not a bad idea to visualize them in R_2. In Fig. 7 the dotted line indicates the partition of the difference set described by 29.3.2. From 29.3.1 and 29.3.2 it follows that if E_1 and E_2 are in $\mathscr{C}$, then $E_1 \cap E_2$ and $E_1 - E_2$ are in $\mathscr{D}$. The result for intersections is easily extended by induction to cover all finite intersections. The relation

$$\left(\bigcup_{k=1}^{n} E_k \right) - \left(\bigcup_{j=1}^{m} F_j \right) = \bigcup_{k=1}^{n} \bigcap_{j=1}^{m} (E_k - F_j)$$

expresses the difference of any two sets in $\mathscr{D}$ as a set in $\mathscr{D}$. The relation

$$E_1 \cup E_2 = E_1 \cup (E_2 - E_1)$$

expresses the union of any two sets from $\mathscr{D}$ as a union of disjoint sets from $\mathscr{D}$, hence as a set in $\mathscr{D}$. Thus, $\mathscr{D}$ is finitely additive. Finally,

$$\bigcup_{n=1}^{\infty} E_n = E_1 \cup \left\{ \bigcup_{n=1}^{\infty} \left[E_{n+1} - \left(\bigcup_{k=1}^{\infty} E_k \right) \right] \right\}$$

expresses any set in $\mathscr{C}_\sigma$ as a union of disjoint sets in $\mathscr{D}$. ∎

By 29.1, the iterated integrals exist and are equal for the characteristic function of every set in $\mathscr{C}$. Since, for disjoint sets, $C_{E_1 \cup E_2} = C_{E_1} + C_{E_2}$, it is obvious that the iterated integrals are equal for characteristic functions of sets in $\mathscr{D}$. If $E \in \mathscr{C}_\sigma$, then C_E is (by 29.3) the limit of a nondecreasing sequence of characteristic functions of sets in $\mathscr{D}$; so by 29.2 the iterated integrals of C_E exist and are equal if and only if the limits of the integrals of the approximating functions are finite. So, we define a function τ_0 over $\mathscr{C}_\sigma$ by setting

$$\tau_0(E) = \int_X \int_Y C_E \, d\beta \, d\alpha = \int_Y \int_X C_E \, d\alpha \, d\beta$$

whenever the integrals exist, and by setting $\tau_0(E) = \infty$ otherwise. It follows from 29.2 that τ_0 satisfies Postulate A–II, though in general its domain is not a completely additive class. It is clear from 25.4.1 that τ_0 is monotone, and these two properties are sufficient to show that τ_0 also satisfies Postulate M–II (see the proof of 10.7). Finally, it follows from 29.1 that τ_0 is an extension of τ.

Theorem 29.4. *Every set in $\mathscr{C}$ is μ^*-measurable.*

Proof. Let Q be any set in $X \times Y$, and let $E \in \mathscr{C}$. Let $\varepsilon > 0$ be given, and let I be a sequence of sets from $\mathscr{C}$ such that

$$\bigcup_{n=1}^{\infty} I_n \supset Q$$

and

$$\sum_{n=1}^{\infty} \tau(I_n) \leq \mu^*(Q) + \varepsilon.$$

For each n for which both $I_n \cap E$ and $I_n - E$ are nonvacuous, we can replace I_n by the three disjoint generalized rectangles of 29.3.1 and 29.3.2; because of the additivity of τ_0 ($= \tau$ here), this does not affect the value of $\sum \tau(I_n)$. Thus, we may assume that part of the sequence I covers $Q \cap E$ and the rest covers $Q - E$. So

$$\mu^*(Q \cap E) + \mu^*(Q - E) \leq \sum_{n=1}^{\infty} \tau(I_n) \leq \mu^*(Q) + \varepsilon.$$

Since this holds for every $\varepsilon > 0$, the result follows from 11.1. ∎

With this result established, we see from 12.3.1 that μ^* is regular whether α^* and β^* are or not.

Theorem 29.5. *If* $E \in \mathscr{C}$, *then* $\tau(E) = \mu(E)$.

Proof. Since E covers itself, we have automatically that $\mu(E) \leq \tau(E)$. Let I be any sequence of sets from $\mathscr{C}$ covering E; then, since τ_0 satisfies M–II,

$$\sum_{n=1}^{\infty} \tau(I_n) = \sum_{n=1}^{\infty} \tau_0(I_n) \geq \tau_0\left(\bigcup_{n=1}^{\infty} I_n\right) \geq \tau_0(E) = \tau(E).$$

Since this holds for every such sequence,

$$\mu(E) \geq \tau(E). \quad \blacksquare$$

Theorem 29.6. *If each of the spaces X and Y is a countable union of disjoint sets of finite measure, then for every μ^*-measurable set $E \subset X \times Y$,*

$$\mu(E) = \int_X \int_Y C_E \, d\beta \, d\alpha = \int_Y \int_X C_E \, d\alpha \, d\beta$$

in the sense that if any one of these quantities is finite, so are the other two, and the equality holds.

Proof. For $E \in \mathscr{C}$, this follows from 29.5 and 29.1. From the complete additivity of both μ and the iterated integrals it follows that if the result holds for each set of a sequence of disjoint sets, then it holds for the union of the sequence. Thus 29.3 enables us to extend the result to every† $E \in \mathscr{C}_\sigma$.

Since $\mathscr{C}$ is closed under the operation of taking finite intersections, the relation

$$\left(\bigcup_{n=1}^{\infty} E_n\right) \cap \left(\bigcup_{k=1}^{\infty} F_k\right) = \bigcup_{n=1}^{\infty} \bigcup_{k=1}^{\infty} (E_n \cap F_k)$$

describes the intersection of two sets from $\mathscr{C}_\sigma$ as a set from $\mathscr{C}_\sigma$, and this may be extended by induction to cover all finite intersections. If $H \in \mathscr{C}_{\sigma\delta}$, then

$$H = \bigcap_{n=1}^{\infty} H_n = \bigcap_{k=1}^{\infty} \bigcap_{n=1}^{k} H_n$$

describes H as the limit of a contracting sequence of sets from $\mathscr{C}_\sigma$; so $-H$ is the limit of an expanding sequence of sets each of whose complements is in $\mathscr{C}_\sigma$. If $\alpha(X) < \infty$ and $\beta(Y) < \infty$, then the integral representation of μ passes to complements because

$$\mu(-E) = \mu(X \times Y) - \mu(E) = \mu(X \times Y) - \int\int C_E$$

$$= \int\int (1 - C_E) = \int\int C_{-E};$$

---

† The reader should note that 29.3 plays a vital role in this step. It is true but not obvious that the iterated integral representation of μ may be extended from any class $\mathscr{E}$ to the class $\mathscr{E}_\sigma$. See, for example, the proof of Fubini's theorem in Saks [27], pp. 82–87.

so the representation is established for sets in $\mathscr{C}_{\sigma\delta}$ by taking complements, applying simultaneously 29.2 to the integrals and 10.3 to μ, then taking complements again. To remove the finiteness condition on $\alpha(X)$ and $\beta(Y)$, we partition $X \times Y$ into disjoint generalized rectangles of finite measure, note that the result holds for each section of the $\mathscr{C}_{\sigma\delta}$ set in question, and use the complete additivity of μ and the integrals to obtain the result for the entire set.

It follows from 12.3 that every μ^*-measurable set E is covered by a set $H \in \mathscr{C}_{\sigma\delta}$ such that $\mu(H) = \mu(E)$. Hence, if $\mu(E) < \infty$, then $\mu(H - E) = 0$; by considering a partition of $X \times Y$ into a countable number of sets of finite measure, we see that in general $H - E$ is a countable union of sets of measure zero. So, the proof of 29.6 is completed if we show that when $\mu(E) = 0$, both iterated integrals of C_E are zero.

Accordingly, let $\mu(E) = 0$. By 12.3 there is a set $H \in \mathscr{C}_{\sigma\delta}$ such that $H \supset E$ and $\mu(H) = 0$. From what has already been proved, we see that

$$\int_X \int_Y C_H \, d\beta \, d\alpha = 0;$$

so, by 25.7,

$$\int_Y C_H(x, \ ) \, d\beta = 0$$

for almost all $x \in X$. Therefore, by 25.7, we have that for almost all $x \in X$,

$$0 \leq C_E(x, \ ) \leq C_H(x, \ ) = 0$$

a. e. in Y. Thus,

$$\int_Y C_E(x, \ ) \, d\beta = 0$$

for almost all $x \in X$; so

$$\int_X \int_Y C_E \, d\beta \, d\alpha = 0.$$

A similar proof applies to the other iterated integral of C_E. ∎

Theorem 29.7 (Fubini). *If each of the spaces X and Y is a countable union of sets of finite measure, if f is a μ^*-measurable function on $X \times Y$, and if any one of the three integrals*

$$\int_{X \times Y} |f| \, d\mu, \quad \int_X \int_Y |f| \, d\beta \, d\alpha, \quad \int_Y \int_X |f| \, d\alpha \, d\beta$$

exists, then the three integrals of f exist, and

$$\int_{X \times Y} f \, d\mu = \int_X \int_Y f \, d\beta \, d\alpha = \int_Y \int_X f \, d\alpha \, d\beta.$$

Proof. For characteristic functions the result has already been established in 29.6. The extension to nonnegative simple functions is trivial. Using the result for simple functions, we extend it to nonnegative measurable functions by applying 26.1 to the μ-integrals and 29.2 to the iterated ones. For the general case, we note that if one of the integrals of $|f|$ exists, then by 25.5 that integral exists for both f^+ and f^-. From what we have already proved it now follows that for each of the functions f^+ and f^- the three integrals exist and are equal. The result for f then follows from 25.2. ∎

Corollary 29.7.1. *If X and Y are any spaces and if f is μ-integrable on $X \times Y$, then the iterated integrals of f exist and*

$$\int_{X \times Y} f \, d\mu = \int_X \int_Y f \, d\beta \, d\alpha = \int_Y \int_X f \, d\alpha \, d\beta.$$

Proof. If f is μ-integrable, then by 21.2,

$$\{(x, y) \mid f(x, y) \neq 0\}$$

is a countable union of sets of finite measure. We may restrict ourselves to this set to satisfy the countability condition in 29.7. If f is μ-integrable, so is $|f|$, and the result follows from 29.7. ∎

In many ways 29.7 is a more useful result than 29.7.1 because it allows us to work from the existence of an iterated integral without worrying about the μ-integral at all. The reader should note, however, that in this case it is integrability of $|f|$ that must be checked. Despite the fact that all our single integrals are absolutely convergent, this does not follow for iterated integrals. The explanation for this lies in the fact that if $g(y) = \int_X f(\ , y) \, d\alpha$, it does not necessarily follow that $g^+(y) = \int_X f^+(\ , y) \, d\alpha$. For an example, see Exercise i below.

Suppose $X = Y = R_1$ so that $X \times Y = R_2$. Let α and β be Lebesgue-Stieltjes measures induced by distribution functions u and v, respectively. Let w be the distribution function (see Exercise 14.w) in R_2 defined by

$$w(x, y) = u(x)v(y).$$

Then (note Exercise 14.v), μ and μ_w are identical on the half-open intervals. Therefore, by 14.5, they are identical on the open sets. By the application of 13.5 to the Lebesgue-Stieltjes measures α and β, it is easily checked that μ and μ_w are identical on the generalized rectangles with measurable sides. Thus, it would appear that μ^* and μ_w^* are identical. This is, indeed, the case, but it does not follow immediately by 11.4 because μ^* and μ_w^* are generated by different covering classes. However, by 14.5, the right continuity of distribution functions, and 13.7.1, any covering by measurable generalized rectangles may be approximated by a covering by open intervals; such an argument would show that $\mu^* = \mu_w^*$.

Thus, 29.7 applies to Lebesgue-Stieltjes product measures in R_2. Therefore, in particular it applies to Lebesgue measure in R_2. Since every Riemann integral is a Lebesgue integral, it follows at once that every absolutely convergent Cauchy-

Riemann integral is a Lebesgue integral. We thus get the following useful corollary to 29.7.

Corollary 29.7.2. *If f is Lebesgue measurable in the plane, if the Cauchy-Riemann integrals*

$$\int_{-\infty}^{\infty} \int_{-\infty}^{\infty} f(x, y) \, dx \, dy \quad \text{and} \quad \int_{-\infty}^{\infty} \int_{-\infty}^{\infty} f(x, y) \, dy \, dx$$

exist, and if one of them is absolutely convergent, then the two are equal.

This is the standard theorem on reversing the order of integration in improper integrals. Sometimes the operation is permissible with conditionally convergent integrals, but this must be checked in each individual case.

Finally, we should remark that though 29.7 is stated in terms of integrals over $X \times Y$, it applies to integrals over other sets in $X \times Y$ as well, because the integrand can always be multiplied by the characteristic function of the set in question.

EXERCISES

a. Let $X = Y = R_1$; let α be Lebesgue measure; and let β be the measure of Exercise 11.*f*. Show that if E is a segment of the line $y = x$, then the iterated integrals of C_E exist but are different. Why does this not contradict 29.7?

b. Consider the example referred to in Exercise 19.*k* to show that the hypothesis that f is μ^*-measurable cannot be dropped in 29.7. That is, it does not follow from the existence of the iterated integrals.

c. Let $X = Y = R_1$; let α and β be Lebesgue measure; let Q be a nonmeasurable set in X; and let Z be a nonvacuous set of measure zero in Y. Show that $C_{Q \times Z}$ is measurable over $X \times Y$, but $C_{Q \times Z}(\ , y)$ is measurable only for almost all $y \in Y$.

d. Discuss the relation between "generalized rectangles with measurable sides" and "measurable generalized rectangles."

e. Let $X = Y = R_1$; let α^* be Lebesgue outer measure; and let β^* be the outer measure of Exercise 11.*h*. Show that the only measurable generalized rectangles of positive measure are those of the form $A \times Y$.

f. Show that every μ^*-measurable set in Exercise *e* is of the form $E - Z$, where E is a generalized rectangle and $\mu^*(Z) = 0$.

g. In Exercise *e*, μ^* must be regular even though β^* is not. What is a measurable cover for the set $A \times B$ where B is a nonvacuous proper subset of Y?

h. Show that

$$\int_0^1 \int_1^\infty (e^{-xy} - 2e^{-2xy}) \, dx \, dy \neq \int_1^\infty \int_0^1 (e^{-xy} - 2e^{-2xy}) \, dy \, dx,$$

though each of these is a convergent Cauchy-Riemann integral. [*Hint*: One of these integrals is positive; the other, negative.]

i. Show that each of the single integration processes in Exercise *h* is actually a Lebesgue integration process; that is, it is absolutely convergent. Note the comment following 29.7.1.

j. Let X and Y be metric spaces with distance functions ρ_1 and ρ_2, respectively. Define a metric in $X \times Y$ by setting†

$$\rho[(a, b), (c, d)] = \max\,[\rho_1(a, c), \rho_2(b, d)].$$

Show that if α^* and β^* are metric outer measures, then μ^* is a metric outer measure.

k. Write out proofs for 29.3.1 and 29.3.2.

l. Write out a proof that if α^* and β^* are Lebesgue-Stieltjes outer measures in R_1, then μ^* is the Lebesgue-Stieltjes outer measure in R_2 induced by the product of their distribution functions.

m. Use the considerations in this section to generalize Exercise 24.*g* to the case of an integral over any space in which a measure is defined.

n. Show that if f and g are integrable on X and Y, respectively, and if $h(x, y) = f(x)g(y)$, then h is integrable on $X \times Y$, and

$$\int_{X \times Y} h\, d\mu = \int_X f\, d\alpha \int_Y g\, d\beta.$$

*30 EXPECTATION OF A RANDOM VARIABLE

In Section 22 we noted that in probability theory a measurable function is called a random variable. If μ is a probability measure in Ω and if f is a random variable over Ω, we define $e(f)$, the *expectation of* f, to be $\int_\Omega f$. If f is not integrable, we say it has no expectation.

In practice a random variable represents a numerical measurement on some chance phenomenon. There may be many ways of representing a given chance phenomenon by a measure space, and for each of these representations of the physical situation there is some random variable which will represent the numerical measurement in question. We should hope that all these random variables would have the same expectation—that is, that the "expectation of the measurement" would be independent of the mathematical representation of the situation. However, it is not obvious from the definition of expectation that this is the case. To show that it is, let us recall from Section 22 that every random variable f generates a probability measure in R_1 in such a way that f may be transformed into the coordinate variable in R_1. Clearly, it suffices to show that expectation is invariant under this transformation.

---

† This is the usual metric in a product of metric spaces.

Theorem 30.1. *Let μ be a probability measure in Ω; let f be a random variable over Ω; let μ_p be the probability measure in R_1 generated by f; and let x be the coordinate variable in R_1. Then,*

$$e(f) = \int_\Omega f \, d\mu = \int_{R_1} x \, d\mu_p$$

in the sense that if either integral exists, so does the other, and the equality holds.

Proof. Clearly, it suffices to prove the theorem for the case in which f is nonnegative. If this is the case, then $\mu_p[(-\infty, 0)] = 0$, so x is nonnegative a. e. Let g be the sequence of simple functions converging to f given by the construction in 21.1, and let y be the sequence constructed from x in the same way. By 23.6.2, it suffices to prove that

$$\lim_n \int_\Omega g_n \, d\mu = \lim_n \int_{R_1} y_n \, d\mu_p.$$

For any n,

$$\int_\Omega g_n \, d\mu = \sum_{i=1}^{n2^n} \frac{i-1}{2^n} \mu\left(g_n^{-1}\left\{\left[\frac{i-1}{2^n}, \frac{i}{2^n}\right)\right\}\right) + n\mu(g_n^{-1}\{[n, \infty]\}),$$

and

$$\int_{R_1} y_n \, d\mu_p = \sum_{i=1}^{n2^n} \frac{i-1}{2^n} \mu_p\left\{\left[\frac{i-1}{2^n}, \frac{i}{2^n}\right)\right\} + n\mu_p\{[n, \infty]\}.$$

These are equal by the definition of μ_p. ∎

Two important theorems concerning expectations are the addition and multiplication theorems. The *addition theorem* says that for any two integrable random variables,

30.2 $e(f + g) = e(f) + e(g).$

The *multiplication theorem* says that for two independent integrable random variables,

30.3 $e(fg) = e(f)e(g).$

One reason for the different hypotheses is that so far as integration theory is concerned these are completely different types of theorem. The addition theorem is 25.2, while the multiplication theorem is Exercise 29.n.

In Section 29 we used measures α and β in spaces X and Y respectively to generate a measure μ in $X \times Y$. In the Lebesgue-Stieltjes case μ, as constructed there, is a product measure; so in the language of probability theory, the considerations of Section 29 apply only to the case in which the coordinate variables are independent. For the study of the general case in probability theory, we reverse the procedure and determine α and β from μ.

Specifically, let X and Y be any two spaces, and let μ^* be any outer measure in $X \times Y$ for which $\mu(X \times Y) = 1$. For each $A \subset X$, let

$$\alpha^*(A) = \mu^*(A \times Y),$$

and for each $B \subset Y$, let

$$\beta^*(B) = \mu^*(X \times B).$$

The measures α and β are called the *marginal probability measures* determined by μ^*.

Though we shall retain the notation of Section 29 in order to indicate without undue bother which measures go with which spaces, the reader should bear in mind that α, β, and μ as used here are not necessarily related as in Section 29. It may seem peculiar that in this apparently more general case it takes only two lines to define the α, β, μ relations, while in Section 29 it took pages. The explanation is that for Fubini's theorem we need a product measure, so μ must be built up from α and β. Furthermore, even if we could somehow define a product measure *a priori* (without knowing product of what), marginal measures can be defined only if $\mu(X \times Y) < \infty$, and we must retain the case of infinite measure in Fubini's theorem in order to get important corollaries such as 29.7.2.

EXERCISES

a. In Exercises 22.b and 22.c compute $e(f)$ from each representation and thereby check 30.1 for these specific examples.

b. The *variance* of a random variable f is defined as

$$\text{var}(f) = e\{[f - e(f)]^2\}.$$

Show that if f and g are independent, then

$$\text{var}(f + g) = \text{var}(f) + \text{var}(g).$$

c. Generalize Exercise b. Show that if f_i and f_j are independent for $i \neq j$, then

$$\text{var}\left(\sum_{i=1}^{n} f_i\right) = \sum_{i=1}^{n} \text{var}(f_i).$$

d. The hypothesis (Exercise c) that a set of random variables are *independent by pairs* is weaker than the hypothesis that they are *totally independent*. Let Ω consist of the eight corner points of the unit cube. Define a measure by assigning the weight $\frac{1}{4}$ to each of the points $(1, 0, 0)$, $(0, 1, 0)$, $(0, 0, 1)$, and $(1, 1, 1)$. Show that this is not a product measure, but that each of the two-dimensional marginal measures is a product measure.

e. Show that if $A \times Y$ is μ^*-measurable, then A is α^*-measurable.

f. Show that if μ^* is regular, then the converse to Exercise e holds.

g. Show that if μ^* is regular, then $A \times B$ is μ^*-measurable whenever A is α^*-measurable and B is β^*-measurable.

h. Show that if μ^* is not regular, Exercises f and g may fail. [*Hint:* Let $X \times Y$ be the four corners of a square, and let μ^* be one-half the outer measure of Exercise 11.h.]

Let $X = Y =$ the set of positive integers; then (note Exercise 10.a) a probability measure μ is determined by a double sequence u such that $u_{ij} \geq 0$ for each i, j and such that

$$\sum_{i=1}^{\infty} \sum_{j=1}^{\infty} u_{ij} = 1.$$

The marginal measures are determined by the sequences p and q defined by

$$p_i = \sum_{j=1}^{\infty} u_{ij},$$

$$q_j = \sum_{i=1}^{\infty} u_{ij}.$$

We may assume that p_i and q_j are never zero, because if $p_i = 0$, then $u_{ij} = 0$ for every j, and this entire column may be removed from the space. Letting x and y be the coordinate variables in X and Y, respectively, we define the *conditional probability that* $y = j$, *given* $x = i$ as

$$q_i(j) = \frac{u_{ij}}{p_i}.$$

Similarly, the *conditional probability that* $x = i$, *given* $y = j$, is defined as

$$p^j(i) = \frac{u_{ij}}{q_j}.$$

We now define the *conditional expectation of* y, *given* $x = i$, as

30.4 $$e_i(y) = \sum_{j=1}^{\infty} j q_i(j).$$

The expectation of y is given in this case by

$$e(y) = \sum_{i=1}^{\infty} \sum_{j=1}^{\infty} j u_{ij},$$

so

30.5 $$e(y) = \sum_{i=1}^{\infty} p_i \sum_{j=1}^{\infty} j \frac{u_{ij}}{p_i} = \sum_{i=1}^{\infty} e_i(y) p_i.$$

In the general case the definitions of conditional expectation and conditional probability are not so straightforward, because we usually have a hypothesis whose probability is zero. For an isolated hypothesis $x = x_0$ of probability zero, the conditional expectation $e_{x_0}(y)$ is meaningless; but we can use a relation analogous to (30.5) to define the aggregate of all such conditional expectations as a measurable function over X.

Let f be any integrable random variable over $X \times Y$. Let σ be defined on the measurable subsets of X by

$$\sigma(A) = \int_{A \times Y} f \, d\mu.$$

Now, if $\alpha(A) = 0$, then $\mu(A \times Y) = 0$, so $\sigma(A) = 0$. Thus, σ is absolutely continuous with respect to α. Therefore, by the Radon-Nikodym theorem, there is a measurable function g defined uniquely a. e. in X by the relation

$$\sigma(A) = \int_A g \, d\alpha.$$

We call the value $g(x)$ of g the *conditional expectation of f, given x*. Our notation for this conditional expectation will be $e_{(x)}(f)$. That is, the conditional expectation of f, given x, is the value at x of the function $e_{()}(f)$, defined a. e. on X by the condition that

30.6
$$\int_{A \times Y} f \, d\mu = \int_A e_{()}(f) \, d\alpha$$

for every measurable $A \subset X$. Similarly, the *conditional expectation of f, given y*, is the value $e^{(y)}(f)$ of the function $e^{()}(f)$ defined a. e. on Y by the condition that

$$\int_{X \times B} f \, d\mu = \int_B e^{()}(f) \, d\beta$$

for every measurable $B \subset Y$.

If C_E is the characteristic function of a measurable set $E \subset X \times Y$, then for each $x \in X$ for which it is defined we set

$$e_{(x)}(C_E) = \mu_{(x)}(E)$$

and call this quantity the *conditional probability of E, given x*. Similarly, we define $\mu^{()}(E)$ over Y by setting

$$e^{(y)}(C_E) = \mu^{(y)}(E).$$

Referring back to 30.6, we may say that the conditional probability of E, given x, is the value at x of the function $\mu_{()}(E)$, defined a. e. on X by the condition that

30.7
$$\int_A \mu_{()}(E) \, d\alpha = \int_{A \times Y} C_E \, d\mu = \mu[E \cap (A \times Y)]$$

for every measurable $A \subset X$.

We want to emphasize that $\mu_{()}(E)$ is a measurable function on X, defined a. e. for each individual E. This is not the same as to say that for almost all $x \in X$, $\mu_{(x)}()$ is defined for all measurable $E \subset X \times Y$. The exceptional set of measure zero may vary from one set E to another with the result that for any given x, $\mu_{(x)}(E)$ may not be defined for any appreciable class of sets E. On the other hand, if we consider only a countable class of sets in any one discussion, the union of the exceptional sets still has measure zero; and we have the following result, which shows the resemblance of $\mu_{(x)}()$ to a measure.

Theorem 30.8. *If E is a sequence of disjoint measurable sets in $X \times Y$, and if*

$$E_0 = \bigcup_{n=1}^{\infty} E_n,$$

then for almost all $x \in X$,

$$\mu_{(x)}(E_0) = \sum_{n=1}^{\infty} \mu_{(x)}(E_n).$$

Proof. From 30.7 we see that for every measurable $A \subset X$,

$$\int_A \mu_{()}(E_0)\, d\alpha = \mu[E_0 \cap (A \times Y)] = \sum_{n=1}^{\infty} \mu[E_n \cap (A \times Y)]$$

$$= \sum_{n=1}^{\infty} \int_A \mu_{()}(E_n)\, d\alpha = \int_A \left[\sum_{n=1}^{\infty} \mu_{()}(E_n)\right] d\alpha.$$

The last step is justified by 26.1, and the result now follows from 25.7.1. ∎

For each measurable $B \subset Y$ we can define the function $\beta_{()}(B)$ a. e. on X by setting

$$\beta_{(x)}(B) = \mu_{(x)}(X \times B).$$

Similarly, for each measurable $A \subset X$, we define $\alpha^{()}(A)$ a. e. on Y by setting

$$\alpha^{(y)}(A) = \mu^{(y)}(A \times Y).$$

Each of these functions has an additivity property induced by 30.8, but again the range of validity of the additivity equation may vary from one sequence of sets to another.

However, it is interesting to note that given a fixed function f on $X \times Y$, we can define an "integral" of f with respect to a conditional probability and limit the discussion to a countable class of sets. This class will, of course, depend on f. That is, we define for nonnegative f,

$$\int_{X \times Y}' f\, d\mu_{(x)} = \lim_n \sum_{i=1}^{k(n)} t_{ni} \mu_{(x)}(E_{ni}),$$

where for each n,

$$\sum_{i=1}^{k(n)} t_{ni} C_{E_{ni}}$$

is one of a monotone sequence of simple functions converging pointwise to f. Clearly, for a given f, this limit exists (at least in the extended number system) for almost all $x \in X$. Similarly, we define

$$\int_Y' f(x,\ )\, d\beta_{(x)} = \lim_n \sum_{i=1}^{k(n)} t_{ni} \beta_{(x)}\{y \mid (x, y) \in E_{ni}\},$$

$$\int_X' f(\ , y)\, d\alpha^{(y)} = \lim_n \sum_{i=1}^{k(n)} t_{ni} \alpha^{(y)}\{x \mid (x, y) \in E_{ni}\}.$$

If $\mu_{(x)}(\ )$ and $\mu^{(y)}(\ )$ happen to be genuine measures for almost all x and almost all y, then these are integrals in the usual sense; but in any case we can prove a sort of generalized Fubini theorem for these pseudo-integrals. For this purpose, let us assume that μ^* is generated by Method I from some function on a class of generalized rectangles whose sides later turn out to be measurable. In particular, this is the case for any (not necessarily product) Lebesgue-Stieltjes probability measure in R_2.

Theorem 30.9. *If μ^* is as described above, if f is a random variable on $X \times Y$, and if $e(f)$ exists, then*

$$\int_Y' f(x,\ )\, d\beta_{(x)}$$

exists for almost all $x \in X$,

$$\int_X' f(\ , y)\, d\alpha^{(y)}$$

exists for almost all $y \in Y$, and

$$\int_X \int_Y' f\, d\beta_{()}\, d\alpha = \int_{X \times Y} f\, d\mu = \int_Y \int_X' f\, d\alpha^{()}\, d\beta.$$

Proof. If f is the characteristic function of $A \times B$, the first equality reduces to

$$\int_A \mu_{()}(A \times B)\, d\alpha = \mu(A \times B),$$

and this is a special case of 30.7. A similar observation may be made about the other equality. The remainder of the argument parallels that used to prove 29.6 and 29.7; it need not be repeated here. The only point to be noted is that for a fixed f, the entire proof uses only a countable class of generalized rectangles; thus in each of the spaces X and Y there is a single set of measure zero outside which all the required pseudo-integrals are defined. ∎

EXERCISES

i. Show that if μ^* is as in 30.9, if f is a random variable on $X \times Y$, and if $e(f)$ exists then for each measurable $A \subset X$,

$$\int_A \int_Y' f\, d\beta_{()}\, d\alpha = \int_{A \times Y} f\, d\beta.$$

j. From Exercise i derive the following parallel to (30.4): For almost all $x \in X$,

$$\int_Y' f(x,\ )\, d\beta_{(x)} = e_{(x)}(f).$$

k. Show that if μ is a product measure, then for each integrable f,

$$\int_Y' f(x, \ ) \, d\beta_{(x)} = \int_Y f(x, \ ) \, d\beta$$

for almost all $x \in X$. [*Hint*: 29.7 and Exercise *i* both apply. Use 25.7.1.]

l. Show that if x and y are independent random variables, then for each measurable $B \subset Y$,

$$\beta_{(x)}(B) = \beta(B)$$

for almost all $x \in X$.

REFERENCES FOR FURTHER STUDY

On integrals over general measure spaces:
  Bartle [2]
  Carathéodory [4], [5]
  Hahn and Rosenthal [9]
  Halmos [11]
  Saks [27]
On integrals with respect to nonmonotone set functions:
  Hahn and Rosenthal [9]
  Halmos [11]
  von Neumann [23]

On Lebesgue integrals:
  Hobson [14]
  Lebesgue [19]
  Natanson [22]
  Royden [25]
  Titchmarsh [28]
On Lebesgue-Stieltjes integrals:
  Lebesgue [19]
  McShane and Botts [20]
  Saks [27]
On random variables and their expectations:
  Halmos [11]
  Kolmogoroff [18]

CHAPTER 6

DIFFERENTIATION

Let f be a function of a single real variable x. The derivative of f at x_0 is usually defined as

$$f'(x_0) = \lim_{x \to x_0} \frac{f(x) - f(x_0)}{x - x_0}. \tag{1}$$

For purposes of generalization, it is convenient to write this in the following form:

$$f'(x_0) = \lim_{\substack{x \to x_0^+ \\ y \to x_0^-}} \frac{f(x) - f(y)}{x - y}, \tag{2}$$

and (contrary to the usual custom) to interpret this double limit to include the possibility that $x = x_0$ or $y = x_0$. Clearly, these last-mentioned special cases of (2) include (1), so (2) implies (1). To go the other way, we note that (1) yields directly the special cases of (2) in which $x = x_0$ or $y = x_0$. For the other cases,

$$\left| \frac{f(x) - f(y)}{x - y} - f'(x_0) \right| \leq \left| \frac{f(x) - f(x_0)}{x - y} - \frac{x - x_0}{x - y} f'(x_0) \right|$$

$$+ \left| \frac{f(x_0) - f(y)}{x - y} - \frac{x_0 - y}{x - y} f'(x_0) \right|;$$

so for $y < x_0 < x$, we multiply the first term on the right by $(x - y)/(x - x_0) > 1$ and the second by $(x - y)/(x_0 - y) > 1$ and have

$$\left| \frac{f(x) - f(y)}{x - y} - f'(x_0) \right| < \left| \frac{f(x) - f(x_0)}{x - x_0} - f'(x_0) \right| + \left| \frac{f(x_0) - f(y)}{x_0 - y} - f'(x_0) \right|.$$

From this it is clear that (1) implies (2) in general.

To see how (2) is generalized, let f be a distribution function on R_1, let μ_f be the Lebesgue-Stieltjes measure induced by f, and let μ be Lebesgue measure in R_1. Then, (2) may be written

$$f'(x_0) = \lim_{\substack{x \to x_0^+ \\ y \to x_0^-}} \frac{\mu_f\{(x, y]\}}{\mu\{(x, y]\}}.$$

In the general case it is customary to refer to this as the derivative of the set function μ_f rather than that of the point function f. Thus, we shall speak of the derivative of a set function σ at a point x_0 and define it by some such relation as

$$\sigma'(x_0) = \lim_{\substack{x_0 \in I \\ d(I) \to 0}} \frac{\sigma(I)}{\mu(I)}.$$

The theory of such derivatives will include as a special case that of derivatives of monotone functions of a real variable; hence, because of the Jordan decomposition theorem, it will include the theory of derivatives of functions of bounded variation.

31 SUMMARY OF THE PROBLEM

Our main purpose in studying differentiation is to determine as precisely as possible the sense in which differentiation and integration are inverse processes. As we remarked at the conclusion of Section 27, this relationship between differentiation and integration is fundamental in many applications of integral calculus. For the case of Lebesgue integrals in R_1 the standard theorems are as follows. (1) *Every function of bounded variation is differentiable* a. e. (2) *Therefore, every indefinite integral is differentiable* a. e. (3) *Furthermore, the derivative of an indefinite integral is* a. e. *equal to the integrand.* (4) *Finally, every absolutely continuous function is the indefinite integral of its derivative.*

A function of bounded variation may be characterized as the difference of two Lebesgue-Stieltjes distribution functions; therefore, in line with our remarks in the introduction to this chapter, we shall obtain (1) above by proving the more general result that every Lebesgue-Stieltjes measure in R_n is differentiable a. e. We shall also prove (3) and (4) above for the case of Lebesgue integrals in R_n. The connecting link (2) is furnished by 31.1 below. Throughout this section and Sections 32 through 35, μ will be the Lebesgue measure function, and all such measure-dependent notions as "$\int$," "a. e.," and "absolutely continuous" will be interpreted with respect to Lebesgue measure.

Theorem 31.1. *If f is nonnegative and Lebesgue integrable on R_n and if $\sigma = \int f$, then σ is a Lebesgue-Stieltjes measure.*

Proof. We define g by setting

$$g(x) = \int_{(-\infty, x]} f;$$

then clearly, $\sigma(I) = \mu_g(I)$ for every half-open interval I. From 14.5 and the complete additivity of μ_g and σ, it follows that $\sigma(G) = \mu_g(G)$ for every open set G. Any set $H \in \mathcal{G}_\delta$ may be described as the limit of a contracting sequence of open sets, and σ is everywhere finite; so it follows from 10.8.3 that $\sigma(H) = \mu_g(H)$ for every $H \in \mathcal{G}_\delta$. If $\mu_g(Z) = 0$, then $Z \subset H \in \mathcal{G}_\delta$ where $\mu_g(H) = 0$; therefore, by what we have already proved $\sigma(H) = \mu_g(H) = 0$, hence $\sigma(Z) = \mu_g(Z)$. Thus, by 13.5.1, $\sigma = \mu_g$ for all sets in the domain of μ_g. However, the domain of σ is that of μ (Lebesgue measure), so we have yet to consider sets for which $\mu = 0$.

If $\mu(Z) = 0$, then $Z \subset H \in \mathscr{G}_\delta$ where $\mu(H) = 0$; therefore by 25.3, $\sigma(H) = 0$. However, $\sigma(H) = \mu_g(H)$, so $\mu_g(Z) = 0$. ∎

The principal complication in the study of differentiation in R_n arises from the different possible ways of defining a derivative. Let us now turn our attention to some of these definitions. We shall say that a sequence I of intervals in R_n *converges to* $x \in R_n$ and write

$$I_k \to x$$

if $x \in I_k$ for each k and if $\lim_k d(I_k) = 0$. We define a *cube* to be an interval with equal, nonzero sides. Let I be a sequence of intervals converging to x; we shall say that I is a *regular sequence* if there is a constant $\alpha > 0$ such that for each k, there is a cube $J_k \supset I_k$ for which

$$\frac{\mu(I_k)}{\mu(J_k)} \geq \alpha.$$

The constant α is called a *parameter of regularity* for the sequence I. Intuitively, a regular sequence is one in which the intervals do not become too "slim."

Let τ be a set function defined at least for the closed intervals in R_n. We define the *upper derivate of τ at x* as

$$\bar{\tau}'(x) = \sup_{J_k \to x} \overline{\lim_k} \frac{\tau(J_k)}{\mu(J_k)},$$

where the sequences J are sequences of closed cubes. Similarly, we define the *lower derivate of τ at x* as

$$\underline{\tau}'(x) = \inf_{J_k \to x} \underline{\lim_k} \frac{\tau(J_k)}{\mu(J_k)},$$

where, again, each J is a sequence of closed cubes. If

$$-\infty < \underline{\tau}'(x) = \bar{\tau}'(x) < \infty,$$

we say that τ is *differentiable at x*, and we define the *derivative of τ at x* as

$$\tau'(x) = \bar{\tau}'(x) = \underline{\tau}'(x).$$

The reader should note that upper and lower derivates are always defined, since we allow them to be infinite. However, we are adopting the convention that derivatives must be finite.

If we relax the restriction that each J be a sequence of cubes to require only that it be a regular sequence of closed intervals, then we define the *regular upper and lower derivates* and the *regular derivative of τ at x*. We shall designate these by $\bar{\tau}'^*(x)$, $\underline{\tau}'^*(x)$, and $\tau'^*(x)$, respectively. Finally, if we relax the restriction of regularity and require only that each J be a sequence of closed intervals such that $\mu(J_k) > 0$ for each k, then we define the *strong upper and lower derivates* and the *strong derivative of τ at* x. We shall designate these by $\bar{D}\tau(x)$, $\underline{D}\tau(x)$, and $D\tau(x)$, respectively.

In R_1 every interval is a cube, so every sequence of intervals is regular; therefore in R_1 these three definitions of a derivative are equivalent. Thus, any theorems concerning any of these types of derivative may be specialized to give theorems about differentiation of functions of a real variable.

In $R_n (n > 1)$, however, the theory differs for different types of derivative. In Sections 32 through 34 we shall prove for derivatives (with respect to cubes) a complete set of theorems paralleling statements (1) through (4) at the beginning of this section. These same theorems hold for regular derivatives, and these proofs are indicated in the exercises. The theory is radically different for strong derivatives, however. The only satisfactory theorem in that case is that for bounded integrands the strong derivative of the indefinite integral exists and equals the integrand a. e. Section 35 is devoted to the proof of this result together with a few other items of interest. Not only do there exist unbounded Lebesgue integrable functions whose indefinite integrals fail to have a finite strong derivative at any point, but indeed, over R_2 the set of all such functions forms a residual set in the space L_1 (see Exercises 35.n through 35.q).

In the light of these results it might seem that the most profitable study of differentiation in R_n is that involving regular derivatives. It is true that the theory is more satisfying in the regular case, but unfortunately the strong derivative is of considerable practical importance. Suppose τ is defined from a Lebesgue-Stieltjes distribution function f by 14.1; then

$$\frac{\tau\{(a, b]\}}{\mu\{(a, b]\}} = \frac{f(b_1, b_2) - f(a_1, b_2) - f(b_1, a_2) + f(a_1, a_2)}{(b_1 - a_1)(b_2 - a_2)}$$

$$= \frac{\dfrac{f(b_1, b_2) - f(a_1, b_2)}{b_1 - a_1} - \dfrac{f(b_1, a_2) - f(a_1, a_2)}{b_1 - a_1}}{b_2 - a_2}.$$

The existence of $D\tau(a)$ will involve the existence of the double limit of this expression as $b_1 \to a_1$ and $b_2 \to a_2$. However, the iterated limit, taken in this order, yields the *cross partial derivative*

$$\frac{\partial^2 f}{\partial x_2\, \partial x_1},$$

and the other iterated limit yields the other cross partial derivative of f. So, these cross partial derivatives of f exist at a and are equal whenever the strong derivative of τ exists at a and the inside limits exist in some neighborhood of a. The inside limits here are the first partial derivatives of f; so with the aid of the strong derivative we can work up from the existence of the first partial derivatives to the existence and equality of the cross partial derivatives.

EXERCISES

a. Show that $\underline{D}\tau \leq \underline{\tau}'^* \leq \underline{\tau}' \leq \bar{\tau}' \leq \bar{\tau}'^* \leq \bar{D}\tau$.

b. Show that the existence of $D\tau(x)$ implies that of $\tau'^*(x)$, which in turn implies that of $\tau'(x)$.

c. Show that if any two of the derivatives $D\tau(x)$, $\tau'^*(x)$, and $\tau'(x)$ exist, then they are equal.

d. Given a mass distribution in R_2 in which the mass is concentrated on a line with the mass of a segment on this line proportional to its length, show that the Lebesgue-Stieltjes measure for this mass distribution is not differentiable at any point on the given line.

e. Given a mass distribution in R_2 in which the mass is concentrated on the x_1 axis with the mass of the segment $[0, x_1]$ equal to x_1^p, show that the Lebesgue-Stieltjes measure for this distribution is not differentiable at any point on the x_1 axis different from the origin.

f. Show that for $p \leq 2$ the Lebesgue-Stieltjes measure in Exercise e is not differentiable at the origin.

g. Show that for $p > 2$ the Lebesgue-Stieltjes measure in Exercise e is regularly but not strongly differentiable at the origin.

h. Let f be defined on R_2 by

$$f(x_1, x_2) = \begin{cases} x_1 - x_2 & \text{for } x_1 \geq 0, x_2 \geq 0, \\ x_2 - x_1 & \text{for } x_1 < 0, x_2 < 0, \\ 0 & \text{otherwise.} \end{cases}$$

Show that f is the difference of two Lebesgue-Stieltjes distribution functions.

i. Let τ be defined by 14.1 from the function f of Exercise h. Show that τ is differentiable but not regularly differentiable at the origin.

j. Show that if a Lebesgue-Stieltjes measure μ_f in R_1 is differentiable at x, then the distribution function f is continuous at x.

k. Show by an example that in R_2 continuity of f does not follow even from strong differentiability of μ_f.

l. Show that if τ_1 and τ_2 are each differentiable at x and if $\tau = \tau_1 + \tau_2$, then τ is differentiable at x and $\tau'(x) = \tau_1'(x) + \tau_2'(x)$. Show that this result also holds for regular and strong derivatives.

m. Show that if $\tau = \tau_1 + \tau_2$ and if $\bar{\tau}_1'(x)$ is finite, then $\bar{\tau}'(x) = \bar{\tau}_1'(x) + \bar{\tau}_2'(x)$. Show that this result holds for regular and strong derivatives.

n. Let f_1 and f_2 be defined on R_1 by

$$f_1(x) = \sqrt[3]{x} + x \sin(1/x),$$
$$f_2(x) = \sqrt[3]{x}.$$

Let τ_1 and τ_2 be defined by $\tau_i([a, b]) = f_i(b) - f_i(a)$ for $i = 1, 2$, and let $\tau = \tau_1 - \tau_2$. Show that

$$\bar{\tau}_1'(0) = \underline{\tau}_1'(0) = \bar{\tau}_2'(0) = \underline{\tau}_2'(0) = \infty,$$

but

$$\bar{\tau}'(0) \neq \underline{\tau}'(0).$$

32 VITALI COVERINGS

A class $\mathscr{J}$ of intervals in R_n is said to *cover a set E in the sense of Vitali*, provided that for every $x \in E$ there is a regular sequence J of intervals from $\mathscr{J}$ such that $J_k \to x$. We shall usually be concerned with coverings by cubes, in which case the regularity of the sequences is automatic. That is, if $\mathscr{J}$ consists of cubes, then $\mathscr{J}$ covers E in the sense of Vitali, provided each point of E is contained in an arbitrarily small cube from $\mathscr{J}$. However, it should be borne in mind that in the more general case the regularity of the sequences plays a vital role.

Theorem 32.1 (Vitali). *If a class $\mathscr{J}$ of closed cubes covers a set $E \subset R_n$ in the sense of Vitali, then there is a sequence J of disjoint cubes from $\mathscr{J}$ such that*

$$\mu\left(E - \bigcup_{k=1}^{\infty} J_k\right) = 0.$$

Proof. We shall first prove the theorem subject to the assumption that E and the entire class $\mathscr{J}$ can be enclosed in some cube I_0.

We define the sequence J inductively. Let J_1 be any cube from $\mathscr{J}$. Given disjoint cubes $J_1, J_2, \ldots, J_k$, we let δ_k be the least upper bound of the diameters of cubes K such that $K \in \mathscr{J}$ and

$$K \cap \left(\bigcup_{i=1}^{k} J_i\right) = \phi.$$

If there is no such cube K, then it follows that $E \subset \bigcup_{i=1}^{k} J_i$, and the theorem is proved. Otherwise, we choose for J_{k+1} any cube from $\mathscr{J}$ such that

$$J_{k+1} \cap \left(\bigcup_{i=1}^{k} J_i\right) = \phi$$

and

$$d(J_{k+1}) > \frac{\delta_k}{2}. \tag{1}$$

To show that the sequence J, thus defined, has the required properties, we introduce the sequence J' defined as follows. For each k, J'_k has the same center as J_k, and

$$d(J'_k) = (4n + 1) d(J_k),$$

where n is the dimension of the space. The effect of this construction is that if $x_0 \in -J'_k$ and $x_1 \in J_k$, then†

$$\rho(x_0, x_1) \geq 2d(J_k) > \delta_{k-1}. \tag{2}$$

---

† If $n = 1$, the factor $4n + 1 = 5$ gives exactly the effect (2). For $n = 2$, the factor $4n + 1 = 9$ is already larger than necessary, but it must increase with n because the diameter of a cube is its main diagonal, and this is $\sqrt{n}$ times the length of a side.

Since the cubes of the sequence J are disjoint and since they are all contained in I_0, we have

$$\sum_{k=1}^{\infty} \mu(J'_k) = (4n + 1)^n \sum_{k=1}^{\infty} \mu(J_k) \leq (4n + 1)^n \mu(I_0). \tag{3}$$

Thus, $\sum \mu(J'_k)$ is convergent. Now, we let

$$A = E - \bigcup_{k=1}^{\infty} J_k, \tag{4}$$

and suppose $\mu^*(A) > 0$. It follows from (3) that there is an integer k_0 such that

$$\sum_{k=k_0}^{\infty} \mu(J'_k) < \mu^*(A).$$

Thus, there is a point $x_0 \in A$ such that $x_0 \in -J'_k$ for each $k \geq k_0$. With $x_0 \in A$, we have from (4) that $x_0 \in -J_k$ for $k = 1, 2, \ldots, k_0$; and since these cubes are closed, there is a cube $K \in \mathscr{J}$ such that $x_0 \in K$ and

$$K \cap J_k = \phi \quad (k = 1, 2, \ldots, k_0). \tag{5}$$

It follows from (1) and (3) that

$$\delta_k \leq 2d(J_{k+1}) = 2\sqrt{n}[\mu(J_{k+1})]^{1/n} \to 0$$

as $k \to \infty$; so it follows from the definition of δ_k that there is a smallest integer k_1 such that

$$K \cap J_{k_1} \neq \phi.$$

It then follows from the definition of δ_k that

$$d(K) \leq \delta_{k_1 - 1}, \tag{6}$$

and it follows from (5) that $k_1 > k_0$. Thus, we get points

$$x_1 \in K \cap J_{k_1} \quad \text{and} \quad x_0 \in K \cap (-J'_{k_1}),$$

and we have from (2) that

$$d(K) \geq \rho(x_0, x_1) > \delta_{k_1 - 1},$$

which contradicts (6). Thus the assumption that $\mu^*(A) > 0$ leads to a contradiction.

We have yet to remove the boundedness assumption made at the beginning of the proof. In the general case, let I be a strictly expanding sequence of cubes such that $\lim_m I_m = R_n$, and let $E_m = E \cap I_m$. We now construct a sequence J of disjoint cubes from $\mathscr{J}$ such that for each m there is an integer k_m such that

$$\mu^*\left(E_m - \bigcup_{k=1}^{k_m} J_k\right) < \frac{1}{m}.$$

Once this is done, we have

$$\mu^*\left(E_m - \bigcup_{k=1}^{\infty} J_k\right) < \frac{1}{m}$$

for each m, and the result follows from 12.1.1. The sequence J may be constructed as follows. The cubes from $\mathscr{J}$ which are contained in I_2 cover E_1 in the sense of Vitali; so the first k_1 cubes come from the fact that the theorem is already proved for E_1. Suppose the first k_m cubes are given. The cubes from $\mathscr{J}$ which are contained in the open set

$$I_{m+2}^0 - \bigcup_{k=1}^{k_m} J_k$$

cover the set

$$A_{m+1} = E_{m+1} - \bigcup_{k=1}^{k_m} J_k$$

in the sense of Vitali. Thus, the results already proved apply to A_{m+1}; and there is a finite set $J_{k_m+1}, \ldots, J_{k_{m+1}}$ of cubes, disjoint from each other and from the preceding k_m cubes, with

$$\frac{1}{m+1} > \mu^*\left(A_{m+1} - \bigcup_{k=k_m+1}^{k_{m+1}} J_k\right) = \mu^*\left(E_{m+1} - \bigcup_{k=1}^{k_{m+1}} J_k\right). \quad\blacksquare$$

As indicated in Exercises a and b below, the Vitali theorem holds not only for coverings by cubes but for any covering by intervals, provided (as specified in the definition of a Vitali covering) each $x \in E$ has a regular sequence of intervals from $\mathscr{J}$ converging to it. The following construction shows that this regularity condition cannot be removed.

Let $[a, b]$ be any closed interval in R_2. Let $\eta > 0$ be given, and let n be the smallest positive integer such that

$$\sum_{k=1}^{n} \frac{1}{k} > \frac{1}{\eta}.$$

We now define subintervals I_k^1 ($k = 1, 2, \ldots, n$) of $[a, b]$ as follows:

$$I_k^1 = \left\{x \mid a_1 < x_1 \leq a_1 + \frac{k(b_1 - a_1)}{n} \ ; \ a_2 < x_2 \leq a_2 + \frac{(b_2 - a_2)}{k}\right\}.$$

For $n = 5$, these are shown in Fig. 8. Each interval I_k^1 in that figure is labeled at its upper right-hand corner, and each has its lower left-hand corner at a.

For each k,

$$\mu(I_k^1) = \frac{(b_1 - a_1)(b_2 - a_2)}{n};$$

and if

$$V^1 = \bigcup_{k=1}^{n} I_k^1,$$

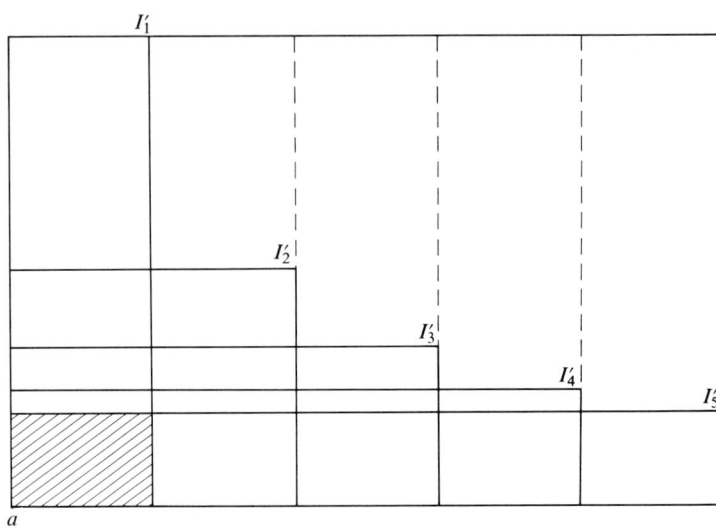

Figure 8

then (add the disjoint areas in the various columns in Fig. 8)

$$\mu(V^1) = \sum_{k=1}^{n} \frac{(b_1 - a_1)(b_2 - a_2)}{nk} > \frac{(b_1 - a_1)(b_2 - a_2)}{\eta n}.$$

Therefore we have

$$\mu(I_1^1) = \mu(I_2^1) = \cdots = \mu(I_n^1) < \eta\mu(V^1),$$

and (see shaded area in Fig. 8)

$$\bigcap_{k=1}^{n} I_k^1 \neq \phi.$$

The set $[a, b] - V^1$ consists of $n - 1$ disjoint intervals (see dotted lines in Fig. 8). We now repeat the above process on each of these to obtain sets $V^2, V^3, \ldots, V^n$, each being a union of n intervals such that

$$\mu(I_1^i) = \mu(I_2^i) = \cdots = \mu(I_n^i) < \eta\mu(V^i)$$

and

$$\bigcap_{k=1}^{n} I_k^i \neq \phi.$$

Now

$$[a, b] - \bigcup_{i=1}^{n} V^i$$

consists of $(n - 1)^2$ disjoint intervals, and the above process may be repeated on each of these. Finally, we may repeat the whole procedure j times to obtain disjoint sets $V^1, V^2, \ldots, V^{N_j}$, where

$$N_j = 1 + (n - 1) + (n - 1)^2 + \cdots + (n - 1)^j.$$

Whenever a set V is constructed in an interval I, we have

$$\mu(I - V) = \frac{\mu(I)}{n} \sum_{k=1}^{n} \frac{k-1}{k} = \theta_\eta \mu(I),$$

where

$$\theta_\eta = \frac{1}{n} \sum_{k=1}^{n} \frac{k-1}{k} < 1.$$

From this it follows that for each j,

$$\mu\left([a, b] - \bigcup_{i=1}^{N_j} V^i\right) = \theta_\eta \mu\left([a, b] - \bigcup_{i=1}^{N_{j-1}} V^i\right);$$

and starting from the fact that $\mu([a, b] - V^1) = \theta_\eta \mu([a, b])$, we have

$$\mu\left([a, b] - \bigcup_{i=1}^{N_j} V^i\right) = \theta_\eta^{j+1} \mu([a, b]).$$

Choosing j so that $\theta_\eta^{j+1} < \eta$, we have $(a, b]$ covered by intervals in such a way that any union of disjoint intervals from this covering has measure less than $2\eta\mu([a, b])$. To see this we note that no two disjoint intervals come from the same set V^i; therefore, of this type of interval we have at most $I_{k_1}^1, I_{k_2}^2, \ldots, I_{k_{N_j}}^{N_j}$ with $\mu(I_{k_i}^i) < \eta\mu(V^i)$, whence

$$\mu\left(\bigcup_{i=1}^{N_j} I_{k_i}^i\right) < \eta \sum_{i=1}^{N_j} \mu(V^i) = \eta\mu\left(\bigcup_{i=1}^{N_j} V^i\right) < \eta\mu([a, b]).$$

All other intervals in a disjoint class must come from

$$[a, b] - \bigcup_{i=1}^{N_j} V^i,$$

and we have just seen that this set has measure

$$\theta_\eta^{j+1} \mu([a, b]) < \eta\mu([a, b]).$$

If each of the intervals I_k^i is closed, we obtain a covering of $[a, b]$ by intervals with the same measures as before, but even fewer of these can be disjoint.

Finally, given $[c, d] \subset R_2$ and given $\varepsilon > 0$, for each positive integer m we subdivide $[c, d]$ into 4^m similar intervals, and for each of these we construct a covering $\mathscr{I}_m$ by closed intervals like that described above with $2\eta = \varepsilon/2^m$. The union of all these coverings is a covering of $[c, d]$ with the property that each $x \in [c, d]$ is contained in an interval from the covering with arbitrarily small diameter. For any sequence J of disjoint intervals from this covering, the ones from $\mathscr{I}_m$ have total measure less than $\varepsilon\mu([c, d])/2^m$; so

$$\mu\left(\bigcup_{r=1}^{\infty} J_r\right) < \sum_{m=1}^{\infty} \frac{\varepsilon}{2^m} \mu([c, d]) = \varepsilon\mu([c, d]).$$

Thus the Vitali theorem fails for this covering.

EXERCISES

a. Show that the Vitali theorem holds provided $\mathscr{J}$ is a class of closed intervals with the property that there is a fixed number $\alpha > 0$ such that for each $x \in E$ there is a sequence J of intervals from $\mathscr{J}$ converging to x and having α as a parameter of regularity.

b. Extend the Vitali theorem to any class $\mathscr{J}$ of closed intervals which covers E in the sense of Vitali. *Hint:* Let E_m be the set of points $x \in E$ such that there is a sequence J of intervals from $\mathscr{J}$ converging to x and having $1/m$ as a parameter of regularity. Use Exercise *a*, and note the argument for removing the boundedness condition in the proof of 32.1.

c. Construct an example to show that 32.1 need not hold unless $\mathscr{J}$ contains arbitrarily small cubes covering each point of E.

d. Construct an example in which the Vitali theorem fails even though each $x \in E$ is the center of an interval from $\mathscr{J}$. *Hint:* Let $\mathscr{J}'$ be the covering constructed at the end of this section. Construct $\mathscr{J}$ by taking, for each $x \in E$, the smallest interval J with center at x for which $J \supset J'$ where $x \in J' \in \mathscr{J}'$.

33 DIFFERENTIATION OF ADDITIVE SET FUNCTIONS

The purpose of this section is to prove that if σ is a completely additive set function in R_n whose domain includes the class of closed cubes, then (in a sense) σ is differentiable a. e. First, we prove some preliminary results.

Theorem 33.1. *If τ is any function defined over the closed cubes in R_n, then $\bar{\tau}'$ and $\underline{\tau}'$ are measurable functions.*

Proof. The proof of this theorem depends on the following lemma, which is of some interest in itself.

Lemma 33.1.1. *Any (countable or noncountable) union of cubes is a measurable set.*

Let $\mathscr{J}$ be any class of cubes, and let

$$A = \bigcup_{I \in \mathscr{J}} I.$$

Let $\mathscr{J}$ be the class of closed cubes J such that $J \subset I$ for some $I \in \mathscr{J}$. Then clearly, $\mathscr{J}$ covers A in the sense of Vitali, and 32.1 gives A as the union of a set of measure zero and a countable class of cubes from $\mathscr{J}$. This proves the lemma.

To complete the proof of the theorem, let a be a real number and let

$$E = \{x \mid \bar{\tau}'(x) > a\}.$$

For each pair of integers h, k, let E_{hk} be the union of all the cubes J such that $d(J) \leq 1/k$ and $\tau(J)/\mu(J) \geq a + 1/h$. Each of these sets is measurable by the lemma, and

$$E = \bigcup_{h=1}^{\infty} \bigcap_{k=1}^{\infty} E_{hk};$$

so E is measurable, and it follows from 19.3 that $\bar{\tau}'$ is measurable. The argument for $\underline{\tau}'$ is completely analogous. ∎

Theorem 33.2. *If σ is a Lebesgue-Stieltjes measure in R_n and if for every $x \in A$,*
$$\bar{\sigma}'(x) \geq a,$$
then
$$\sigma(E) \geq a\mu^*(A)$$
for every measurable set $E \supset A$.

Proof. Given $\varepsilon > 0$, we have by 13.7.1 an open set $G \supset E$ such that
$$\sigma(E) \geq \sigma(G) - \varepsilon.$$
Given $b < a$, we denote by $\mathscr{J}$ the class of closed cubes K such that $K \subset G$ and
$$\sigma(K) \geq b\mu(K).$$
It follows at once from the fact that $b < a \leq \bar{\sigma}'(x)$ for all $x \in A$ that $\mathscr{J}$ covers A in the sense of Vitali; thus by 32.1 there is a sequence J of disjoint cubes from $\mathscr{J}$ such that
$$\sum_{k=1}^{\infty} \mu(J_k) = \mu\left(\bigcup_{k=1}^{\infty} J_k\right) \geq \mu^*(A).$$
So we have
$$\sigma(E) \geq \sigma(G) - \varepsilon \geq \sigma\left(\bigcup_{k=1}^{\infty} J_k\right) - \varepsilon = \sum_{k=1}^{\infty} \sigma(J_k) - \varepsilon$$
$$\geq b \sum_{k=1}^{\infty} \mu(J_k) - \varepsilon \geq b\mu^*(A) - \varepsilon.$$
Since this holds for arbitrary $b < a$ and $\varepsilon > 0$, we have the desired result. ∎

Theorem 33.3 (Lebesgue). *If σ is a Lebesgue-Stieltjes measure in R_n, then σ is differentiable a. e.*

Proof. For each pair of positive integers h, k, we set
$$A_{hk} = \left\{ x \mid \bar{\sigma}'(x) > \frac{h+1}{k} > \frac{h}{k} > \underline{\sigma}'(x) \right\}. \tag{1}$$

Each of these sets is measurable by 33.1; so by 13.7.1, given $\varepsilon > 0$, there is for each h, k an open set G_{hk} such that $G_{hk} \supset A_{hk}$ and
$$\mu(G_{hk}) \leq \mu(A_{hk}) + \varepsilon. \tag{2}$$

For a fixed, but arbitrary, choice of integers h, k let $\mathscr{J}$ be the class of all closed cubes $K \subset G_{hk}$ such that
$$\sigma(K) \leq \frac{h}{k}\mu(K). \tag{3}$$

It then follows from (1) that $\mathscr{J}$ covers A_{hk} in the sense of Vitali; so by 32.1 there is a sequence J of disjoint cubes from $\mathscr{J}$ such that

$$\mu\left(A_{hk} - \bigcup_{i=1}^{\infty} J_i\right) = 0, \tag{4}$$

whence

$$\mu\left(\bigcup_{i=1}^{\infty} J_i \cap A_{hk}\right) = \mu(A_{hk}). \tag{5}$$

Using (3) and (2), we have

$$\sigma\left(\bigcup_{i=1}^{\infty} J_i\right) = \sum_{i=1}^{\infty} \sigma(J_i) \leq \frac{h}{k} \sum_{i=1}^{\infty} \mu(J_i) \leq \frac{h}{k} \mu(G_{hk}) \leq \frac{h}{k}[\mu(A_{hk}) + \varepsilon].$$

On the other hand, using (1) with 33.2 and also using (5), we have

$$\sigma\left(\bigcup_{i=1}^{\infty} J_i\right) \geq \sigma\left(\bigcup_{i=1}^{\infty} J_i \cap A_{hk}\right) \geq \frac{h+1}{k} \mu\left(\bigcup_{i=1}^{\infty} J_i \cap A_{hk}\right) = \frac{h+1}{k} \mu(A_{hk}).$$

Putting these results together, we get

$$(h + 1)\mu(A_{hk}) \leq h[\mu(A_{hk}) + \varepsilon],$$

whence $\mu(A_{hk}) < h\varepsilon$ for every $\varepsilon > 0$; so for each h, k we have

$$\mu(A_{hk}) = 0. \tag{6}$$

It follows immediately from (1) that

$$A = \{x \mid \bar{\sigma}'(x) > \underline{\sigma}'(x)\} = \bigcup_{h=1}^{\infty} \bigcup_{k=1}^{\infty} A_{hk};$$

so from (6) we get $\mu(A) = 0$. That is, $\bar{\sigma}' = \underline{\sigma}'$ a. e.

To show that the derivative is finite a. e., we partition R_n into a countable class of intervals. If I is one of these intervals and if $\bar{\sigma}'(x) = \infty$ for every $x \in A \subset I$ where $\mu(A) > 0$, then we have from 33.2 that $\sigma(I) = \infty$. However, a Lebesgue-Stieltjes measure is finite on every interval; so $\bar{\sigma}' < \infty$ a. e. in I, and it follows at once that $\bar{\sigma}' < \infty$ a. e. in R_n. A similar argument may be used to show that $\underline{\sigma}' > -\infty$ a. e. ∎

This gives the result announced at the beginning of this section for the most important class of completely additive set functions in R_n. The case of non-monotone σ is easily handled (see Exercise 31.m), provided the variations of σ are Lebesgue-Stieltjes measures. In other cases, we may have $\bar{\sigma}' = \infty$ on a set of positive measure, but at least we can get $\bar{\sigma}' = \underline{\sigma}'$ a. e. This does not follow from the proofs already given because in both 33.2 and 33.3 we applied 13.7.1 to σ, and this essentially requires that σ be a Lebesgue-Stieltjes measure. However, the general case may be treated as follows.

Theorem 33.4. *If σ is any completely additive set function in R_n whose domain includes the class of all closed cubes, then $\bar{\sigma}' = \underline{\sigma}'$ a. e.*

Proof. First, we observe that every half-open interval is a countable union of closed cubes; hence the domain of σ includes the class of all half-open intervals, and hence, by 14.5, the class of all open sets.

Let us consider first the case in which σ is a measure function. Let E be the set of all points x such that there is a closed cube I containing x for which $\sigma(I) < \infty$. Clearly, E is covered in the sense of Vitali by the class of all closed cubes for which σ is finite; therefore by 32.1, it is covered except for a set of measure zero by a countable class of such cubes. Since the faces of a cube have Lebesgue measure zero, E is covered except for a set of measure zero by a countable class of open cubes for each of which σ is finite. Let (a, b) be one of these open cubes. We define a function f on $[a, b]$ by setting $f(x) = 0$ for each x on one of the closed faces of $[a, b]$ and setting $f(x) = \sigma\{(a, x]\}$ for each $x \in (a, b)$. Clearly, the Lebesgue-Stieltjes measure μ_f is equal to σ on the half-open intervals in (a, b); therefore by 14.5, on the open sets in (a, b); and so, by taking complements, on the closed cubes in (a, b). For any $x \in (a, b)$, $\bar{\sigma}'(x)$ and $\underline{\sigma}'(x)$ are determined by the values of σ on closed cubes in (a, b); so it follows from 33.3 that σ' exists a. e. in (a, b), hence a. e. in E. If $x \in -E$, then for any closed cube J such that $x \in J$, $\sigma(J) = \infty$; so $\bar{\sigma}'(x) = \underline{\sigma}'(x) = \infty$ for every $x \in -E$. This proves the result for the case in which σ is a measure function.

Let σ be any completely additive set function and let $\sigma = \sigma_1 - \sigma_2$ be its Jordan decomposition. By 10.5, σ_2 must be finite whenever σ_1 is infinite; so it follows from the above argument that $\sigma_2'(x)$ exists for almost all x for which $\bar{\sigma}_1'(x) = \infty$. Similarly, $\sigma_1'(x)$ exists for almost all x for which $\bar{\sigma}_2'(x) = \infty$, and the result follows at once. ∎

EXERCISES

a. Show that the results of this section hold for regular derivatives.

b. Show that if f is defined on R_1 and is of bounded variation, then f is differentiable a. e.

c. Let f be defined on R_1 as follows. If x is rational and different from zero, write $x = r/s$ where the fraction r/s is in lowest terms, and set $f(x) = 1/r^2s^2$. For all other x, set $f(x) = 0$. Show that f is of bounded variation, hence differentiable a. e. However, f is nondifferentiable on a dense set.

d. Show that if σ is Hausdorff one-dimensional measure in R_2, then $\bar{\sigma}'(x) = \underline{\sigma}'(x) = \infty$ for every $x \in R_2$.

e. Construct an example of a completely additive set function whose Jordan decomposition $\sigma = \sigma_1 - \sigma_2$ has the property that at some point x, $\bar{\sigma}_1'(x) = \bar{\sigma}_2'(x) = \infty$.

f. In the light of Exercise e, discuss the proof of 33.4. The Jordan decomposition theorem cannot be invoked at the beginning of that proof.

g. Prove 33.2 for the case in which σ is any bounded measure function on the Lebesgue measurable sets in R_n. *Hint:* Follow the proof of 33.4 to show that σ is equivalent on the closed sets to a Lebesgue-Stieltjes measure. Then apply 13.7.2 to the Lebesgue measure function to show that for every $\varepsilon > 0$, $\sigma(E) \geq a\mu^*(A) - \varepsilon$.

34 THE LEBESGUE DECOMPOSITION

We know from 33.1 that derivatives, when they exist, are automatically measurable. From 33.3 we see that the derivative of a Lebesgue-Stieltjes measure exists a. e. We now want to show that if the Lebesgue-Stieltjes measure is everywhere finite, then this derivative is not only measurable, it is integrable. Furthermore, if the Lebesgue-Stieltjes measure is a Lebesgue integral (note 31.1), then its derivative is equal to the integrand a. e. Thus any absolutely continuous Lebesgue-Stieltjes measure (necessarily the integral of something according to the Radon-Nikodym theorem) may be expressed as the integral of its derivative. This will complete the results summarized at the beginning of Section 31. Finally, these results may be used to obtain an important decomposition theorem for Lebesgue-Stieltjes measures.

Theorem 34.1. *If σ is a nondecreasing sequence of Lebesgue-Stieltjes measures on R_n such that for each measurable $E \subset R_n$,*

$$\lim_k \sigma_k(E) = \sigma_0(E),$$

where σ_0 is also a Lebesgue-Stieltjes measure, then

$$\lim_k \sigma'_k = \sigma'_0 \ [\text{a. e.}].$$

Proof. Let $\tau_k = \sigma_0 - \sigma_k$, and for each positive integer m, let

$$E_m = \left\{ x \mid \lim_k \bar{\tau}'_k(x) \geq \frac{1}{m} \right\}.$$

Since τ is a nonincreasing sequence, so is $\bar{\tau}'$; therefore for each $x \in E_m$,

$$\bar{\tau}'_k(x) \geq \frac{1}{m}$$

for every k. Each E_m is measurable by 33.1 and 20.3.2, so we have from 33.2 that

$$m\tau_k(E_m) \geq \mu(E_m)$$

for every k; hence

$$\mu(E_m) \leq m \lim_k \tau_k(E_m) = 0.$$

Since

$$E = \{x \mid \lim_k \bar{\tau}'_k(x) > 0\} = \bigcup_{m=1}^{\infty} E_m,$$

it now follows that $\mu(E) = 0$; the proof is completed once we note that $\lim_k \underline{\tau}'_k(x) \geq 0$ for every x, inasmuch as $\tau_k(A) \geq 0$ for each k and all measurable A. ∎

Theorem 34.2. *If σ is a Lebesgue-Stieltjes measure function in R_n, then on each measurable set E for which $\sigma(E)$ is finite, σ' is integrable and*

$$\sigma(E) \geq \int_E \sigma'.$$

Proof. We know from 33.1 that σ' is measurable; let f be a monotone sequence of simple functions converging to σ'. For an arbitrary positive integer k, let the simple function f_k have the form

$$f_k = \sum_{i=1}^m a_i C_{E_i}$$

where the sets E_i are disjoint. Then, for each $x \in E_i$,

$$\sigma'(x) \geq f_k(x) = a_i;$$

so by 33.2, for any measurable E,

$$\sigma(E \cap E_i) \geq a_i \mu(E \cap E_i) = \int_{E \cap E_i} f_k,$$

and it follows that

$$\sigma(E) = \sum_{i=1}^m \sigma(E \cap E_i) \geq \sum_{i=1}^m \int_{E \cap E_i} f_k = \int_E f_k.$$

Since this is true for each k, we have

$$\sigma(E) \geq \lim_k \int_E f_k = \int_E \sigma'. \quad \blacksquare$$

Theorem 34.3. *If f is Lebesgue integrable on R_n and if $\sigma = \int f$, then $\sigma' = f$ a. e.*

Proof. If f is the characteristic function of an open set G and if $x \in G$, then there is a number $\delta > 0$ such that if a closed cube J contains x and has diameter less than δ, then $J \subset G$; hence for all such J, $\sigma(J) = \mu(J)$, and it follows that

$$\sigma'(x) = f(x) = 1$$

for every $x \in G$.

Now let A be any measurable set. Using 13.7.1 we can find a contracting sequence G of open sets such that $A \subset G_n$ for each n and

$$\mu(\lim_n G_n - A) = 0.$$

Thus for any measurable E,

$$\int_E C_{G_n}$$

tends monotonically to
$$\int_E C_A;$$
so by 34.1
$$\left(\int C_A\right)' = \lim_n \left(\int C_{G_n}\right)' \text{ [a. e.]}.$$
However, we have already established that
$$\left(\int C_{G_n}\right)'(x) = 1$$
for every $x \in G_n$; so
$$\left(\int C_A\right)' = 1 \text{ a. e. in } A. \tag{1}$$
Since $\int_E C_A + \int_E C_{-A} = \mu(E)$, it is clear that
$$\left(\int C_A + \int C_{-A}\right)' = 1;$$
so whenever they exist (that is, a. e.)
$$\left(\int C_A\right)' + \left(\int C_{-A}\right)' = 1.$$
This together with (1) establishes the result for characteristic functions. The extension to simple functions is obvious, and the further extension to arbitrary integrable functions is accomplished with the aid of 34.1. ∎

Corollary 34.3.1. *If σ is an everywhere finite, absolutely continuous, completely additive set function on the Lebesgue measurable sets in R_n, then $\sigma = \int \sigma'$.*

Proof. By the Radon-Nikodym theorem (27.3), there is an integrable function f such that $\sigma = \int f$; and by 34.3, $f = \sigma'$ a. e.; so the result follows from 25.5.3. ∎

We often think of a Lebesgue-Stieltjes measure σ as describing a mass distribution. If σ is absolutely continuous, then the mass of any set is given by the integral over that set of some fixed function f. This function f is called the *mass density function*. This terminology is motivated by 34.3.1. We see there that $f = \sigma'$ a. e.; that is, for almost all x, $f(x)$ is the limiting value of the mass per unit volume in cubes containing x. Not every mass distribution can be described by a density function, but we turn now to the interesting theorem that every mass distribution is the sum of two mass distributions, one of which has a mass density function and the other of which consists solely of mass concentrated in a set of measure zero.

To describe this latter type, let us introduce the notion of a *singular function*. A set function σ is singular if there is a set E_0 of measure zero such that σ vanishes on every measurable subset of $-E_0$.

Lemma 34.4. *If σ is additive, absolutely continuous and singular, then $\sigma \equiv 0$.*

Proof. Let E_0 be the set outside which σ vanishes because of being singular, and let E be any measurable set; then

$$\sigma(E) = \sigma(E \cap E_0) + \sigma(E - E_0).$$

Since $\mu(E_0) = 0$ and since σ is absolutely continuous, we have $\sigma(E \cap E_0) = 0$. By the definition of E_0, $\sigma(E - E_0) = 0$. ∎

Lemma 34.5. *A necessary and sufficient condition that a Lebesgue-Stieltjes measure σ be singular is that $\sigma' = 0$ a. e.*

Proof. If σ is singular, we see from 34.2 that

$$\int_{-E_0} \sigma' \leq \sigma(-E_0) = 0,$$

and it follows from 25.7 that $\sigma' = 0$ a. e.

If $\sigma' = 0$ a. e., let $E = \{x \mid \sigma'(x) = 0\}$; then $(-\sigma)' = 0$ on E, and by 33.2, $-\sigma(E) \geq 0$. Since $\sigma(E) \geq 0$ and since $\mu(-E) = 0$, it follows that σ is singular. ∎

Theorem 34.6 (Lebesgue Decomposition Theorem). *If σ is an everywhere finite Lebesgue-Stieltjes measure in R_n, then σ has a unique decomposition $\sigma = \alpha + \beta$ where α is absolutely continuous and β is singular.*

Proof. By 33.3, σ is differentiable a. e., and by 34.2, σ' is integrable; so we set $\alpha = \int \sigma'$. By 27.1.1, α is absolutely continuous, and by 34.3, $\sigma' = \alpha'$ a. e. Therefore, by 34.5, $\beta = \sigma - \alpha$ is singular. To prove the uniqueness of this decomposition, suppose $\alpha_1 + \beta_1 = \alpha_2 + \beta_2$; then $\alpha_1 - \alpha_2 = \beta_2 - \beta_1$. One side of this last equation is absolutely continuous, the other singular; so by 34.4, each side vanishes identically. ∎

EXERCISES

a. Show that all the results in this section hold with derivatives replaced by regular derivatives.

b. Let σ be any completely additive set function on the Lebesgue measurable sets in R_n such that $V(\sigma, R_n) < \infty$. Show that σ has a Lebesgue decomposition. [*Hint:* Use Exercise 33.g to generalize 34.2 and 34.5; then take a Jordan decomposition of σ.]

c. Let σ be a convergent sequence of completely additive set functions for which the limit function is also completely additive. Show that if for each n, σ_n is absolutely continuous [singular], then the limit function is absolutely continuous [singular].

d. Show that if σ is any completely additive set function on the Lebesgue measurable sets in R_n for which $V(\sigma, \ )$ is finite on bounded sets, then σ has a Lebesgue decomposition.

e. Show that Hausdorff one-dimensional measure in R_2 does not have a Lebesgue decomposition.

f. A function f of a real variable is called *singular* if $f' = 0$ a. e. Show that if g is a function of bounded variation, then $g = g_1 + g_2$ where g_1 is absolutely continuous and g_2 is singular. Show that this decomposition is unique except for an additive constant.

g. Show that any step function is singular, and thus a singular function need not be constant.

h. Show that the Cantor function (Section 27) is singular, and thus a continuous singular function need not be constant.

i. Show that the function of Exercise 33.c is singular, and thus a singular function need not be constant on any interval.

j. Let f be the characteristic function of the rationals, and let $\sigma = \int f$. Show that $\sigma' \equiv 0$; so while 34.3 applies, it is not true that $\sigma'(x) = f(x)$ whenever $\sigma'(x)$ exists.

*35 METRIC DENSITY AND APPROXIMATE CONTINUITY

Let A be any set (measurable or not) in R_n, and let σ_A be the completely additive (see 11.2.5) set function defined for measurable E by

$$\sigma_A(E) = \mu^*(A \cap E).$$

Let x be any point of R_n. The upper and lower strong derivates $\bar{D}\sigma_A(x)$ and $\underline{D}\sigma_A(x)$ are called, respectively, the *upper and lower outer density of A at x*. If $D\sigma_A(x)$ exists, we call it the *outer density of A at x*. If A is measurable, we delete the word outer in each of these expressions. A point x is called a *point of density for a set A* if the outer density of A at x is unity; x is called a *point of dispersion for A* if the outer density of A at x is zero. It is easily seen that if A is measurable, then any point of density for A is a point of dispersion for $-A$.

Our first project in this section is to show that for any set A, almost all points of A are points of density for A; and if A is measurable, then almost all points of $-A$ are points of dispersion for A. First, let us observe that if A is measurable, we may write

$$\sigma_A(E) = \int_E C_A;$$

so if strong derivatives are replaced by derivatives with respect to cubes, then the result for measurable sets is a special case of 34.3. In R_1 derivatives and strong derivatives are equivalent, so in that case the result is already proved. We shall want to use this fact in the general development given below.

Lemma 35.1. *If F is a bounded, closed set in R_n, then almost all points of F are points of density for F.*

Proof. For simplicity of notation, let us consider the case of R_2. The argument for R_n is essentially the same. Let us represent R_2 as $X \times Y$ where $X = Y = R_1$, and let α, β, and μ be Lebesgue measures in X, Y, and $X \times Y$, respectively. Given $E \subset X \times Y$ and $y \in Y$, let us write

$$E^{(y)} = \{x \mid (x, y) \in E\}.$$

Let $\varepsilon > 0$ be given. For each positive integer n, let E_n be the set of all points $(x, y) \in F$ such that

$$\alpha(F^{(y)} \cap I) \geq (1 - \varepsilon)\alpha(I)$$

for every closed interval $I \subset X$ such that $x \in I$ and $\alpha(I) < 1/n$.

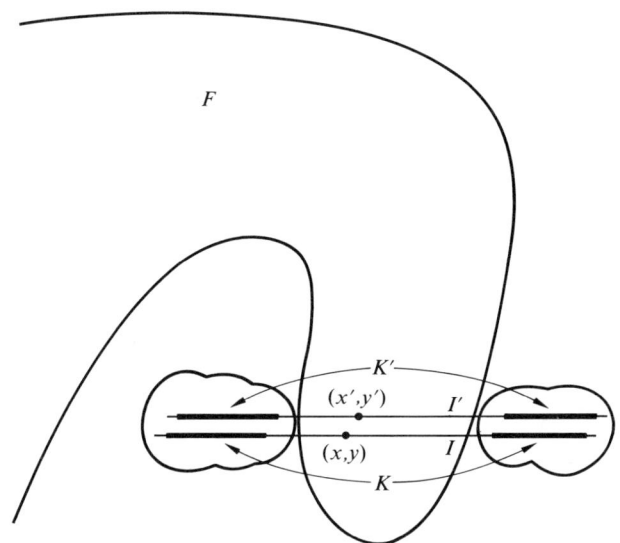

Figure 9

First, we want to show that each set E_n is closed. Perhaps it is easier to see that $-E_n$ is open. If $(x, y) \in -E_n$, then there is an interval $I \subset X$ such that $x \in I$, $\alpha(I) < 1/n$, and $\alpha(F^{(y)} \cap I) < (1 - \varepsilon)\alpha(I)$. That is, if $G = -F$, then the open linear set $G^{(y)} \cap I^0$ has measure greater than $\varepsilon \alpha(I)$; hence there is a closed set $K \subset G^{(y)} \cap I^0$ such that

$$\alpha(K) > \varepsilon \alpha(I).$$

In the space $X \times Y$ the set K may be covered by neighborhoods each of which is contained in the open set G; so it is covered by a finite number of these. Any sufficiently small translation of I (in any direction in $X \times Y$) will leave K in the union of the sets from this finite covering (see Fig. 9); that is, it will leave K inside G. So, any such translation I' of I will have the property that

$$\alpha(F^{(y')} \cap I') < (1 - \varepsilon)\alpha(I');$$

thus all points (x', y') sufficiently close to (x, y) are also in $-E_n$, and it follows that $-E_n$ is open.

Clearly E is an expanding sequence of sets. We set

$$N = F - \lim_n E_n.$$

Since the sets E_n are closed, it follows that N is measurable. Let y be any fixed point in Y, and let $x \in N^{(y)}$. Then for each n there is an interval I with $x \in I$ such that $\alpha(I) < 1/n$ and such that

$$\alpha(N^{(y)} \cap I) < (1 - \varepsilon)\alpha(I);$$

therefore, the lower density of $N^{(y)}$ at x is at most $1 - \varepsilon$. However, by 34.3, the density of $N^{(y)}$ at x is unity for almost all $x \in N^{(y)}$; so it follows that

$$\alpha(N^{(y)}) = 0.$$

Since N is measurable, we can apply Fubini's theorem (29.7) to C_N to see that

$$\mu(N) = \int_Y \alpha(N^{(\,)}) \, d\beta = 0.$$

Since F is bounded, it follows by 10.8.2 that, given $\eta > 0$, there is an integer n_0 such that

$$\mu(F - E_{n_0}) < \eta.$$

We now repeat the above procedure, interchanging the roles of X and Y and using the closed set E_{n_0} in place of F. We thus obtain a set $H_{m_0} \subset E_{n_0} \subset F$ with the following properties:

$$\mu(F - H_{m_0}) < 2\eta;$$

if $(x, y) \in H_{m_0}$, then for each interval I_1 with $x \in I_1 \subset X$ and $\alpha(I_1) < 1/n_0$,

$$\alpha(F^{(y)} \cap I_1) \geq (1 - \varepsilon)\alpha(I_1);$$

also for each interval I_2 with $y \in I_2 \subset Y$ and $\beta(I_2) < 1/m_0$,

$$\beta(E_{n_0}^{(x)} \cap I_2) \geq (1 - \varepsilon)\beta(I_2).$$

Let $J = I_1 \times I_2$ with I_1 and I_2 as above. By Fubini's theorem,

$$\begin{aligned}
\mu(F \cap J) &= \int_{I_2} \alpha(F^{(\,)} \cap I_1) \, d\beta \\
&\geq \int_{E_{n_0}^{(x)} \cap I_2} \alpha(F^{(\,)} \cap I_1) \, d\beta \\
&\geq (1 - \varepsilon)\alpha(I_1)\beta(E_{n_0} \cap I_2) \\
&\geq (1 - \varepsilon)^2 \alpha(I_1)\beta(I_2) \\
&= (1 - \varepsilon)^2 \mu(J).
\end{aligned}$$

Thus, the lower density of F is at least $(1 - \varepsilon)^2$ at each point of H_{m_0} where $\mu(F - H_{m_0}) < 2\eta$. Since ε and η are arbitrary, it follows that the density of F is unity a. e. in F. ∎

Theorem 35.2. *If A is any set (measurable or not) in R_n, then x is a point of density for A for almost all $x \in A$. A necessary and sufficient condition that A be measurable is that x be a point of dispersion for A for almost all $x \in -A$.*

Proof. If E is measurable, then by 13.5.2,

$$E = Z \cup \left(\bigcup_{k=1}^{\infty} F_k \right)$$

where $\mu(Z) = 0$ and each F_k is closed. Replacing each F_k by its intersection with the closed sphere with center at the origin and radius k, we may assume that each set F_k is also bounded. By 35.1, for each k, F_k (and therefore E) has unit density a. e. in F_k; thus E has unit density a. e. in E. If A is any set, let E be a measurable cover for A. For any interval J,

$$\mu(E \cap J) = \mu(E) - \mu(E - J) = \mu^*(A) - \mu(E - J)$$
$$\leq \mu^*(A) - \mu^*(A - J) \leq \mu^*(A \cap J),$$

but obviously

$$\mu(E \cap J) \geq \mu^*(A \cap J);$$

so it follows that the outer density of A is always equal to the density of E. Thus, the outer density of A is unity a. e. in A.

To prove the second statement, we note that if A is measurable, then by what we have already proved, almost all points of $-A$ are points of density for $-A$, hence points of dispersion for A. If A is nonmeasurable and E is a measurable cover for A, then almost all points of $E - A$ are points of density for E and so for A. However, $\mu^*(E - A) > 0$, so the set of points of $-A$ which are not points of dispersion for A is a set with positive outer measure. ∎

EXERCISES

a. Show that if I is any interval in R_n, then at any point on the boundary of I, the upper density of I is 1 and the lower density is zero.

b. Let E be the set in R_1 consisting of the intervals

$$\left[\frac{1}{2^{2k+1}}, \frac{1}{2^{2k}} \right] \quad (k = 1, 2, 3, \dots)$$

and their reflections through the origin. Show that at 0, E has upper density 2/3 and lower density 1/3.

c. Construct a set in R_1 which has density 1/2 at the origin.

d. Show that if E is any set and if x is a point of dispersion for E, then x is a point of density for $-E$.

e. Show that the converse to Exercise d holds only if E is measurable. Indeed, if E is nonmeasurable, then there must be points which are points of density for both E and $-E$.

f. Complete the proof of 35.1 for the case of n dimensions.

From 35.2 we see that if E is a measurable set and if $\sigma = \int C_E$, then $D\sigma = C_E$ a. e. This is the beginning of an analog to 34.3 for strong derivatives. As we pointed out in Section 31, this analogy extends only to the case of bounded integrands. In addition to proving this theorem on strong differentiability of integrals, we want to investigate the possibility of a characterization in terms of properties of the integrand of the points at which the strong derivative of an integral is equal to the integrand. That is, just because an integral is differentiable (in any sense) a. e. and its derivative equals the integrand a. e., it does not follow (see Exercise 34.j) that the equality holds everywhere that the derivative in question exists.

If f is a point function over R_n, we say that f is *approximately continuous at* x_0 provided that for every $\varepsilon > 0$, x_0 is a point of dispersion for the set

$$\{x \mid |f(x) - f(x_0)| \geq \varepsilon\}.$$

A continuous function is obviously approximately continuous, but the converse is certainly not true. For example, the characteristic function of the rationals is discontinuous everywhere, but it is approximately continuous at every irrational point. Indeed, the following theorem indicates the latitude allowed in the behavior of an approximately continuous function.

Theorem 35.3. *A necessary and sufficient condition that a point function over R_n be measurable is that it be approximately continuous a. e.*

Proof. Suppose f is measurable; then by Lusin's theorem (21.4; note also Exercise 21.o), given $\varepsilon > 0$, there is a closed set F such that $\mu(-F) < \varepsilon$ and the restriction of f to F is continuous. If x is a point of dispersion for $-F$, then it is clear that f is approximately continuous at x; thus by 35.2, f is approximately continuous a. e. in F. So, the set on which f is not approximately continuous has measure less than ε. Since ε is arbitrary, this set has measure zero, and the proof of necessity is complete.

Suppose f is approximately continuous a. e. Let a be any real number, and let

$$E = \{x \mid f(x) > a\}.$$

If $x_0 \in E$ and is a point of approximate continuity of f, we set $\varepsilon = f(x_0) - a > 0$, and it follows that x_0 is a point of dispersion for $-E$. Since this is true for almost all $x_0 \in E$, it follows from 35.2 that $-E$ is measurable, so f is measurable by 19.3. ∎

It is interesting to compare 35.3 and 24.4. For bounded functions on a bounded interval, a necessary and sufficient condition for Riemann integrability is continuity a. e., while for Lebesgue integrability a necessary and sufficient condition is approximate continuity a. e.

Theorem 35.4. *If f is nonnegative and integrable over R_n, if $\sigma = \int f$, and if $D\sigma(x) = f(x)$, then x is a point of approximate continuity for f.*

Proof. There is no loss of generality in assuming that $f(x) = 0$; subtracting the constant $f(x)$ from each value of f in some neighborhood of x does not affect approximate continuity of f or strong differentiability of σ. Let $\varepsilon > 0$ be given, and let
$$E = \{y \mid f(y) \geq \varepsilon\}.$$
Then, for any interval J,
$$\frac{\sigma(J)}{\mu(J)} \geq \frac{\sigma(J \cap E)}{\mu(J)} \geq \varepsilon \frac{\mu(J \cap E)}{\mu(J)};$$
so if $D\sigma(x) = 0$, it follows that the density of E at x is zero. In a similar manner we can show that the density at x of
$$\{y \mid f(y) \leq -\varepsilon\}$$
is zero. Thus, x is a point of dispersion for the union of these two sets, so f is approximately continuous at x. ∎

Theorem 35.5. *If f is bounded and integrable on R_n, if $\sigma = \int f$, and if x is a point of approximate continuity for f, then $D\sigma(x) = f(x)$.*

Proof. As in the proof of 35.4, assume $f(x) = 0$. Let M be an upper bound for $|f|$, and for an arbitrary $\varepsilon > 0$ let
$$A = \{y \mid |f(y)| < \varepsilon\};$$
then x is a point of density for A and a point of dispersion for $-A$, so
$$\left|\frac{\sigma(J)}{\mu(J)}\right| = \left|\frac{\sigma(J \cap A) + \sigma(J - A)}{\mu(J)}\right| \leq \varepsilon \frac{\mu(J \cap A)}{\mu(J)} + M \frac{\mu(J - A)}{\mu(J)} \to \varepsilon$$
as $J \to x$. Since ε is arbitrary, it follows that $D\sigma(x) = 0$. ∎

Corollary 35.5.1. *If f is bounded and integrable on R_n, and if $\sigma = \int f$, then $D\sigma = f$ a. e.*

Proof. This follows from 35.5 and 35.3. ∎

A precise characterization in terms of f alone of those points x at which $D(\int f)(x) = f(x)$ does not seem to be known. We see in 35.4 that for nonnegative f approximate continuity is necessary and in 35.5 that for bounded f approximate continuity is sufficient. Thus for bounded, nonnegative f approximate continuity does indeed characterize the points at which the integral differentiates back to the integrand. However, the restrictions on f make this a very limited result. Exercise g shows that nonnegativity is essential in 35.4 and Exercise h shows that boundedness is essential in 35.5. Both of these examples are in R_1; so the question of dealing with strong rather than regular derivatives is not an issue. Note, however, Exercises n through q which outline a proof that boundedness is essential in 35.5.1, whereas the parallel for regular derivatives, 34.3, is not restricted in this way.

EXERCISES

g. Let E be a set with metric density $1/2$ at x and let $f = C_E - C_{-E}$. Show that $D(\int f)(x) = 0$ but that f is not approximately continuous at x.

h. Let
$$f(x) = \begin{cases} 2^n & \text{for } (1/2^n) - (1 - 2^{2n+1}) \le x \le 1/2^n; \\ 0 & \text{otherwise,} \end{cases} \quad n = 0, 1, 2, \ldots,$$

and let $\sigma = \int f$. Show that f is approximately continuous at 0 but that σ is not differentiable at 0.

i. Let f be a bounded integrable function on R_2 and let g be defined by

$$g(x, y) = \int_{((-\infty, -\infty),(x,y)]} f.$$

Show that if $\partial g/\partial x$ and $\partial g/\partial y$ exist everywhere, then

$$\frac{\partial^2 g}{\partial x \, \partial y} = \frac{\partial^2 g}{\partial y \, \partial x} = f \text{ a. e.}$$

j. Define g as in Exercise i. Show that for each fixed y, $\partial g/\partial x$ exists a. e. (with respect to Lebesgue measure in R_1) on the horizontal line with ordinate y. State and prove a similar result for $\partial g/\partial y$.

k. Define g as in Exercise i. Show that the set in R_2 on which $\partial g/\partial x$ exists is a measurable set. *Hint:* Let

$$L_n(x, y) = \overline{\lim_k} \frac{1}{r_k} [g(x + r_k, y) - g(x, y)],$$

$$l_n(x, y) = \underline{\lim_k} \frac{1}{r_k} [g(x + r_k, y) - g(x, y)]$$

where r is a sequence assuming all rational values less in absolute value than $1/n$. L and l are monotone as functions of n; let $L_0(x, y) = \lim_n L_n(x, y)$ and $l_0(x, y) = \lim_n l_n(x, y)$. Show that L_0 and l_0 are measurable and that the set on which they are equal is the set on which $\partial g/\partial x$ exists.

l. Using Fubini's theorem and Exercises j and k, show that with g defined as in Exercise i, $\partial g/\partial x$ and $\partial g/\partial y$ each exist a. e.

m. Formulate a compromise definition of cross partial derivative and prove from 35.5.1 and Exercise l that this pseudo-cross partial of the integral of a bounded function exists and equals the integrand a. e. *Hint:* The difficulty is that in the usual definition of

$$\frac{\partial}{\partial y}\left(\frac{\partial g}{\partial x}\right)$$

it is required that $\partial g/\partial x$ exist everywhere on some vertical linear interval through (x, y). Show that for almost all (x, y), $\partial g/\partial x$ will exist on a dense set in such a vertical linear interval, and define the second partial from these values.

n. Let m be any fixed positive integer, and let the unit square K in R_2 be subdivided into m^2 equal squares. In each of these small squares S follow the construction of a non-Vitali covering given at the end of Section 32 and let $\eta = 1/m^2$. Define f_m over each S (and hence over K) as follows:

$$f_m(x) = \begin{cases} \dfrac{\mu(V^i)}{m\mu\left(\bigcap_{k=1}^{n} I_k^i\right)} & \text{for } x \in \bigcap_{k=1}^{n} I_k^i; \quad i = 1, 2, \ldots, N_j, \\ m & \text{for } x \in S - \bigcup_{i=1}^{N_j} V^i, \\ 0 & \text{otherwise.} \end{cases}$$

Show that

$$\int_K f_m \leq \frac{2}{m},$$

and for each $x \in K$ there is an interval J such that $x \in J$, $d(J) \leq \sqrt{2}/m$, and

$$\int_J f_m \geq m\mu(J).$$

o. Let K be the unit square in R_2 and let L_1 be the space of all integrable functions on K. With $\rho(f, g) = \int_K |f - g|$, L_1 is a pseudo-metric space. For each positive integer r, let $A_r \subset L_1$ be the set of all functions g with the property that there exists $x \in K$ such that for every interval J with $x \in J \subset K$ and $d(J) \leq 1/r$,

$$\int_J g \leq r\mu(J).$$

Show that A_r is nowhere dense in L_1. *Hint*: Given $\varepsilon > 0$, let $g \in A_r$ and let f be a bounded function such that $\int_K |g - f| < \varepsilon/2$. Let M be an upper bound for $|f|$ and choose m in Exercise n so that $m > \sqrt{2}(M + r)$ and $2/m < \varepsilon/2$. Letting f_m be the function constructed in Exercise n, show that $\int_K |g - (f + f_m)| < \varepsilon$ but $f + f_m$ is not in $\bar{A}_r$.

p. Show from Exercise o that those functions on K whose integrals have at any point a finite strong derivative form a set of Cat. I in L_1.

q. In Section 42 it will be shown that the space L_1 introduced in Exercise o is complete. Assuming this, what does Exercise p say about the boundedness restriction in 35.5.1?

r. An alternate definition of approximate continuity is that f is approximately continuous at x provided there is a measurable set E such that x is a point of density for E and such that the restriction of f to E is continuous at x. Show that this definition is equivalent to the one given in the text.

*36 DIFFERENTIATION WITH RESPECT TO NETS

The arguments given in Sections 33 and 34 are standard ones (due essentially to Lebesgue) for showing that differentiation and integration are inverse processes. In this section we want to formulate the problem in an abstract space in such a way that these same arguments will carry over. The key to the entire discussion in Sections 33 and 34 is the result in 33.2 to the effect that if $\bar{\sigma}'(x) \geq a$ for every $x \in E$, then $\sigma(E) \geq a\mu(E)$. Thus, we seek a definition of derivate in an abstract space for which this result holds. The remainder of the theory can then be developed as in Sections 33 and 34.

The proof of 33.2 is based on two things: (1) the Vitali covering theorem and (2) the fact that each measurable E is contained in an open set with only slightly larger measure, and the Vitali covering of E can be contained in this open set.

In the abstract case we circumvent the Vitali theorem altogether by our method of defining a derivative. Let Ω and μ^* be given, and let Ω be the union of a countable class of measurable sets of finite measure. A *net* in Ω is defined as a countable class of disjoint measurable sets of finite measure whose union is Ω. We shall say that a sequence $\mathcal{N}$ of nets is *monotone* if for each positive integer n, every set of $\mathcal{N}_{n+1}$ is a subset of some set of $\mathcal{N}_n$.

Given a monotone sequence $\mathcal{N}$ of nets and a set function σ on the class of measurable sets, we define for each positive integer n the point function d_n by choosing for each $x \in \Omega$ the (unique) set $E \in \mathcal{N}_n$ such that $x \in E$ and setting

$$d_n(x) = \begin{cases} \sigma(E)/\mu(E) & \text{for } \mu(E) > 0, \\ \infty & \text{for } \mu(E) = 0, \ \sigma(E) \geq 0, \\ -\infty & \text{for } \mu(E) = 0, \ \sigma(E) < 0. \end{cases}$$

We now define, respectively, the *upper* and *lower derivates of σ with respect to $\mathcal{N}$ and μ* as

$$\bar{D}_{\mathcal{N}\mu}\sigma(x) = \overline{\lim_n} \, d_n(x),$$

$$\underline{D}_{\mathcal{N}\mu}\sigma(x) = \underline{\lim_n} \, d_n(x).$$

As usual, we say that σ is *differentiable at x with respect to $\mathcal{N}$ and μ* if

$$-\infty < \underline{D}_{\mathcal{N}\mu}\sigma(x) = \bar{D}_{\mathcal{N}\mu}\sigma(x) < \infty,$$

and we define this common value to be the *derivative of σ at x with respect to $\mathcal{N}$ and μ* and designate it by

$$D_{\mathcal{N}\mu}\sigma(x).$$

In the theory of differentiation with respect to nets the Vitali theorem may be replaced by the following.

Theorem 36.1. *If $\mathcal{N}$ is a monotone sequence of nets, if*

$$\mathcal{N}_0 = \bigcup_{n=1}^{\infty} \mathcal{N}_n,$$

and if $\mathcal{K} \subset \mathcal{N}_0$, then there is a sequence N of disjoint sets from $\mathcal{K}$ such that

$$\bigcup_{i=1}^{\infty} N_i = \bigcup_{K \in \mathcal{K}} K.$$

Proof. Let $\mathscr{K}_1$ be the (countable) class of all those sets in $\mathscr{K}$ which are also in $\mathscr{N}_1$. For each $n > 1$, let $\mathscr{K}_n$ be the class of all those sets of $\mathscr{K}$ which are in $\mathscr{N}_n$ and are not subsets of any set from

$$\bigcup_{j=1}^{n-1} \mathscr{N}_j.$$

Then,

$$\bigcup_{n=1}^{\infty} \mathscr{K}_n$$

is a countable class of disjoint sets, and on arranging these sets in a sequence, we have the desired result. ∎

In order to obtain the other feature needed to prove a theorem such as 33.2, we introduce the following notion. Let $\mathscr{N}$ be a sequence of nets and let

$$\mathscr{N}_0 = \bigcup_{n=1}^{\infty} \mathscr{N}_n.$$

We say that $\mathscr{N}$ is *regular on the measurable set E with respect to the measure μ* if, given $\varepsilon > 0$, there is a sequence K of sets from $\mathscr{N}_0$ such that

$$\mu\left(E - \bigcup_{i=1}^{\infty} K_i\right) = 0 \quad \text{and} \quad \mu\left(\bigcup_{i=1}^{\infty} K_i\right) \leq \mu(E) + \varepsilon.$$

Lemma 36.2. *If $\mathscr{N}$ is regular on E with respect to μ, if $\mu(E) < \infty$, and if the measure σ is absolutely continuous with respect to μ, then $\mathscr{N}$ is regular on E with respect to σ.*

Proof. Given $\varepsilon > 0$, there is a $\delta > 0$ such that $\sigma(A) \leq \varepsilon$ whenever $\mu(A) \leq \delta$. Let K be chosen so that

$$\mu\left(E - \bigcup_{i=1}^{\infty} K_i\right) = 0 \quad \text{and} \quad \mu\left(\bigcup_{i=1}^{\infty} K_i\right) \leq \mu(E) + \delta.$$

Clearly,

$$\sigma\left(E - \bigcup_{i=1}^{\infty} K_i\right) = 0,$$

and since $\mu(E) < \infty$, we have

$$\mu\left(\bigcup_{i=1}^{\infty} K_i - E\right) \leq \delta,$$

whence

$$\sigma\left(\bigcup_{i=1}^{\infty} K_i - E\right) \leq \varepsilon;$$

so

$$\sigma\left(\bigcup_{i=1}^{\infty} K_i\right) \leq \sigma(E) + \varepsilon. \quad \blacksquare$$

Theorem 36.3. *Let $\mathcal{N}$ be a monotone sequence of nets, regular on E with respect to μ, and let the measure σ be absolutely continuous with respect to μ. If*

$$\overline{D}_{\mathcal{N}\mu}\sigma(x) \geq a$$

for every $x \in E$, then

$$\sigma(E) \geq a\mu(E).$$

Proof. First, let us suppose $\mu(E) < \infty$; then by 36.2 there is for every $\varepsilon > 0$ a sequence K of sets from the nets of $\mathcal{N}$ such that

$$\mu\left[\bigcup_{i=1}^{\infty} (K_i \cap E)\right] = \mu(E) \quad \text{and} \quad \sigma\left(\bigcup_{i=1}^{\infty} K_i\right) \leq \sigma(E) + \varepsilon.$$

Given $b < a$, it follows from the fact that $\overline{D}_{\mathcal{N}\mu}\sigma \geq a$ on E that each set $K_i \cap E$ is covered by a sequence J_i of sets from the nets of $\mathcal{N}$ such that for each k,

$$J_{ik} \subset K_i \quad \text{and} \quad \sigma(J_{ik}) \geq b\mu(J_{ik}).$$

By 36.1, the double sequence J may be replaced by a simple sequence I of disjoint sets taken from the sets in J; so we have

$$\sigma(E) + \varepsilon \geq \sigma\left(\bigcup_{i=1}^{\infty} K_i\right) \geq \sigma\left(\bigcup_{j=1}^{\infty} I_j\right) = \sum_{j=1}^{\infty} \sigma(I_j) \geq b \sum_{j=1}^{\infty} \mu(I_j)$$

$$= b\mu\left(\bigcup_{j=1}^{\infty} I_j\right) \geq b\mu\left[\bigcup_{i=1}^{\infty} (K_i \cap E)\right] = b\mu(E).$$

Since $\varepsilon > 0$ and $b < a$ were arbitrary, the result follows at once.

If $\mu(E) = \infty$, we can express E as the limit of an expanding sequence A of sets of finite measure; then

$$\sigma(E) = \lim_n \sigma(A_n) \geq \lim_n a\mu(A_n) = \infty. \quad \blacksquare$$

Theorem 36.4. *Let $\mathcal{N}$ be a monotone sequence of nets, regular with respect to μ on each measurable set. If f is μ-integrable and $\sigma = \int f\, d\mu$, then*

$$D_{\mathcal{N}\mu}\sigma = f \text{ a. e.}$$

If σ is everywhere finite, completely additive, and absolutely continuous with respect to μ, then $D_{\mathcal{N}\mu}\sigma$ exists a. e., and

$$\sigma = \int D_{\mathcal{N}\mu}\sigma\, d\mu.$$

Proof. Measurability of $D_{\mathcal{N}\mu}\sigma$ follows at once from the obvious measurability of the functions d_n used to define it. The remainder of the proof follows that of 34.3 and 34.3.1. That is, we prove the theorem for the case in which f is the characteristic function of a measurable set; then we employ a monotone convergence theorem. The proof of the required monotone convergence theorem is essentially the same as that of 34.1, the only variation being that we substitute 36.3 for 33.2.

To prove that $\underline{D}_{\mathcal{N}\mu}(\int C_E) = C_E$, let $\varepsilon > 0$ be given and let

$$A = \left\{x \mid x \in E;\ \underline{D}_{\mathcal{N}\mu}\left[\int C_E\right](x) < 1 - 2\varepsilon\right\}.$$

Since $\underline{D}_{\mathcal{N}\mu}(\int C_E)$ is a measurable function, A is measurable; and since we need consider only the case $\mu(E) < \infty$, we may take $\mu(A) < \infty$. This being the case, the regularity of $\mathcal{N}$ yields a sequence K of sets from the nets of $\mathcal{N}$ such that

$$\mu\left(A - \bigcup_{i=1}^{\infty} K_i\right) = 0$$

and

$$\mu\left(\bigcup_{i=1}^{\infty} K_i - A\right) < \varepsilon^2. \tag{1}$$

Letting $K_0 = \bigcup_{i=1}^{\infty} K_i$, we note that because of the definition of A, each $x \in A \cap K_0$ is contained in a set J_x from one of the nets of $\mathcal{N}$ such that $J_x \subset K_0$ and

$$\mu[J_x \cap (K_0 - E)] > \varepsilon\mu(J_x). \tag{2}$$

By 36.1 the class $\mathscr{J} = \{J_x \mid x \in A \cap K_0\}$ may be replaced by a sequence I of disjoint sets from $\mathscr{J}$. From (2) and (1) we have

$$\varepsilon \sum_{k=1}^{\infty} \mu(I_k) < \sum_{k=1}^{\infty} \mu[I_k \cap (K_0 - E)] = \mu(K_0 - E) \leq \mu(K_0 - A) < \varepsilon^2,$$

whence

$$\mu(A) \leq \mu(K_0) = \sum_{k=1}^{\infty} \mu(I_k) < \varepsilon.$$

Thus, $\underline{D}_{\mathcal{N}\mu}(\int C_E) \geq 1$ a. e. in E, and it is easily seen that $\overline{D}_{\mathcal{N}\mu}(\int C_E) \leq 1$ a. e.; so $D_{\mathcal{N}\mu}(\int C_E) = 1$ a. e. in E. Replacing E by $-E$ and subtracting, we have $D_{\mathcal{N}\mu}(\int C_E) = 0$ a. e. in $-E$. ∎

In conclusion, let us look at some criteria for regularity of sequences of nets. As a first step in that direction, we have the following.

Lemma 36.5. *If $\mathcal{N}$ is regular with respect to μ on each set of a sequence E, then it is regular on*

$$E_0 = \bigcup_{n=1}^{\infty} E_n.$$

Proof. Let $\varepsilon > 0$ be given. For each n, let K_n be a sequence of sets from the nets of $\mathcal{N}$ such that

$$\mu\left(E_n - \bigcup_{k=1}^{\infty} K_{nk}\right) = 0 \quad \text{and} \quad \mu\left(\bigcup_{k=1}^{\infty} K_{nk}\right) \leq \mu(E_n) + \frac{\varepsilon}{2^n}.$$

Then
$$\mu\left(\bigcup_{n=1}^{\infty} E_n - \bigcup_{n=1}^{\infty}\bigcup_{k=1}^{\infty} K_{nk}\right) \leq \sum_{n=1}^{\infty} \mu\left(E_n - \bigcup_{k=1}^{\infty} K_{nk}\right) = 0,$$
and
$$\mu\left(\bigcup_{n=1}^{\infty}\bigcup_{k=1}^{\infty} K_{nk}\right) \leq \sum_{n=1}^{\infty} \mu\left(\bigcup_{k=1}^{\infty} K_{nk}\right) \leq \sum_{n=1}^{\infty} \left[\mu(E_n) + \frac{\varepsilon}{2^n}\right] \leq \mu(E_0) + \varepsilon. \quad \blacksquare$$

First, let us investigate regularity in an abstract space.

Theorem 36.6. *If μ^* is constructed from $\mathscr{C}$ and τ by Method I and if $\mathscr{N}$ is regular with respect to μ on each set in $\mathscr{C}$, then $\mathscr{N}$ is regular with respect to μ on every measurable set.*

Proof. By 36.5, $\mathscr{N}$ is regular on every set in $\mathscr{C}_\sigma$. Given any measurable set E and any $\varepsilon > 0$, we cover E by a sequence H of sets from $\mathscr{C}$ such that if

$$H_0 = \bigcup_{n=1}^{\infty} H_n,$$

then

$$\mu(H_0) \leq \sum_{n=1}^{\infty} \mu(H_n) \leq \sum_{n=1}^{\infty} \tau(H_n) \leq \mu(E) + \frac{\varepsilon}{2}.$$

The result follows at once from the regularity of $\mathscr{N}$ on H_0. $\blacksquare$

Finally, we investigate the situation in a metric space. In that case it is natural to require that for each $x \in \Omega$ and each $\varepsilon > 0$ there be a set from one of the nets of $\mathscr{N}$ which contains x and has diameter less than ε. If the sequence $\mathscr{N}$ of nets has this property, we shall call it an *indefinitely fine* sequence.

Theorem 36.7. *If μ^* satisfies the conditions of 13.5 and if $\mathscr{N}$ is a monotone, indefinitely fine sequence of nets, then $\mathscr{N}$ is regular with respect to μ on every measurable set.*

Proof. Because of 13.5.2, it suffices to prove that $\mathscr{N}$ is regular on every set in $\mathscr{F}_\sigma$; so by 36.5 it suffices to prove that $\mathscr{N}$ is regular on each closed set. Accordingly, let F be closed. Since $\mathscr{N}$ is indefinitely fine, it follows that for each positive integer n, F is covered by a sequence K of sets from the nets of $\mathscr{N}$ such that

$$\bigcup_{i=1}^{\infty} K_i \subset \bigcup_{x \in F} N\left(x, \frac{1}{n}\right) = G_n.$$

G is a contracting sequence with $\lim_n G_n = F$; so the result follows by 10.8.2 if $\mu(G_n) < \infty$ for any n. Otherwise, we partition G_n by means of the net $\mathscr{N}_1$. Since $\mathscr{N}$ is monotone, the covering of F is partitioned into coverings of the resulting subsets of F. We obtain the result for each of these and then apply 36.5. $\blacksquare$

EXERCISES

a. Write out a complete proof of 36.4.

b. Show that if $\mathscr{M}$ and $\mathscr{N}$ are two monotone sequences of nets, each regular with respect to μ on every measurable set, and if σ is absolutely continuous with respect to μ, then
$$D_{\mathscr{M}\mu}\sigma = D_{\mathscr{N}\mu}\sigma \text{ a. e.}$$

c. Complete the theorem on change of variable in an abstract integral: If σ is absolutely continuous with respect to μ, if f is σ-integrable, and if $\mathscr{N}$ is a monotone sequence of nets, regular with respect to μ on every measurable set, then for every measurable E,
$$\int_E f \, d\sigma = \int_E f D_{\mathscr{N}\mu}\sigma \, d\mu.$$
Note Exercise 27.q.

d. Show that if σ is defined on the Lebesgue measurable sets in R_n and is differentiable (in the sense of Section 31) at x, then σ is differentiable at x with respect to every indefinitely fine, monotone sequence of nets.

e. Prove the converse to Exercise *d*.

f. Construct an example of an indefinitely fine, monotone sequence $\mathscr{N}$ of nets in R_1 and a function σ such that σ is differentiable at some point x with respect to $\mathscr{N}$ but not differentiable at x.

g. Show that in R_n every sequence of nets that is regular with respect to Lebesgue measure on every measurable set must be indefinitely fine.

h. Show by an example that the result in Exercise *g* need not hold if we substitute a Lebesgue-Stieltjes measure for Lebesgue measure.

i. Let μ^* be a metric outer measure in Ω such that neighborhoods have finite, nonzero measure. Show that in order that there exist an indefinitely fine sequence of nets in Ω it is necessary and sufficient that Ω be separable.

REFERENCES FOR FURTHER STUDY

On derivatives in R_1:
Hewitt and Stromberg [13]
Hobson [14]
Titchmarsh [28]

On derivatives in R_n:
Carathéodory [4]
Hahn and Rosenthal [9]
Saks [27]

On derivatives in abstract spaces:
Hahn and Rosenthal [9]

CHAPTER 7

CONVERGENCE THEOREMS

In this chapter we want to use the concepts of measure and integration to give several useful definitions of the notion of limit of a sequence of functions. We shall write

$$\lim_n f_n = f_0 \, [\quad]$$

and place in the brackets an abbreviation which will indicate the type of convergence we have in mind. For example, we have already used this notation in connection with uniform convergence and convergence a. e. Thus, the statements

$$\lim_n f_n = f_0 \, [\text{a. e.}] \quad \text{and} \quad \lim_n f_n = f_0 \, [\text{unif.}]$$

should already be intelligible to the reader. We shall add other types of convergence to our repertory as the chapter progresses.

For any type of convergence, designated for the moment by a blank in the brackets, we shall say that f is a *Cauchy sequence* [ ] provided

$$\lim_{\substack{m \to \infty \\ n \to \infty}} |f_m - f_n| = \theta \, [\quad],$$

where $\theta(x) \equiv 0$.

Each of the notions of convergence discussed here applies to a sequence of functions on a set, not to a sequence of function values at a point. We hope that the functional notation we are employing will serve to emphasize this fact. We always write

$$\lim_n f_n = f_0 \, [\quad],$$

never

$$\lim_n f_n(x) = f_0(x) \, [\quad].$$

The latter notation would indicate that the notion of convergence [ ] applied to a sequence of numbers, and this is not true for any of the types of convergence we discuss here.

The reader will see from our definitions that each type of convergence could be made to read

$$\lim_n f_n = f_0 \, [\quad] \text{ on } E.$$

198

The role of the set E is not too important, and we simplify the notation without detracting from the theory by letting $E = \Omega$ in all cases. Most of our notions of convergence will depend on a measure μ. As usual, we shall tacitly assume in each discussion that Ω and μ are given.

We also want to make the blanket assumption that *each function f_n or f_0 in this chapter is finite a. e.* This will simplify several discussions and will rule out few, if any, cases of any importance.

37 UNIFORM AND ALMOST EVERYWHERE CONVERGENCE

For the sake of having our discussion of convergence complete, let us review briefly the principal facts about uniform convergence and convergence a. e. We say that the sequence f *converges uniformly* to f_0 and write

$$\lim_n f_n = f_0 \text{ [unif.]},$$

provided that to every $\varepsilon > 0$ there corresponds an integer n_0 such that if $n > n_0$, then $|f_n(x) - f_0(x)| < \varepsilon$ for every $x \in \Omega$. We say that the sequence f *converges almost everywhere* to f_0 and write

$$\lim_n f_n = f_0 \text{ [a. e.]},$$

provided there exists a set E such that $\mu(-E) = 0$ and such that to every ordered pair (ε, x) where $\varepsilon > 0$ and $x \in E$ there corresponds an integer n_0 such that if $n > n_0$, then $|f_n(x) - f_0(x)| < \varepsilon$. Motivated by the conclusion to Egoroff's theorem (21.3), we say that a sequence f *converges almost uniformly* to f_0 and write

$$\lim_n f_n = f_0 \text{ [a. un.]},$$

provided that to every $\varepsilon > 0$ there corresponds a set E such that $\mu(-E) < \varepsilon$ and such that $\lim_n f_n = f_0$ [unif.] on E.

In these terms Egoroff's theorem states that if $\mu(\Omega) < \infty$, then convergence [a. e.] implies convergence [a. un.]. The implication is not valid (Exercise 21.*i*) if the finiteness restriction on μ is removed. However, the following theorem shows that a dominant function (as used in Section 28) is sufficient to restore the [a. e.] to [a. un.] implication.

Theorem 37.1. *If f_n is a sequence of measurable functions such that $\lim_n f_n = f_0$ [a. e.] and if there is an integrable function g_0 such that $|f_n| \leq g_0$ a. e., then $\lim_n f_n = f_0$ [a. un.].*

Proof. For this modification of Egoroff's theorem (21.3), let us use a portion of the proof of that theorem. The condition $\mu(\Omega) < \infty$ is used in the proof of 21.3 only to establish equation (1) in that proof; so we set

$$E_{kn} = \bigcap_{m=n}^{\infty} \left\{ x \mid |f_m(x) - f_0(x)| < \frac{1}{k} \right\},$$

and it suffices to prove from our present hypotheses that for each k,
$$\lim_n \mu(-E_{kn}) = 0. \tag{1}$$

It follows from the definition of convergence that for each k,
$$\{x \mid \lim_m f_m(x) = f_0(x)\} \subset \bigcup_{n=1}^{\infty} E_{kn};$$
so
$$\mu\left[\bigcap_{n=1}^{\infty} (-E_{kn})\right] = 0.$$

Now, $-E_k$ is a contracting sequence; so (1) follows by 10.3.1 if we show that for some n, $\mu(-E_{kn}) < \infty$.

From the present hypotheses it is clear that $|f_0| \leq g_0$ a. e.; so for each m, $|f_m - f_0| \leq 2g_0$ a. e. Thus
$$-E_{kn} = \bigcup_{m=n}^{\infty} \left\{x \mid |f_m(x) - f_0(x)| \geq \frac{1}{k}\right\} \subset \left\{x \mid g_0(x) \geq \frac{1}{2k}\right\} \cup Z,$$
where $\mu(Z) = 0$. Since g_0 is integrable, it follows that $\mu(-E_{kn}) < \infty$ for each k and each n. ∎

EXERCISES

a. Discuss the relation between convergence [a. un.] and convergence [unif.] on a set E for which $\mu(-E) = 0$.

b. Given $\varepsilon > 0$, there exists a set E such that $\mu(-E) < \varepsilon$ and also such that for every $\eta > 0$ there is an integer n_0 such that for all $n > n_0$, $|f_n(x) - f_0(x)| < \eta$ for every $x \in E$. Show that this statement describes convergence [a. un.].

c. Given $\varepsilon > 0$, there exists a set E and an integer n_0 such that $\mu(-E) = 0$ and such that for all $n > n_0$, $|f_n(x) - f_0(x)| < \varepsilon$ for every $x \in E$. Compare Exercise b. Show that this statement describes the other concept mentioned in Exercise a.

d. Given $\varepsilon > 0$, there exists a set E such that $\mu(-E) < \varepsilon$ and such that for every $x \in E$ and every $\eta > 0$ there is an integer n_0 such that for all $n > n_0$, $|f_n(x) - f_0(x)| < \eta$. Show that this statement describes convergence [a. e.].

38 CONVERGENCE IN MEASURE AND IN MEAN

We say that a sequence f of measurable functions *converges in measure* to f_0 and write
$$\lim_n f_n = f_0 \text{ [meas.]},$$
provided that to every ordered pair (ε, η) of positive numbers there corresponds an integer n_0 such that if $n > n_0$, then
$$\mu\{x \mid |f_n(x) - f_0(x)| \geq \varepsilon\} < \eta.$$

Another way of saying the same thing is to say that for each fixed $\varepsilon > 0$,
$$\lim_n \mu\{x \mid |f_n(x) - f_0(x)| \geq \varepsilon\} = 0.$$

Before investigating the relation of convergence [meas.] to other types of convergence, let us stop to note that this type of convergence has some of the familiar properties that we expect of the notion of a limit.

Theorem 38.1. *If* $\lim_n f_n = f_0$ [meas.] *and if* $\lim_n g_n = g_0$ [meas.], *then* $\lim_n (f_n + g_n) = f_0 + g_0$ [meas.].

Proof. Given $\varepsilon > 0$, it follows from the inequality
$$|f_n(x) + g_n(x) - f_0(x) - g_0(x)| \leq |f_n(x) - f_0(x)| + |g_n(x) - g_0(x)|$$
that
$$\{x \mid |f_n(x) + g_n(x) - f_0(x) - g_0(x)| \geq \varepsilon\}$$
$$\subset \left\{x \mid |f_n(x) - f_0(x)| \geq \frac{\varepsilon}{2}\right\} \cup \left\{x \mid |g_n(x) - g_0(x)| \geq \frac{\varepsilon}{2}\right\}.$$

The measure of each of these last two sets tends to zero; therefore so does that of the first set. ∎

Theorem 38.2. *If* $\lim_n f_n = f_0$ [meas.], *if* $\lim_n g_n = g_0$ [meas.], *and if* $\mu(\Omega) < \infty$, *then* $\lim_n f_n g_n = f_0 g_0$ [meas.].

Proof. We break this proof up into two lemmas. Let θ be the function which is identically zero.

Lemma 38.2.1. *If* $\lim_n f_n = \theta$ [meas.] *and if* $\lim_n g_n = \theta$ [meas.], *then* $\lim_n f_n g_n = \theta$ [meas.].

This follows immediately from the fact that
$$\{x \mid |f_n(x) g_n(x)| \geq \varepsilon\} \subset \{x \mid |f_n(x)| \geq \sqrt{\varepsilon}\} \cup \{x \mid |g_n(x)| \geq \sqrt{\varepsilon}\}.$$

Lemma 38.2.2. *If* $\lim_n f_n = \theta$ [meas.], *if* g_0 *is measurable, and if* $\mu(\Omega) < \infty$, *then* $\lim_n f_n g_0 = \theta$ [meas.].

Since we are tacitly assuming that g_0 is finite a. e. and since $\mu(\Omega) < \infty$, it follows from 10.8 that
$$\lim_{k \to \infty} \mu\{x \mid |g_0(x)| \geq k\} = 0;$$
so, given $\eta > 0$, there is a k_0 such that $\mu\{x \mid |g_0(x)| \geq k_0\} < \eta/2$. The result now follows when we note that given $\varepsilon > 0$,
$$\{x \mid |f_n(x) g_0(x)| \geq \varepsilon\} \subset \left\{x \mid |f_n(x)| \geq \frac{\varepsilon}{k_0}\right\} \cup \{x \mid |g_0(x)| \geq k_0\}.$$

To complete the proof of 38.2, we note that
$$f_n g_n - f_0 g_0 = (f_n - f_0)(g_n - g_0) + f_0(g_n - g_0) + g_0(f_n - f_0)$$
and apply 38.2.1, 38.2.2, and 38.1. ∎

We turn now to the relation between convergence [meas.] and convergence [a. e.].

Theorem 38.3. *If f is a sequence of measurable functions, if $\mu(\Omega) < \infty$, and if $\lim_n f_n = f_0$ [a. e.], then $\lim_n f_n = f_0$ [meas.].*

Proof. It is obvious that convergence [a. un.] implies convergence [meas.]; so the result follows from Egoroff's theorem (21.3). ∎

Theorem 38.4. *If f is a sequence of measurable functions, if there is an integrable function g_0 such that $|f_n| \leq g_0$ for each n, and if $\lim_n f_n = f_0$ [a. e.], then $\lim_n f_n = f_0$ [meas.].*

Proof. This follows from 37.1 in the same way that 38.3 follows from 21.3. ∎

Theorem 38.5. *If f is a Cauchy sequence [meas.], then there is a subsequence of f which is a Cauchy sequence [a. un.].*

Proof. For each positive integer k, there is a positive integer m_k such that
$$\mu\left\{x \mid |f_m(x) - f_n(x)| \geq \frac{1}{2^k}\right\} < \frac{1}{2^k} \tag{1}$$
for all m, n such that $m \geq m_k$ and $n \geq m_k$. If we set
$$n_1 = m_1, \quad n_2 = \max[n_1 + 1, m_2], \quad n_3 = \max[n_2 + 1, m_3], \quad \ldots,$$
then the indices n_k determine an infinite subsequence of f, and (1) holds, provided $m \geq n_k$ and $n \geq n_k$. For each k, we let
$$E_k = \left\{x \mid |f_{n_{k+1}}(x) - f_{n_k}(x)| \geq \frac{1}{2^k}\right\}.$$

Now, for each positive integer i, if $\varepsilon > 0$ is given, if $r > s \geq i$, and if $1/2^{s-1} < \varepsilon$, then
$$|f_{n_r}(x) - f_{n_s}(x)| \leq \sum_{k=s}^{r-1} |f_{n_{k+1}}(x) - f_{n_k}(x)| \leq \sum_{k=s}^{r-1} \frac{1}{2^k} \leq \frac{1}{2^{s-1}} < \varepsilon$$
for every $x \in \bigcap_{k=i}^{\infty}(-E_k) = -\bigcup_{k=i}^{\infty} E_k$. That is, the indices n_k determine a subsequence which is a Cauchy sequence [unif.] on $-\bigcup_{k=i}^{\infty} E_k$ for each i. However, given $\eta > 0$, we have from (1) that if $1/2^{i-1} < \eta$, then
$$\mu\left(\bigcup_{k=i}^{\infty} E_k\right) \leq \sum_{k=i}^{\infty} \mu(E_k) \leq \sum_{k=i}^{\infty} \frac{1}{2^k} = \frac{1}{2^{i-1}} < \eta. \quad \blacksquare$$

Corollary 38.5.1. *If f is a Cauchy sequence* [meas.], *then there is a subsequence of f which is a Cauchy sequence* [a. e.].

Proof. This follows immediately from 38.5 and the fact that convergence [a. un.] always implies convergence [a. e.]. ∎

Finally, we have a Cauchy theorem for convergence [meas.].

Theorem 38.6. *If f is a Cauchy sequence* [meas.] *and if $\mu(\Omega) < \infty$, then there exists a function f_0 such that $\lim_n f_n = f_0$* [meas.].

Proof. The subsequence of f given by 38.5.1 determines a function f_0 such that
$$\lim_k f_{n_k} = f_0 \text{ [a. e.]};$$
hence by 38.3,
$$\lim_k f_{n_k} = f_0 \text{ [meas.]}. \tag{1}$$
For any positive integers m and k and any $\varepsilon > 0$,
$$\{x \mid |f_m(x) - f_0(x)| \geq \varepsilon\}$$
$$\subset \left\{x \mid |f_m(x) - f_{n_k}(x)| \geq \frac{\varepsilon}{2}\right\} \cup \left\{x \mid |f_{n_k}(x) - f_0(x)| \geq \frac{\varepsilon}{2}\right\}.$$
The first set on the right may be given arbitrarily small measure because f is a Cauchy sequence [meas.]; the measure of the second may be made small by (1). ∎

The notion of convergence in measure is a very important one in probability theory. In that connection it is usually called *convergence in probability*. If μ is a probability measure, then the statement that
$$\lim_n f_n = f_0 \text{ [meas.]}$$
may be interpreted as follows: for each $\varepsilon > 0$, the probability that $|f_n(x) - f_0(x)| < \varepsilon$ tends to unity as $n \to \infty$. The statement that
$$\lim_n f_n = f_0 \text{ [a. e.]}$$
may be interpreted as follows: the probability that $\lim_n f_n(x) = f_0(x)$ is equal to unity.

These are the ideas involved in the laws of large numbers. If f is a sequence of integrable random variables, we form the variables
$$g_n = \frac{1}{n} \sum_{k=1}^{n} [f_k - e(f_k)].$$

That is, g_n is the arithmetic mean of the deviations of the first n variables f_k from their respective expectations. The sequence f obeys the *weak law of large numbers* provided
$$\lim_n g_n = 0 \, [\text{meas.}].$$
The sequence f obeys the *strong law of large numbers* provided
$$\lim_n g_n = 0 \, [\text{a. e.}].$$

EXERCISES

a. Let μ be a probability measure and let f be a sequence of random variables. Show that
$$\lim_n f_n = 0 \, [\text{meas.}]$$
if and only if, given $\varepsilon > 0$ and $\eta > 0$, there exists an integer n_0 such that
$$\mu\{x \mid |f_n(x)| < \varepsilon\} > 1 - \eta$$
for all $n > n_0$. Note: The weak law of large numbers is frequently stated in this form.

b. Let μ and f be as in Exercise a. Show that
$$\lim_n f_n = 0 \, [\text{a. e.}]$$
if and only if, given $\varepsilon > 0$ and $\eta > 0$, there exists an integer n_0 such that
$$\mu\{x \mid |f_n(x)| < \varepsilon \quad \text{for all} \quad n > n_0\} > 1 - \eta.$$
Note: The strong law of large numbers is frequently stated in this form.

c. Let μ and f be as in Exercise a. Suppose that given $\varepsilon > 0$, there exists an integer n_0 such that
$$\mu\{x \mid |f_n(x)| < \varepsilon \quad \text{for all} \quad n > n_0\} = 1.$$
Show that this is equivalent to the statement that there is a set E such that $\mu(-E) = 0$ and
$$\lim_n f_n = 0 \, [\text{unif.}]$$
on E.

d. Let μ and f be as in Exercise a. Suppose that given $\varepsilon > 0$, there exists an integer n_0 such that
$$\mu\{x \mid |f_n(x)| < \varepsilon\} = 1$$
for all $n > n_0$. Show that this is equivalent to the statement in Exercise c.

e. Given $\varepsilon > 0$ and $\eta > 0$, there exist a set E and an integer n_0 such that $\mu(-E) < \varepsilon$ and for all $n > n_0$, $|f_n(x) - f_0(x)| < \eta$ for every $x \in E$. This statement describes what type of convergence?

f. Prove that if $\mu(\Omega) < \infty$ and if $\lim_n f_n = f_0$ [meas.] with $f_n \neq 0$ a. e. for each n and $f_0 \neq 0$ a. e., then
$$\lim_n \frac{1}{f_n} = \frac{1}{f_0} \, [\text{meas.}].$$

g. Show that if $\lim_n f_n = f_0$ [meas.] and $\lim_n f_n = g_0$ [meas.], then $f_0 = g_0$ a. e.
h. Show that if $\lim_n f_n = f_0$ [meas.], then f is a Cauchy sequence [meas.].
i. Show that if $\lim_n f_n = f_0$ [meas.], then there is a subsequence of f such that $\lim_k f_{n_k} = f_0$ [a. un.].
j. Show that if μ is the measure of Exercise 11.f, then convergence [meas.] is equivalent to convergence [unif.].
k. Let $g_0(x) = x$ and $f_n(x) = 1/n$ to show that 38.2.2 cannot be extended to the case of infinite measure.

We say that a sequence f *converges in the mean* to f_0 and write
$$\lim_n f_n = f_0 \text{ [mean]}$$
provided that
$$\lim_n \int_\Omega |f_n - f_0| = 0.$$

The results of Section 28 may be rewarded in terms of the definitions given here. Specifically, 28.1 says that given an integrable dominant function, convergence [a. e.] implies convergence [mean]; 28.3 says that given an integrable dominant function, convergence [meas.] implies convergence [mean]. The sophisticated theorems on mean convergence appear in Section 40.

Theorem 38.7. *If* $\lim_n f_n = f_0$ [mean], *then* $\lim_n f_n = f_0$ [meas.].

Proof. Suppose convergence in measure fails; there exist positive numbers ε and η such that
$$\mu\{x \mid |f_n(x) - f_0(x)| \geq \varepsilon\} \geq \eta$$
for an infinite number of values of n. Thus
$$\int_\Omega |f_n - f_0| \geq \varepsilon\eta$$
for an infinite number of values of n, and mean convergence fails. ∎

EXERCISES

l. Show that if $\lim_n f_n = f_0$ [mean] and if g_0 is bounded and measurable, then $\lim_n f_n g_0 = f_0 g_0$ [mean].
m. Halmos [11] defines an integral as follows. A function f_0 is integrable provided there is a sequence f of integrable simple functions such that $\lim_n f_n = f_0$ [meas.] and such that f is a Cauchy sequence [mean]. If this is the case, he defines
$$\int_E f_0 = \lim_n \int_E f_n.$$
Show that this definition is equivalent to that given in Section 24.

39 RELATIONS AMONG CONVERGENCE TYPES

We have been concerned here with five types of convergence for sequences of measurable functions: [unif.], [a. un.], [a. e.], [meas.], and [mean]. It is the purpose of this section to summarize the implications among these convergence types. There are certain intrinsic implications and others are obtained by imposing side conditions. We shall consider two side conditions: (i) finite measure for Ω and (ii) domination of the sequence by an integrable function. The charts in Fig. 10 show the results. There $[A] \to [B]$ means, "If $\lim_n f_n = f_0[A]$, then $\lim_n f_n = f_0[B]$."

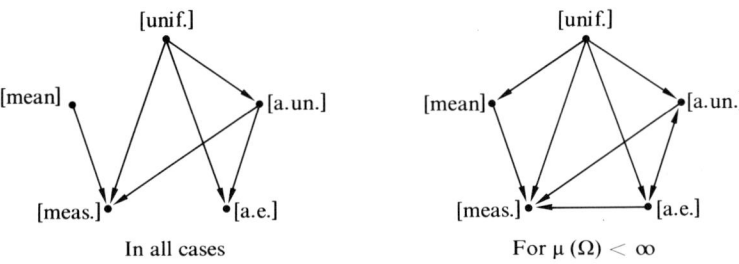

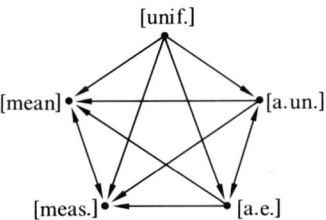

For g_0 integrable and $|f_n| \leq g_0$ a.e.

Figure 10

We now want to show that these charts are correct and complete. That is, every implication listed in Fig. 10 is valid, and for every situation in which an implication is not shown there is a counterexample. The necessary list of counterexamples is surprisingly short.

Example 39.1. $f_n = (1/n)C_{[0,n]}$; μ is Lebesgue measure on $[0, \infty)$.

Example 39.2. $f_n = nC_{[0,1/n]}$; μ is Lebesgue measure on $[0, 1]$.

Example 39.3. $f_n(x) = x^n$; μ is Lebesgue measure on $[0, 1]$.

Example 39.4. μ is Lebesgue measure on $[0, 1]$; E is a sequence of intervals with $\mu(E_n) = 1/n$. These intervals are laid end to end, lapping back when they reach the end of the unit interval (Fig. 11) $f_n = C_{E_n}$.

Example 39.5. $f_n = C_{[n,n+1]}$; μ is Lebesgue measure on $[0, \infty)$.

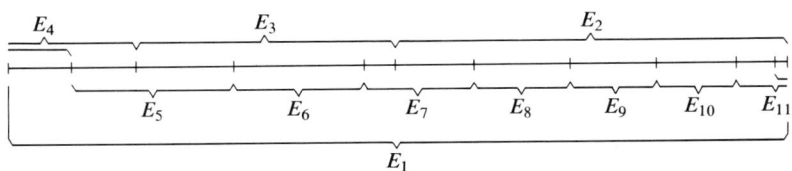

Figure 11

There follows a summary of the proof that the charts in Fig. 10 are complete and correct. Where we list something as an exercise we mean that the proof is relatively simple (often trivial) but has not actually been given here. Other simple but worthwhile exercises consist of verifying that the examples cited below do indeed show what we claim they show.

In all cases

[unif.]:
    → [a. un.] exercise   → [meas.] exercise
    → [a. e.] exercise   ↛ [mean] 39.1

[a. un.]:
    ↛ [unif.] 39.3   → [meas.] exercise
    → [a. e.] exercise   ↛ [mean] 39.1

[a. e.]:
    ↛ [unif.] 39.3   ↛ [meas.] 39.5
    ↛ [a. un.] 39.5   ↛ [mean] 39.2

[meas.]:
    ↛ [unif.] 39.3   ↛ [a. e.] 39.4
    ↛ [a. un.] 39.4   ↛ [mean] 39.2

[mean]:
    ↛ [unif.] 39.3   ↛ [a. e.] 39.4
    ↛ [a. un.] 39.4   → [meas.] 38.6

$\mu(\Omega) < \infty$

[unif.]:
    → [a. un.] exercise   → [meas.] exercise
    → [a. e.] exercise   → [mean] exercise

[a. un.]:
    ↛ [unif.] 39.3   → [meas.] exercise
    → [a. e.] exercise   ↛ [mean] 39.2

[a. e.]:
    ↛ [unif.] 39.3   → [meas.] 21.3
    → [a. un.] 21.3   ↛ [mean] 39.2

[meas.]:
    ↛ [unif.] 39.3   ↛ [a. e.] 39.4
    ↛ [a. un.] 39.4   ↛ [mean] 39.2

[mean]:
    ↛ [unif.] 39.3   ↛ [a. e.] 39.4
    ↛ [a. un.] 39.4   → [meas.] 38.6

$$g_0 \text{ integrable and } |f_n| \leq g_0 \text{ [a. e.]}$$

[unif.]:
    → [a. un.] exercise   → [meas.] exercise
    → [a. e.] exercise   → [mean] 28.1

[a. un.]:
    ↛ [unif.] 39.3   → [meas.] exercise
    → [a. e.] exercise   → [mean] 28.1

[a. e.]:
    ↛ [unif.] 39.3   → [meas.] 37.1
    → [a. un.] 37.1   → [mean] 28.1

[meas.]:
    ↛ [unif.] 39.3   ↛ [a. e.] 39.4
    ↛ [a. un.] 39.4   → [mean] 28.3

[mean]:
    ↛ [unif.] 39.3   ↛ [a. e.] 39.4
    ↛ [a. un.] 39.4   → [meas.] 38.6

40 CONVERGENCE OF MEASURES AND INTEGRALS

Let μ be a finite measure on a class $\mathcal{M}$ of measurable sets. Some very interesting convergence theorems emerge if we make $\mathcal{M}$ into a pseudo-metric space. For $A, B \in \mathcal{M}$ we define

40.1 $$\rho(A, B) = \mu(A - B) + \mu(B - A).$$

Obviously $\rho(A, A) = 0$; the triangle inequality follows from the fact that

$$(A - C) \cup (C - A) \subset (A - B) \cup (B - A) \cup (B - C) \cup (C - B).$$

Theorem 40.2. *The pseudo-metric space $\mathcal{M}$ with distance defined by* 40.1 *is complete.*

Proof. Let A be a Cauchy sequence in $\mathcal{M}$. For each i there is an m_i such that

$$\rho(A_m, A_{m_i}) < \frac{1}{2^i} \tag{1}$$

for $m > m_i$; also we may assume that $m_i > m_j$ for $i > j$. It follows from (1) and the definition of ρ that for $m > m_i$

$$\mu(A_m - A_{m_i}) < \frac{1}{2^i} \quad \text{and} \quad \mu(A_{m_i} - A_m) < \frac{1}{2^i}. \tag{2}$$

Now let

$$B = \varlimsup_i A_{m_i} = \bigcap_{k=1}^{\infty} \bigcup_{i=k}^{\infty} A_{m_i},$$

$$C = \varliminf_i A_{m_i} = \bigcup_{k=1}^{\infty} \bigcap_{i=k}^{\infty} A_{m_i}.$$

Since

$$B - A_{m_k} \subset \left(\bigcup_{i=k}^{\infty} A_{m_i}\right) - A_{m_k} \subset \bigcup_{i=k}^{\infty} (A_{m_{i+1}} - A_{m_i})$$

we have by (2) that

$$\mu(B - A_{m_k}) \leq \sum_{i=k}^{\infty} \mu(A_{m_{i+1}} - A_{m_i}) < \sum_{i=k}^{\infty} \frac{1}{2^i} = \frac{1}{2^{k-1}}. \tag{3}$$

On the other hand, $B \supset C$; so

$$A_{m_k} - B \subset A_{m_k} - C \subset A_{m_k} - \left(\bigcap_{i=k}^{\infty} A_{m_i}\right) \subset \bigcup_{i=k}^{\infty} (A_{m_i} - A_{m_{i+1}});$$

so it follows by (2) that

$$\mu(A_{m_k} - B) \leq \sum_{i=k}^{\infty} \mu(A_{m_i} - A_{m_{i+1}}) < \sum_{i=k}^{\infty} \frac{1}{2^i} = \frac{1}{2^{k-1}}. \tag{4}$$

By (3) and (4), $\rho(A_{m_k}, B) < 1/2^{k-2}$, and the result now follows by 4.7. ∎

The idea is to apply the theorems of Moore and Osgood (Section 6) to sequences of functions on $\mathcal{M}$. This means we will be concerned with continuity and equicontinuity. On the other hand, the interesting functions on $\mathcal{M}$ are additive; and with this feature present, continuity and equicontinuity can be "moved around."

Theorem 40.3. *For each n, let σ_n be a finitely additive function on $\mathcal{M}$. Then, the σ_n are equicontinuous at $A \in \mathcal{M}$ if and only if they are equicontinuous at ϕ with respect to the metric in $\mathcal{M}$.*

Proof. Since $\rho(\phi, E) = \mu(E)$, equicontinuity at ϕ means that given $\varepsilon > 0$, there exists $\delta > 0$ such that $\mu(E) < \delta$ implies $|\sigma_n(E)| < \varepsilon$ for all n. Suppose this holds and suppose $\rho(A, B) < \delta$; then

$$|\sigma_n(A - B)| < \varepsilon \quad \text{and} \quad |\sigma_n(B - A)| < \varepsilon.$$

However,

$$|\sigma_n(A) - \sigma_n(B)| = |\sigma_n(A) - \sigma_n(A \cup B) + \sigma_n(A \cup B) - \sigma_n(B)|$$
$$\leq |\sigma_n(A \cup B) - \sigma_n(A)| + |\sigma_n(A \cup B) - \sigma_n(B)|$$
$$= |\sigma_n(A - B)| + |\sigma_n(B - A)| < 2\varepsilon;$$

so the σ_n are equicontinuous at A. Conversely, suppose $\rho(A, B) < \delta$ implies $|\sigma_n(A) - \sigma_n(B)| < \varepsilon$, and let $\mu(E) < \delta$. Then,

$$\rho(A, A \cup E) < \delta \quad \text{and} \quad \rho(A, A - E) < \delta;$$

so

$$|\sigma_n(E)| = |\sigma_n(A \cup E) - \sigma_n(A - E)|$$
$$\leq |\sigma_n(A \cup E) - \sigma_n(A)| + |\sigma_n(A) - \sigma_n(A - E)| < 2\varepsilon. \quad \blacksquare$$

We shall frequently employ 40.3 to say that a single additive function on $\mathcal{M}$ is continuous if and only if it is continuous at ϕ; this is merely the special case in which all σ_n are the same.

So far it would seem that we take a finite measure μ, define a metric in $\mathcal{M}$, then look at other functions on $\mathcal{M}$ and hope they are continuous. Actually we can do better than this. We take a sequence of functions that we want to study and let this sequence generate μ in such a way that the individual functions come out continuous on the pseudo-metric space $\mathcal{M}$. The mechanism for this is as follows.

Theorem 40.4. *Let σ be a sequence of finite, completely additive functions on a class $\mathcal{M}$, and let μ be defined by*

$$\mu(E) = \sum_{n=1}^{\infty} \frac{V(\sigma_n, E)}{2^n[1 + V(\sigma_n, \Omega)]}.$$

Then μ is a finite measure on $\mathcal{M}$, and each σ_n is continuous with respect to the metric generated in $\mathcal{M}$ by μ.

Proof. Clearly μ is finitely additive. If E_k tends monotonically to ϕ, then for each n, $V(\sigma_n, E_k) \to 0$ and by 7.1 $\mu(E_k) \to 0$; so μ is continuous at ϕ, hence completely additive by 10.9. On the other hand, by 7.1, $\mu(E) \to 0$ implies $V(\sigma_n, E) \to 0$ for each n; so each σ_n is continuous at ϕ with respect to the metric generated by μ and by 40.3 each σ_n is thus continuous on $\mathcal{M}$. $\blacksquare$

Theorem 40.5. *Let σ be a sequence of finite, completely additive functions on a class $\mathcal{M}$ and let*

$$-\infty < \sigma_0(E) = \lim_n \sigma_n(E) < \infty$$

for each $E \in \mathcal{M}$; then σ_0 is completely additive.

Proof. It is clear that σ_0 is finitely additive. We apply 40.4 and σ becomes a pointwise convergent sequence of continuous functions on a complete pseudo-metric space. By 6.5.1, σ_0 is continuous somewhere and by 40.3 it is continuous at ϕ with respect to the metric in $\mathcal{M}$. If E_k tends monotonically to ϕ, $\mu(E_k) \to 0$ and so $\sigma_0(E_k) \to 0$; thus σ_0 is continuous at ϕ, hence completely additive by 10.9. $\blacksquare$

EXERCISES

a. Let μ be a sequence of finite measures with
$$\mu_0(E) = \lim_n \mu_n(E)$$
finite for each measurable E. Show that if f is a simple function, then
$$\int_E f \, d\mu_0 = \lim_n \int_E f \, d\mu_n.$$

b. Extend the result of Exercise a to the case where f is bounded and measurable. [*Hint*: Approximate f by simple functions f_k and show that $\lim_n \int_E f_k \, d\mu_n = \int_E f_k \, d\mu_0$ uniformly in k. Use 6.4.]

c. Show that boundedness of f is essential in Exercise b. *Hint*: Let λ be Lebesgue measure in R_1; set
$$\mu_n(E) = \lambda[E \cap (\tfrac{1}{2}^{n+1}, \tfrac{1}{2}^n)]$$
and let $f(x) = 1/x$.

d. Show that finiteness of μ_0 is essential in Exercises a and b. [*Hint*: Let $\mu_n(E) = \lambda[E \cap (0, n)]$, $f(x) = 1$.]

e. Show that if $\mu_{n+1}(E) \geq \mu_n(E)$ for each n and each E, then
$$\int_E f \, d\mu_0 = \lim_n \int_E f \, d\mu_n$$
given only that f is μ_0-integrable.

f. Prove the following integration by parts theorem for Lebesgue-Stieltjes integrals in R_1. Let f and g be nondecreasing and right continuous and assume that f and g have no common discontinuities; then
$$\int_{(a,\,b]} f \, d\mu_g + \int_{(a,\,b]} g \, d\mu_f = f(b)g(b) - f(a)g(a).$$
[*Hint*: Prove by direct computation for simple f and g; use Exercise b.]

g. Let $f(x) = x$ and let g be the Cantor function (Section 27). Use Exercise f to compute
$$\int_{[0,\,1]} f \, d\mu_g.$$

h. Show that the hypothesis that f and g have no common discontinuities is essential in Exercise f.

i. Give another proof of 40.2 in which the set C defined there appears as the limit of the sequence A.

Given a pointwise convergent sequence of continuous functions on a complete pseudo-metric space, Theorems 6.4 and 6.5 yield two major conclusions: (i) the limit function is continuous somewhere and (ii) the sequence is equicontinuous somewhere. Theorem 40.5 applies the first of these conclusions to additive set functions; application of the second conclusion yields a significant result in the study of mean convergence.

First, we need to clarify the meaning of "equicontinuity at ϕ" for a set of additive set functions. Following the definition of continuity at ϕ given in Section 10, we will say that additive set functions σ_n are *equicontinuous at* ϕ if for every contracting sequence E with $E_k \to \phi$ we have $\lim_k \sigma_n(E_k) = 0$ uniformly in n. On the other hand, given a measure μ (finite or not), we will say that the σ_n are *uniformly absolutely continuous* with respect to μ provided that given $\varepsilon > 0$, there exists $\delta > 0$ such that $\mu(E) < \delta$ implies $|\sigma_n(E)| < \varepsilon$ for every n. If μ happens to be finite and generates a metric on $\mathcal{M}$, then uniform absolute continuity with respect to μ is just equicontinuity at the point $\phi \in \mathcal{M}$ in the usual sense of equicontinuity on pseudo-metric spaces. These concepts are not synonymous, and the following theorem gives some basic relations.

Theorem 40.6. *Let μ be a finite measure on $\mathcal{M}$ and let σ be a sequence of finite, completely additive functions on $\mathcal{M}$ such that the σ_n are uniformly absolutely continuous with respect to μ.*

40.6.1 *The σ_n are equicontinuous at ϕ.*

40.6.2 *The $V(\sigma_n,\ )$ are uniformly absolutely continuous with respect to μ.*

Proof. If E_k tends monotonically to ϕ, then $\mu(E_k) \to 0$; so $\lim_k \sigma_n(E_k) = 0$ uniformly in n. This proves 40.6.1. To prove 40.6.2, let $\varepsilon > 0$ be given and let δ be such that $\mu(E) < \delta$ implies $|\sigma_n(E)| < \varepsilon/2$ for all n. Given E satisfying $\mu(E) < \delta$, note that $A \subset E$ implies $\mu(A) < \delta$; so

$$\sigma_n(A) < \frac{\varepsilon}{2} \quad \text{for all } n;$$

and we have

$$0 \leq \overline{V}(\sigma_n, E) \leq \frac{\varepsilon}{2} \quad \text{for all } n.$$

Similarly,

$$0 \geq \underline{V}(\sigma_n, E) \geq -\frac{\varepsilon}{2} \quad \text{for all } n,$$

and the result follows because $V(\sigma_n, E) = \overline{V}(\sigma_n, E) - \underline{V}(\sigma_n, E)$. ∎

We are now ready to study mean convergence.

Theorem 40.7. *Let f be a sequence of integrable functions and let f_0 be integrable; let*

$$\lim_n f_n = f_0 \text{ [meas.]}$$

and let the set functions $\int |f_n|$ be equicontinuous at ϕ. Then

$$\lim_n f_n = f_0 \text{ [mean]}.$$

Proof. The proof is based on two lemmas.

Lemma 40.7.1. *The conclusion follows, given the added hypothesis that the set functions $\int |f_n|$ are uniformly absolutely continuous with respect to the measure that generates the integrals.*

Proof of 40.7.1. Since

$$\int |f_n - f_0| \leq \int |f_n| + \int |f_0|$$

and since $\int |f_0|$ is continuous at ϕ and absolutely continuous, we now have the following: $\int |f_n - f_0|$ equicontinuous at ϕ (this is 28.2.1); $\int |f_n - f_0|$ uniformly absolutely continuous (this is 28.2.2); convergence in measure. The lemma now follows by 28.2. ∎

Lemma 40.7.2. *If each of the set functions σ_n is nonnegative, completely additive, and absolutely continuous, and if the σ_n are equicontinuous at ϕ, then they are uniformly absolutely continuous.*

Proof. Suppose the σ_n are not uniformly absolutely continuous. Then there exist an $\varepsilon > 0$, a sequence E of measurable sets, and a sequence n of positive integers such that for each k,

$$\mu(E_k) \leq \frac{1}{2^k}$$

and

$$\sigma_{n_k}(E_k) \geq \varepsilon.$$

Let A be the sequence defined by

$$A_i = \bigcup_{k=i}^{\infty} E_k.$$

Then A is a contracting sequence, and

$$\mu(A_i) \leq \sum_{k=i}^{\infty} \frac{1}{2^k} = \frac{1}{2^{i-1}};$$

so if $A_0 = \lim_i A_i$, we have

$$\mu(A_0) = \lim_i \mu(A_i) = 0.$$

Now, $\lim_i (A_i - A_0) = \phi$; and since each σ_n is absolutely continuous, we have for each i,

$$\sigma_{n_i}(A_i - A_0) = \sigma_{n_i}(A_i) \geq \sigma_{n_i}(E_i) \geq \varepsilon,$$

thus contradicting the assumption of equicontinuity at ϕ.

The theorem now follows because by 40.7.2 the added hypothesis of 40.7.1 is implied by one of the original hypotheses. ∎

Theorem 40.8. Let f be a sequence of integrable functions and let f_0 be integrable; let
$$\lim_n f_n = f_0 \text{ [meas.]}$$
and for each measurable set E let
$$\lim_n \int_E f_n = \int_E f_0.$$
Then
$$\lim_n f_n = f_0 \text{ [mean]}.$$

Proof. The measure generating the integrals may be infinite; it will not be used here. Instead, we define a finite measure v by
$$v(E) = \sum_{n=1}^{\infty} \frac{\int_E |f_n|}{2^n [1 + \int_\Omega |f_n|]}.$$
By 40.4 the functions $\int f_n$ are continuous on the pseudo-metric space $\mathscr{M}$ with metric generated by v. Our second hypothesis here is that these continuous functions converge "pointwise" on $\mathscr{M}$. Thus by 6.5.1 they are equicontinuous somewhere on $\mathscr{M}$, and by 40.3 they are therefore uniformly absolutely continuous with respect to v. By 27.2.2 and 40.6.2 we now have the functions $\int |f_n|$ uniformly absolutely continuous with respect to v; so by 40.6.1 these are equicontinuous at ϕ. The result now follows by 40.7. ∎

Theorem 40.9. *In each of Theorems 40.7 and 40.8 the two sufficient conditions given are also necessary.*

Proof. Convergence in mean always implies convergence in measure (Section 39). It is obvious that mean convergence implies the second hypothesis of 40.8, and the proof of 40.8 shows that this in turn implies the second hypothesis of 40.7. ∎

EXERCISES

j. Let μ be Lebesgue measure on $[0, 1]$ and let
$$f_n(x) = \begin{cases} 1 & \text{for } 2i/2n \leq x < (2i+1)/2n; \quad i = 1, 2, \ldots, n-1; \\ -1 & \text{for } (2i+1)/2n \leq x < (2i+2)/2n; \quad i = 1, 2, \ldots, n-1. \end{cases}$$
Show that f satisfies the second hypothesis of 40.8 but not the first, and that the conclusion fails.

k. In 40.6 drop the requirement that μ be finite. Show that 40.6.1 may fail but 40.6.2 is still true.

l. Prove the following modification of 40.7. Assume $\mu(\Omega) < \infty$ and replace the second hypothesis by the requirement that the set functions $\int f_n$ be uniformly absolutely continuous.

m. Show that the assumption $\mu(\Omega) < \infty$ is essential in Exercise *l*.

n. In Exercise *l* the uniformity condition is imposed on the $\int f_n$ while in 40.7 it is imposed on the $\int |f_n|$. Can the absolute value signs be removed in the second hypothesis of 40.7? So far as we can determine, the answer to this question is not known.

41 THE L_p SPACES

For $p \geq 1$ we define L_p as the space of all measurable functions x on Ω such that $|x|^p$ is integrable. Note that as we turn now to a discussion of function spaces, we shall (as in Section 7) use x, y, and z to denote elements of the function space and normally use t to denote an element of the underlying space Ω.

Theorem 41.1. *Let $x \in L_p$ and $y \in L_p$ and let c be a constant; then $cx \in L_p$ and $x + y \in L_p$.*

Proof. That $cx \in L_p$ follows from the fact that $\int_\Omega |cx|^p = |c|^p \int_\Omega |x|^p$. The other conclusion is established as follows. Let

$$E_1 = \{t \mid |x(t)| \geq |y(t)|\},$$
$$E_2 = \{t \mid |x(t)| < |y(t)|\};$$

then

$$|x(t) + y(t)|^p \leq \begin{cases} |2x(t)|^p & \text{for } t \in E_1, \\ |2y(t)|^p & \text{for } t \in E_2. \end{cases}$$

So, let

$$z(t) = \begin{cases} |2x(t)|^p & \text{for } t \in E_1, \\ |2y(t)|^p & \text{for } t \in E_2; \end{cases}$$

then z is integrable and $x + y \in L_p$ by 25.5.2. ∎

We define distance in L_p by setting

$$\rho(x, y) = \left(\int_\Omega |x - y|^p \right)^{1/p}.$$

The primary purpose of this section is to show that each L_p is a pseudo-metric space. Clearly, $\rho(x, y) = 0$. Equally clearly, we do not get a metric space if Ω contains nonvacuous sets of measure zero because $\rho(x, y) = 0$ if and only if $x = y$ [a. e.].

Verification of the triangle postulate in L_p is a major project. We establish it by looking at two classical inequalities, each of which has many applications in analysis.

The first of these, Hölder's inequality, follows from an algebraic inequality. If $a > 0$, $b > 0$, $\alpha > 0$, $\beta > 0$, and $\alpha + \beta = 1$, then

41.2 $$a^\alpha b^\beta \leq \alpha a + \beta b.$$

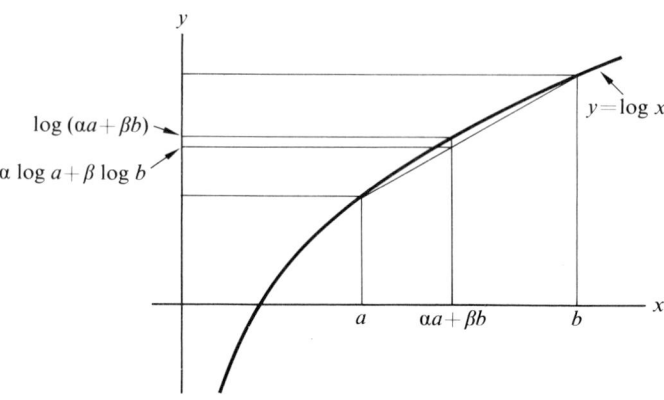

Figure 12

This can be proved by purely algebraic methods,† but one of the easiest ways of reducing it to something familiar is to take logarithms:

$$\alpha \log a + \beta \log b \leq \log(\alpha a + \beta b).$$

From the conditions given here we have $a \leq \alpha a + \beta b \leq b$ or $b \leq \alpha a + \beta b \leq a$; so in logarithmic form, 41.2 merely describes the concavity of the logarithm curve (see Fig. 12).

Theorem 41.3 (Hölder's inequality). *If $x \in L_p$ and $y \in L_q$ where $p > 1$ and*

$$\frac{1}{p} + \frac{1}{q} = 1,$$

then $xy \in L_1$ and for every measurable $E \subset \Omega$,

$$\int_E |xy| \leq \left(\int_E |x|^p\right)^{1/p} \left(\int_E |y|^q\right)^{1/q}.$$

Proof. Integrability of $|xy|$ follows from the fact that if

$$|y(t)| \leq |x(t)|^{p-1},$$

then

$$|x(t)y(t)| \leq |x(t)|^p;$$

otherwise we have

$$|x(t)| < |y(t)|^{1/(p-1)} = |y(t)|^{q-1},$$

and

$$|x(t)y(t)| \leq |y(t)|^q.$$

To prove the inequality, we note that if $xy = 0$ a. e. in E, then the left side vanishes; so we may assume that for every $t \in E$, $x(t)y(t) \neq 0$. In this case we apply 41.2 with

$$\alpha = \frac{1}{p}, \quad \beta = \frac{1}{q}, \quad a = \frac{|x(t)|^p}{\int_E |x|^p}, \quad b = \frac{|y(t)|^q}{\int_E |y|^q};$$

---

† See, for example, Titchmarsh [28], p. 383.

thus for each $t \in E$,

$$\frac{|x(t)\,y(t)|}{(\int_E |x|^p)^{1/p}(\int_E |y|^q)^{1/q}} \leq \frac{|x(t)|^p}{p \int_E |x|^p} + \frac{|y(t)|^q}{q \int_E |y|^q}.$$

Integrating over E, we have

$$\frac{\int_E |xy|}{(\int_E |x|^p)^{1/p}(\int_E |y|^q)^{1/q}} \leq \frac{1}{p} + \frac{1}{q} = 1. \quad \blacksquare$$

Theorem 41.4 (Minkowski's inequality). *If $p \geq 1$, $x \in L_p$ and $y \in L_p$, then for each measurable $E \subset \Omega$,*

$$\left(\int_E |x+y|^p\right)^{1/p} \leq \left(\int_E |x|^p\right)^{1/p} + \left(\int_E |y|^p\right)^{1/p}.$$

Proof. Existence of the left-hand integral follows from 41.1. If $p = 1$ the inequality comes from integrating over E the inequality

$$|x(t) + y(t)| \leq |x(t)| + |y(t)|,$$

valid for every $t \in E$ by the triangle inequality for real numbers. Therefore for $p > 1$ we have

$$\int_E |x+y|^p = \int_E |x+y|\,|x+y|^{p-1} \leq \int_E |x|\,|x+y|^{p-1} + \int_E |y|\,|x+y|^{p-1}.$$

Applying Hölder's inequality to each of these last two integrals and noting that $(p-1)q = p$, we have

$$\int_E |x+y|^p \leq \left(\int_E |x|^p\right)^{1/p}\left(\int_E |x+y|^p\right)^{1/q} + \left(\int_E |y|^p\right)^{1/p}\left(\int_E |x+y|^p\right)^{1/q}.$$

If $\int_E |x+y|^p = 0$, the result is obvious to begin with; otherwise we divide the above inequality by $(\int_E |x+y|^p)^{1/q}$ and note that $1 - (1/q) = 1/p$. $\blacksquare$

EXERCISES

a. Show that if $x_1, x_2, \ldots, x_n$ and $y_1, y_2, \ldots, y_n$ are two sets of n real numbers each, and if $p \geq 1$, then

$$\left(\sum_{i=1}^n |x_i + y_i|^p\right)^{1/p} \leq \left(\sum_{i=1}^n |x_i|^p\right)^{1/p} + \left(\sum_{i=1}^n |y_i|^p\right)^{1/p}.$$

Note: For $p = 2$, this gives the triangle inequality in R_n.

b. State and prove Minkowski's inequality for infinite series. Note: For $p = 2$, this will give the triangle inequality for the space l_2 (Hilbert space), introduced in Section 7.

c. State and prove Hölder's inequality for finite sums and for infinite series.

d. Hölder's inequality for $p = 2$ is usually called Schwarz's inequality. Give a separate proof for this case. *Hint:* For any a,

$$0 \leq \int_E (|x| - a|y|)^2 = \int_E |x|^2 - 2a \int_E |xy| + a^2 \int_E |y|^2;$$

discuss the case $\int_E |y|^2 = 0$ separately; then set

$$a = \frac{\int_E |xy|}{\int_E |y|^2}.$$

e. Show by an example that Minkowski's inequality cannot be extended to the case $p < 1$.

f. The *coefficient of linear correlation* of two random variables f and g is defined as

$$r = \frac{e(fg) - e(f)e(g)}{[\operatorname{var}(f)]^{1/2}[\operatorname{var}(g)]^{1/2}}.$$

Show that $|r| \leq 1$.

g. Show that if in Exercise f, $f - e(f) = a[g - e(g)]$, where a is a constant, then $|r| = 1$.

h. Show that if f and g are independent random variables, then $r = 0$ in Exercise f.

If in Minkowski's inequality we set $E = \Omega$ and replace x by $x - z$ and y by $z - y$, we have the triangle inequality for L_p. Thus for $p \geq 1$, L_p is a pseudometric space. In this connection, note carefully Exercise e above.

The L_p spaces introduce a new mode of convergence for sequences of measurable functions. Specifically, $\lim_n x_n = x_0$ in L_p means

41.5
$$\lim_n \int_\Omega |x_n - x_0|^p = 0.$$

When 41.5 holds, we say that x_n tends to x_0 in the mean of order p. We abbreviate this

$$\lim_n x_n = x_0 \ [\operatorname{mean}^p].$$

Hölder's inequality yields very easily some quite useful information about mean^p convergence.

Theorem 41.6. *If $\mu(\Omega) < \infty$, if $\lim_n x_n = x_0 \ [\operatorname{mean}^{p_1}]$ and if $1 \leq p_2 < p_1$, then $\lim_n x_n = x_0 \ [\operatorname{mean}^{p_2}]$.*

Proof. In Hölder's inequality, let $x = |x_n - x_0|^{p_2}$; let $p = p_1/p_2$; let $y(t) \equiv 1$. Then

$$\int_\Omega |x_n - x_0|^{p_2} \leq \left(\int_\Omega |x_n - x_0|^{p_1} \right)^{p_2/p_1} [\mu(\Omega)]^{p_1/(p_1 - p_2)}. \ \blacksquare$$

Theorem 41.7. *If $\lim_n x_n = x_0 \ [\operatorname{mean}^p]$ and if $y \in L_q$ where $(1/p) + (1/q) = 1$, then $\lim_n x_n y = x_0 y \ [\operatorname{mean}]$.*

Proof. By Hölder's inequality,

$$\int_\Omega |x_n y - x_0 y| \le \left(\int_\Omega |x_n - x_0|^p\right)^{1/p} \left(\int_\Omega |y|^q\right)^{1/q}. \quad \blacksquare$$

This last theorem is particularly important when applied to series. We say that

$$\sum_{k=1}^\infty x_k = x_0 \, [\text{mean}^p]$$

provided that

$$\lim_n \sum_{k=1}^n x_k = x_0 \, [\text{mean}^p].$$

Corollary 41.7.1. *If*

$$\sum_{k=1}^\infty x_k = x_0 \, [\text{mean}^p],$$

and if $y \in L_q$, *then*

$$\sum_{k=1}^\infty x_k y = x_0 y \, [\text{mean}].$$

Proof. In 41.7, replace x_n by $\sum_{k=1}^n x_k$. ∎

EXERCISES

i. Show by an example that 41.6 cannot be extended to the case $\mu(\Omega) = \infty$.

j. Let $\Omega = [0, 1]$ and let μ be Lebesgue measure. Show that L_p is not locally compact.

k. Show that if $\Omega = R_n$ and if μ is a Lebesgue-Stieltjes measure, then L_p is separable. *Hint:* Consider half-open intervals whose corner points have rational coordinates. Use linear combinations of the characteristic functions of these with rational coefficients to obtain a countable dense set.

l. Show that if $\Omega = [0, 1]$ and if μ is the measure of Exercise 11.*f*, then L_p is non-separable.

m. Prove that if $\lim_n x_n = x_0$ [meas.] and if there is a $z \in L_1$ such that for each n, $|x_n|^p \le z$ [a. e.], then $\lim_n x_n = x_0$ [mean^p]. [*Hint:* See Section 28 for the case $p = 1$.]

n. Prove that if $\lim_n x_n = x_0$ [meas.] and if the set functions $\int |x_n|^p$ are equicontinuous at ϕ, then $\lim_n x_n = x_0$ [mean^p]. [*Hint:* See Section 40.]

o. Prove that if $\lim_n x_n = x_0$ [mean^p], then $\lim_n x_n = x_0$ [meas.] and the set functions $\int |x_n|^p$ are equicontinuous at ϕ. Note: This is proved in Section 40 for $p = 1$, but the methods used there will not apply here because there is no parallel to 40.8 for $p > 1$. For a direct proof of equicontinuity, let A_i tend monotonically to ϕ and note that

$$\left(\int_{A_i} |x_n|^p\right)^{1/p} \le \left(\int_{A_i} |x_{n_0}|^p\right)^{1/p} + \left(\int_\Omega |x_n - x_{n_0}|^p\right)^{1/p};$$

for sufficiently large n_0 the last integral on the right is small for $n > n_0$; for $n \le n_0$ each of the other integrals is small for i sufficiently large.

p. Show that if f is a sequence of random variables such that

$$\lim_n \frac{1}{n^2} \operatorname{var}\left(\sum_{i=1}^n f_i\right) = 0,$$

then f obeys the weak law of large numbers.

q. Show that if f is a sequence of independent random variables with $\operatorname{var}(f_n) = b$ for each n, then f obeys the weak law of large numbers. [*Hint:* See Exercise 30.c.]

42 CAUCHY THEOREMS

In Section 7 we introduced the pseudo-metric function spaces c_0, c, m, s, C and M. In Section 41 we have just introduced L_p ($p \geq 1$). We mentioned l_2 (Hilbert space) in Section 7; in the light of Section 41 we should generalize this to l_p ($p \geq 1$). Specifically, l_p is the space of sequences x such that

$$\sum_{i=1}^\infty |x_i|^p$$

is convergent. In l_p we define

$$\rho(x, y) = \left(\sum_{i=1}^\infty |x_i - y_i|^p\right)^{1/p}.$$

Clearly, l_p is a special case of an L_p space in which Ω is the set of positive integers with each point having measure one.

There are two other function spaces that we shall encounter in Chapter 8. If x is a measurable function on Ω, we define the *essential supremum* of x:

$$\operatorname{essup} x = \sup \{M \mid \mu[\{t \mid x(t) < M\}] > 0\}.$$

The function x is *essentially bounded* if $\operatorname{essup} |x| < \infty$. The space L_∞ is the set of all essentially bounded functions on Ω with

$$\rho(x, y) = \operatorname{essup} |x - y|.$$

The second space we want to call attention to is the space S of all measurable functions on a space of finite measure. In S we define

$$\rho(x, y) = \int_\Omega \frac{|x - y|}{1 + |x - y|}.$$

An important item in functional analysis is the fact that all of these function spaces are complete. Completeness of L_p is one of the major items of measure theory and the principal purpose of this section is to give the proof of the completeness theorem for L_p. First, however, we should like to point out that completeness proofs for function spaces follow a fairly standard pattern. A Cauchy sequence x in the space means a sequence of functions that is a Cauchy sequence [ ], where the blank brackets mean, "Fill in your own mode of convergence, depending on how distance is defined in the space under consideration."

To get a limit function we need to go from convergence [ ] to pointwise convergence or to a. e. convergence if the metric is such that sets of measure zero can be ignored. The point here is that by the completeness of the real number system a pointwise Cauchy sequence yields a limit number for each point and thus defines a (pointwise) limit function. In some cases convergence [ ] implies pointwise convergence and this step is easy. In other cases we must turn to a theorem in which convergence [ ] guarantees a. e. convergence of a subsequence. In any case, we get at least a subsequence at least a. e. convergent to a function x_0. The remaining steps are as follows. In the sense determined by convergence [ ] we have $|x_{n_k} - x_{n_j}|$ "small." Hopefully, we can take an a. e. limit on j and get $|x_{n_k} - x_0|$ "small" in the same sense. If the space is right, this will put x_0 in it. Furthermore, it shows the subsequence convergent [ ] to x_0 and so by 4.7 x converges [ ] to x_0.

In short, a completeness proof for a function space involves three principal steps. (i) Construct a limit function x_0. (ii) Obtain convergence to x_0 in the appropriate sense. (iii) Show that x_0 is in the space. It is interesting to note (Theorem 42.3 and Exercises f through j) that the relative complexity of these three steps varies considerably from one space to another.

One of the major motivations for the development of measure theory was the fact that the Riemann integrable functions do not form a complete space with respect to mean convergence. To see this, let Ω be $[0, 1]$ and let μ be Lebesgue measure. Let K be a nowhere dense perfect set with positive measure, and let E be a sequence of open intervals such that

$$\bigcup_{k=1}^{\infty} E_k = -K.$$

Define the sequence x by setting x_n equal to the characteristic function of $\bigcup_{k=1}^{n} E_k$. For each n, x_n is a Riemann integrable function, and it is clear that

$$\lim_n x_n = C_{-K} \text{ [a. e.]}.$$

The functions x_n are uniformly bounded; so by 28.1,

$$\lim_n x_n = C_{-K} \text{ [mean]}.$$

Thus, x is a Cauchy sequence [mean]; and since each x_n is Riemann integrable, it follows that x is a mean Cauchy sequence with respect to Riemann integrals. However, a mean limit is essentially unique, so any limit function must be equal a. e. to C_{-K}.

It follows from the construction of K that, given any $t_0 \in [0, 1]$ and any $\varepsilon > 0$, the set $(t_0 - \varepsilon, t_0 + \varepsilon) - K$ contains an interval; therefore if $x_0 = C_{-K}$ a. e., then $(t_0 - \varepsilon, t_0 + \varepsilon)$ certainly contains points t for which $x_0(t) = 1$. Thus, for every $t_0 \in [0, 1]$,

$$\varlimsup_{t \to t_0} x_0(t) \geq 1;$$

however, since $\mu(K) > 0$, we have $x_0 = 0$ on a set of positive measure. On this set x_0 is obviously discontinuous; so by 24.4 it is not Riemann integrable. Thus, no mean limit of the sequence x can be Riemann integrable.

The Cauchy theorem for mean convergence is one of the great improvements of the Lebesgue theory of integration over the Riemann theory. We turn now to the proof of this fundamental theorem.

Lemma 42.1. *If x is a Cauchy sequence* [mean^p], *then x is a Cauchy sequence* [meas.].

Proof. This is just like 38.6. If
$$\mu\{t \mid |x_m(t) - x_n(t)| \geq \varepsilon^{1/(2p)}\} > \varepsilon^{1/2},$$
then
$$\int_\Omega |x_m - x_n|^p \geq \varepsilon. \quad \blacksquare$$

Lemma 42.2 (Fatou). *Let x be a sequence of nonnegative, integrable functions with $x_0 = \varliminf_n x_n$* [a. e.]; *and let $\varliminf_n \int_\Omega x_n$ be finite. Then x_0 is integrable, and*
$$\int_\Omega x_0 \leq \varliminf_n \int_\Omega x_n.$$

Proof For each $t \in \Omega$, we let
$$y_k(t) = \inf_{n \geq k} x_n(t);$$
then $0 \leq y_k \leq x_k$; so each y_k is integrable, and
$$\lim_k \int_\Omega y_k \leq \varliminf_k \int_\Omega x_k.$$
However, y is a nondecreasing sequence with
$$\lim_k y_k = \varliminf_n x_n = x_0 \text{ [a. e.]},$$
so the result follows by 26.1. $\blacksquare$

Theorem 42.3 (Riesz-Fischer). *For each $p \geq 1$, L_p is a complete pseudo-metric space.*

Proof. Let x be a Cauchy sequence in L_p; then x is a Cauchy sequence [mean^p] of functions on Ω. By 42.1, x is a Cauchy sequence [meas.], so by 38.5.1 there is a subsequence of x which is a Cauchy sequence [a. e.]. This subsequence determines a. e. a function x_0 such that
$$\lim_k x_{n_k} = x_0 \text{ [a. e.]}.$$
For a fixed index i, we then have
$$\lim_k |x_{n_i} - x_{n_k}|^p = |x_{n_i} - x_0|^p \text{ [a. e.]};$$

so by Fatou's lemma (42.2),

$$\varliminf_{k} \int_\Omega |x_{n_i} - x_{n_k}|^p \geq \int_\Omega |x_{n_i} - x_0|^p. \tag{1}$$

Let $\varepsilon > 0$ be given. Since x is a Cauchy sequence [mean^p], it follows that there is an integer k_0 such that if $i > k_0$ and $k > k_0$, then

$$\int_\Omega |x_{n_i} - x_{n_k}|^p < \varepsilon.$$

Therefore, we have from (1) that for $i > k_0$,

$$\int_\Omega |x_{n_i} - x_0|^p \leq \varepsilon.$$

Thus $x_0 \in L_p$ by 41.1 because $x_{n_i} \in L_p$. Furthermore,

$$\lim_n x_{n_i} = x_0 \; [\text{mean}^p]$$

and the result follows by 4.7. ∎

EXERCISES

a. Give an example in which the definite inequality holds in Fatou's lemma.

b. Given an example to show that the restriction that the functions x_n be nonnegative cannot be removed in Fatou's lemma.

c. Show that with the integral defined in terms of a product measure (Exercise 24.g), Fatou's lemma becomes a special case of 10.8.

d. Show that $\lim_n x_n = x_0$ in L_∞ if and only if there exists a set A with $\mu(A) = 0$ and $\lim_n x_n = x_0$ uniformly on $-A$. *Hint:* Let

$$E_{kn} = \{t \mid |x_n(t) - x_0(t)| > 1/k\};$$

for each k there is an n_k such that $n \geq n_k$ implies $\mu(E_{kn}) = 0$; let

$$A = \bigcup_{k=1}^\infty \bigcup_{n=n_k}^\infty E_{kn}.$$

e. Show that $\lim_n x_n = x_0$ in S if and only if $\lim_n x_n = x_0$ [meas.]. [*Hint:* $|x|/(1 + |x|) \leq 1$; use 28.3.]

f. Prove completeness of c_0, c, and C. In these proofs pointwise convergence is obtained trivially; uniform convergence to the limit function is provable; then something like 6.4.4 is needed to show that the limit function is in the space.

g. Prove completeness of m and M. This is easier than Exercise f because boundedness of the limit function follows trivially from the uniform approximation.

h. Prove completeness of L_∞. In this proof a. e. convergence emerges trivially; after that it is a modification of Exercise g.

i. Prove completeness of S. Here one must go to a subsequence to get a. e. convergence; then the remaining steps are trivial.

j. Prove completeness of s. This is the easiest of all; each step is trivial.

It is interesting to note that this discussion of function spaces involves many of the modes of convergence looked at earlier in this chapter. In c_0, c, C, m, and M we have uniform convergence; in S we have convergence in measure; in L_p we have mean^p convergence. Convergence in L_∞ is uniform on the complement of a set of measure zero. Note (Exercise 21.*j*) that this is different from almost uniform convergence. Pointwise convergence on the integers appears in s. Pointwise convergence on R_1 is not describable by a metric but it is generated by the product topology (Section 8) in $R_1{}^{R_1}$. Conspicuously absent from the list are [a. e.] and [a. un.], and they must remain absent. There is no topology on the space of measurable functions for which convergence is a. e. convergence. A topology would determine neighborhoods of the zero function. Suppose there were one and consider a sequence x (see 39.4) that converges to zero in measure but not a. e. This would yield a neighborhood N of zero and a subsequence y of x ($y_i = x_{n_i}$) such that

$$y_i \notin N \qquad \text{for every } i. \tag{1}$$

However, y still converges to zero in measure and by 38.5 it has a subsequence convergent to zero a. e., but this contradicts (1).

*43 ORTHOGONAL EXPANSIONS IN HILBERT SPACE

In Section 7 we defined Hilbert space as the sequence space l_2. If we let Ω be the set of positive integers and define μ by giving each point unit weight, then l_2 appears as the space L_2 over Ω. In this chapter we shall refer to an arbitrary L_2 space as *Hilbert space*. This would appear to be a generalization of the definition in Section 7, but actually it is not as much of a generalization as it appears to be. The definition given here introduces nonseparable Hilbert spaces (see Exercise 41.*l*); so in that sense it is a generalization. However, the separable cases are by far the most important (see Exercise 41.*k*), and it can be shown (see Exercises *n* to *p* below) that all such L_2 spaces are equivalent to l_2. Thus, for the separable cases, we have introduced here only a varied representation of Hilbert space.

If $x \in L_2$ and $y \in L_2$, then by 41.3, xy is integrable. We call the integral of this product the *scalar product* of x and y:

$$x \cdot y = \int_\Omega xy.$$

The term scalar product is borrowed from vector analysis. If $A = (A_1, A_2, A_3)$ and $B = (B_1, B_2, B_3)$ are two vectors in 3-space, then

$$A \cdot B = A_1 B_1 + A_2 B_2 + A_3 B_3$$

defines their scalar product. The definition in L_2 is an obvious generalization of this.

It is a simple matter of trigonometry to show that two vectors (in 3-space) are perpendicular, or orthogonal, if and only if their scalar product is zero. We adopt the obvious analogy to define orthogonal functions. Elements x and y of L_2 are called *orthogonal* when $x \cdot y = 0$. We say that x is an *orthogonal sequence* in L_2 when $x_m \cdot x_n = 0$ for $m \neq n$. If, in addition, $x_n \cdot x_n = 1$ for each n, we say that x is an *orthonormal sequence*. It is easily checked that if x is any orthogonal sequence, then the functions $x_n/\sqrt{x_n \cdot x_n}$ form an orthonormal sequence.

If Ω is the interval $[-\pi, \pi]$ and if μ is Lebesgue measure, then the sequence in L_2 defined by

43.1
$$x_0(t) = \frac{1}{2\pi},$$
$$x_{2n-1}(t) = \frac{1}{\pi} \cos nt,$$
$$x_{2n}(t) = \frac{1}{\pi} \sin nt$$

is an important example of an orthonormal sequence. Other examples are described in Exercises *d* through *g* below.

EXERCISES

a. Prove the following properties of scalar products:
$$x \cdot y = \overline{y \cdot x},$$
$$x \cdot (ay + bz) = a(x \cdot y) + b(x \cdot z),$$
$$(x - y) \cdot (x - y) = [\rho(x, y)]^2.$$

b. Show that the functions defined by 43.1 are orthogonal on $[-\pi, \pi]$ with respect to Lebesgue measure.

c. The *Sturm-Liouville differential operator* L is defined by
$$L(u) = (pu')' - qu,$$
where p and q are functions of the real variable t and primes indicate differentiation with respect to t. Show that if u and v are functions such that all the formal operations called for are valid and if
$$v(b)p(b)u'(b) - v(a)p(a)u'(a) = u(b)p(b)v'(b) - u(a)p(a)v'(a) = 0,$$
then
$$\int_{[a,b]} uL(v) = \int_{[a,b]} vL(u),$$
where the integrals are with respect to Lebesgue measure. [*Hint*: Integrate by parts.]

d. A *Sturm-Liouville differential equation* is one of the form

$$L(u) + \lambda wu = 0,$$

where λ is a constant and w is a bounded measurable function of t. Show that if u_m and u_n are solutions of this equation for different values λ_m and λ_n of the parameter λ, if u_m and u_n are both in L_2, and if Exercise c applies, then

$$\int_{[a,b]} wu_m u_n = 0,$$

where the integral is with respect to Lebesgue measure.

e. *Legendre polynomials* are defined by the differential equation

$$(1 - t^2)P_n''(t) - 2tP_n'(t) + n(n + 1)P_n(t) = 0.$$

Show that for $m \neq n$, P_m is orthogonal to P_n on $[-1, 1]$ with respect to Lebesgue measure.

f. *Laguerre polynomials* are defined by the differential equation

$$tL_n''(t) + (1 - t)L_n'(t) + nL_n(t) = 0.$$

Show that for $m \neq n$, L_m is orthogonal to L_n on $[0, \infty]$ with respect to the Lebesgue-Stieltjes measure induced by the distribution function f where $f(t) = 1 - e^{-t}$. *Hint:* Multiply the differential equation by e^{-t} and use Exercise d. Note also Exercise 27.r.

g. *Hermite polynomials* are defined by the differential equation

$$H_n''(t) - 2tH_n'(t) + 2nH_n(t) = 0.$$

Show that for $m \neq n$, H_m is orthogonal to H_n on $(-\infty, \infty)$ with respect to the Lebesgue-Stieltjes measure induced by a distribution function f where $f'(t) = e^{-t^2}$. [*Hint:* Multiply the differential equation by e^{-t^2}.]

h. Let x be a sequence of linearly independent elements of L_2, and let z be the sequence defined by †

$$z_1 = x_1,$$

$$z_n = x_n - \sum_{i=1}^{n-1} (x_n \cdot z_i) z_i.$$

Show that z is an orthogonal sequence.

A very useful line of investigation in connection with orthogonal sequences is concerned with series expansions of functions in terms of such sequences. That is, we look for relations of the form

$$y = \sum_{n=1}^{\infty} a_n x_n,$$

where x is an orthogonal sequence of functions and a is a sequence of numbers.

---

† This is known as the *Schmidt orthogonalization process*.

The subject of orthogonal expansions is one that can be pursued at great length. We give here only a few sample theorems.

We did not specify what type of convergence is expected of the series above, and indeed, the theory differs for different types of convergence. However, since orthogonality appears in connection with Hilbert space, it is certainly natural to consider convergence [mean^2] for orthogonal expansions, and it is to this that we turn our attention. Our theorems will be concerned with the following questions. (1) Given that

$$y = \sum_{n=1}^{\infty} a_n x_n \text{ [mean}^2\text{]},$$

is this expansion unique, and if so, what are the coefficients a_n? (2) Under what conditions on the sequence a does $\sum a_n x_n$ converge [mean^2] to a function $y \in L_2$ for each orthogonal sequence x? (3) Under what conditions on the orthogonal sequence x is every $y \in L_2$ the sum [mean^2] of an expansion in terms of x?

Lemma 43.2. *If x is any sequence in L_2, if $z \in L_2$, and if*

$$y = \sum_{n=1}^{\infty} x_n \text{ [mean}^2\text{]},$$

then

$$y \cdot z = \sum_{n=1}^{\infty} x_n \cdot z.$$

Proof. This follows at once from 41.7.1, the definition of a scalar product, and the fact that a mean convergent series may be integrated term by term. ∎

Theorem 43.3. *If $y \in L_2$, if x is an orthonormal sequence in L_2, and if*

$$y = \sum_{n=1}^{\infty} a_n x_n \text{ [mean}^2\text{]},$$

then the coefficients a_n are given by the formula

$$a_n = x_n \cdot y = \int_\Omega x_n y.$$

Proof. By 43.2, for each k,

$$x_k \cdot y = \sum_{n=1}^{\infty} a_n x_k \cdot x_n = a_k,$$

because by the orthogonality condition on x, every term vanishes except one. ∎

Theorem 43.4 (Riesz-Fischer). *If a is a sequence of numbers such that*

$$\sum_{n=1}^{\infty} a_n^2 < \infty,$$

then for every orthonormal sequence x in L_2 there exists $y \in L_2$ such that

$$y = \sum_{n=1}^{\infty} a_n x_n \; [\text{mean}^2].$$

Proof. We note that for any integers n and m with $n > m$,

$$\left[\rho\left(\sum_{k=1}^{n} a_k x_k, \sum_{k=1}^{m} a_k x_k\right)\right]^2 = \left(\sum_{k=m}^{n} a_k x_k\right) \cdot \left(\sum_{k=m}^{n} a_k x_k\right)$$

$$= \sum_{k} a_k^2 x_k \cdot x_k + \sum_{j \neq k} a_j a_k x_j \cdot x_k$$

$$= \sum_{k=m}^{n} a_k^2;$$

so from the convergence of $\sum a_k^2$ it follows that the sequence of partial sums of $\sum a_k x_k$ is a Cauchy sequence in L_2. The result now follows from 42.3. ∎

If the orthonormal sequence x is the one given in 43.1, the expansion is called a *Fourier expansion*. The Riesz-Fischer theorem as it originally appeared involved not only the manipulation shown above for the special case of Fourier expansions, but also a proof of the fundamental theorem 42.3. Indeed, the search for a theorem of this sort for Fourier expansions was one of the chief motivating forces behind the development of the Lebesgue integral. As we pointed out in Section 42, there is no such theorem available on the basis of the Riemann theory of integration.

To answer the last of the three questions raised in the discussion preceding 43.2 we need the following result.

Theorem 43.5 (Bessel's inequality). *If x is an orthonormal sequence in L_2, if $y \in L_2$, and if a is the sequence of numbers defined by*

$$a_n = y \cdot x_n,$$

then

$$\sum_{n=1}^{\infty} a_n^2 \leq y \cdot y.$$

Proof. For each positive integer k, we have

$$0 \leq \left(y - \sum_{n=1}^{k} a_n x_n\right) \cdot \left(y - \sum_{n=1}^{k} a_n x_n\right)$$

$$= y \cdot y - 2 \sum_{n} a_n y \cdot x_n + \sum_{n} a_n^2 x_n \cdot x_n + \sum_{m \neq n} a_m a_n x_m \cdot x_n$$

$$= y \cdot y - \sum_{n=1}^{k} a_n^2;$$

thus for each k,

$$\sum_{n=1}^{k} a_n^2 \leq y \cdot y,$$

and the result follows if we let $k \to \infty$. ∎

Theorem 43.6. *If the orthonormal sequence x has the property that $x_n \cdot w = 0$ for each n implies $w = \theta$, then for every $y \in L_2$,*

$$y = \sum_{n=1}^{\infty} (x_n \cdot y) x_n \ [\text{mean}^2].$$

Proof. As usual, we let $x_n \cdot y = a_n$. It follows from 43.5 and 43.4 that

$$\sum_{n=1}^{\infty} a_n x_n$$

is always convergent [mean^2]. Let

$$z = \sum_{n=1}^{\infty} a_n x_n \ [\text{mean}^2];$$

then for each k, we have from 43.2 that

$$(y - z) \cdot x_k = y \cdot x_k - \sum_{n=1}^{\infty} a_n x_n \cdot x_k = y \cdot x_k - a_k = 0,$$

so according to our present hypotheses, $y - z = \theta$. ∎

EXERCISES

i. Prove that if x is an orthonormal sequence and if

$$y = \sum_{n=1}^{\infty} a_n x_n \ [\text{mean}^2],$$

then

$$y \cdot y = \sum_{n=1}^{\infty} a_n^2.$$

j. Prove the converse to Exercise *i*. If x is an orthonormal sequence and if

$$y \cdot y = \sum_{n=1}^{\infty} (x_n \cdot y)^2,$$

then

$$y = \sum_{n=1}^{\infty} (x_n \cdot y) x_n \ [\text{mean}^2].$$

Hint: Let z be the sum of the series, and show that $y \cdot y = y \cdot z = z \cdot z$; hence $\rho(y, z) = 0$. Note: The results in Exercises *i* and *j* constitute *Parseval's theorem*.

k. Show that in the Schmidt orthogonalization process (Exercise *h*),

$$x_n = \sum_{i=1}^{n} (x_n \cdot z_i) z_i / \sqrt{z_i \cdot z_i}$$

for each n.

l. Show that if x is any sequence in L_2, then there is an orthonormal sequence z such that for each n, x_n is a linear combination of the z_i.

m. Show that if L_2 is separable, then there is an orthonormal sequence z such that for every $y \in L_2$ there is a sequence a such that

$$y = \sum_{i=1}^{\infty} a_i z_i \, [\text{mean}^2].$$

n. Show that if L_2 is separable, then there is a one-to-one transformation $T(x) = y$ from L_2 onto the sequence space l_2. *Hint:* Let z be a fixed orthonormal sequence in L_2 such that (Exercise m) every $x \in L_2$ is the sum [mean^2] of an expansion in terms of z. Let y be the sequence of coefficients in the expansion of x.

o. Show that the transformation in Exercise n preserves the algebraic operations; that is, $T(ax_1 + bx_2) = aT(x_1) + bT(x_2)$.

p. Show that the transformation in Exercise n preserves distances.

REFERENCES FOR FURTHER STUDY

On convergence theorems:
  Hahn and Rosenthal [9]
  Halmos [11]
  Zaanen [31]
On probability limit theorems:
  Halmos [11]

On L_p spaces:
  Bartle [2]
  Hewitt and Stromberg [13]
  McShane and Botts [20]
  Zaanen [31]
On Hilbert space:
  Hewitt and Stromberg [13]

CHAPTER 8

FUNCTIONAL ANALYSIS

Let x's with subscripts be elements of L_p and let $y \in L_q$. The following facts are easily verified:

1) $ax_1 + bx_2 \in L_p$ by Minkowski's inequality.

2) $x_n y$ is integrable by Hölder's inequality.

3) $\int_\Omega (ax_1 + bx_2) y = a \int_\Omega x_1 y + b \int_\Omega x_2 y$, using elementary properties of the integral.

4) If $\lim_n x_n = x_0$ [mean^p], then $\lim_n \int_\Omega x_n y = \int_\Omega x_0 y$ by Hölder's inequality.

Item (1) points out that L_p admits the operations of a linear space, addition, and multiplication by scalars. Items (2) and (3) point out that the operations of multiplication by y followed by integration over Ω define a linear function on the linear space L_p. Item (4) points out that when L_p is metricized by mean^p convergence, this linear function is continuous.

In functional analysis we abstract the ideas introduced in this example. That is, we study the general theory of linear spaces with appropriate metrics and continuous linear functions thereon. In this study the principal tools used come from topology and linear algebra; however, integration theory is never too far away for the following reason. The above example of a function f on L_p defined by setting $f(x) = \int_\Omega xy$ is not just an isolated example of the object of study in functional analysis; it is essentially a typical example. That is, the prime examples of linear spaces are function spaces, and (see Section 46) on these the continuous linear functions are generated by integration or a reasonable facsimile thereof. Thus, in a sense, functional analysis is advanced integration theory without integral signs.

On the other hand, functional analysis is an interesting and extensive branch of mathematics in its own right, and the present chapter is a bare introduction to the subject. However, in the current state of the art there are supposed to be three "big theorems" in functional analysis, and these are given here. The Banach-Steinhaus theorem is in Section 44; the Hahn-Banach theorem is in Section 45; and the closed-graph theorem is in Section 50.

44 BANACH SPACES

Suppose B is a set for which addition of elements and multiplication of elements by real numbers is defined and that B is closed under these operations. We say that B is a *linear space* provided the following postulates are satisfied (x, y, and z are arbitrary elements of B and a and b are arbitrary real numbers):

L–I. $x + y = y + x$;

L–II. $x + (y + z) = (x + y) + z$;

L–III.   if $x + y = x + z$, then $y = z$;

L–IV. $a(x + y) = ax + ay$;

L–V. $(a + b)x = ax + bx$;

L–VI. $a(bx) = (ab)x$;

L–VII. $1 \cdot x = x$.

In any linear space we can define subtraction of elements by setting

$$x - y = x + (-1) \cdot y.$$

A linear space B has a zero element θ such that for every $x \in B$,

$$x + \theta = x.$$

To see this, take $y \in B$ and set $\theta = 0 \cdot y$. Then,

$$y + \theta = (1 + 0) \cdot y = y,$$

and for any $x \in B$,

$$x + y = x + (y + \theta) = (x + \theta) + y;$$

so by L–III, $x = x + \theta$. It follows immediately from L–III that θ is unique.

If addition of functions and multiplication of functions by numbers is defined "pointwise," that is

$$(x + y)(t) = x(t) + y(t),$$

$$(ax)(t) = ax(t),$$

then each of the standard function spaces becomes a linear space. In particular, C, c, c_0, l_2, m, M (Section 7) all qualify, as do L_p (Section 41) and l_p (Section 42).

We say that a linear space B is a *normed linear space* if for each $x \in B$ there is a number $\|x\|$ (norm of x) such that the following postulates are satisfied:

N–I. $\|x\| \geq 0$ for each $x \in B$;

N–II. $\|\theta\| = 0$;

N–III. $\|x + y\| \leq \|x\| + \|y\|$ for every $x, y \in B$;

N–IV. $\|ax\| = |a|\, \|x\|$ for each real a and each $x \in B$.

It is easily verified that a norm, as thus defined, generates a pseudo-metric if we set $\rho(x, y) = \|x - y\|$. One can try to go the other way by setting $\|x\| = \rho(x, \theta)$, but postulate N–IV is not always satisfied. For example, s (defined in Section 7 and used in Section 15) is a linear space with a metric which does not come from a norm. In s,

$$\rho(ax, \theta) = \sum_{i=1}^{\infty} \frac{|ax_i|}{2^i(1 + |ax_i|)}$$

and $|a|$ does not factor out of this expression. On the other hand, we do have the following function space norms:

c_0, c, m: $\|x\| = \sup_i |x_i|$.

C, M: $\|x\| = \sup_{0 \leq t \leq 1} |x(t)|$.

l_p: $\|x\| = \left[\sum_{i=1}^{\infty} |x_i|^p \right]^{1/p}$.

L_p: $\|x\| = \left[\int_\Omega |x|^p \right]^{1/p}$.

A *Banach space* is a complete normed linear space. The function spaces whose norms are listed above are Banach spaces. The completeness questions were discussed in Section 42.

We must turn now to a complicated question of terminology. As noted above, our definition of a norm is such that our norms generate pseudo-metrics which are not necessarily metrics. Thus many authors would insist that what we have defined is a "pseudo-norm" with the word "norm" reserved for the case in which $x \neq \theta$ implies $\|x\| > 0$. Note that the principal offender here is L_p; $\|x\| = 0$ in L_p provided $x = \theta$ [a. e.]. Usually the distinction between the "pseudo" and "true" cases is based on the question, "Does your postulate read $\|x\| = 0$ *if* $x = \theta$, or *if and only if* $x = \theta$?" However, there is another approach to the matter. It is readily verified that in a normed linear space (as defined here) the following three statements are equivalent:

1) $\|x\| > 0$ if $x \neq \theta$.

2) $\{\theta\}$ is a closed set.

3) Every singleton set is closed.

Now statement (3) is a standard item from general topology known as the T_1 property.

So our terminology will be as follows. Banach space will mean complete normed linear space with the understanding that statements (1), (2), and (3) above may not hold. If we want conditions (1), (2), and (3) imposed, we shall refer to a T_1-Banach space. Our primary reason for adopting this (nonstandard) terminology is that most of the standard theory of Banach spaces is completely independent of the T_1 property. We want to develop the core of this theory with the name Banach attached but with emphasis on the fact that the T_1 postulate is seldom needed.

Let A and B be Banach spaces. A function f from A to B is called *linear* if $f(ax + by) = af(x) + bf(y)$ for every $x, y \in A$ and all real a, b. Such a function is *continuous* at $y \in A$ if for every $\varepsilon > 0$ there exists $\delta > 0$ such that $\|x - y\| < \delta$ implies $\|f(x) - f(y)\| < \varepsilon$.

Theorem 44.1. *A linear function from a Banach space A to a Banach space B is continuous at an arbitrary $y \in A$ if and only if it is continuous at 0.*

Proof. Suppose f is continuous at 0. Given $\varepsilon > 0$, there exists $\delta > 0$ such that $\|z\| < \delta$ implies $\|f(z) - f(0)\| < \varepsilon$. However, $f(0) = 0$ because

$$f(x) = f(x + 0) = f(x) + f(0)$$

for every $x \in A$. Thus, $\|z\| < \delta$ implies $\|f(z)\| < \varepsilon$; so given $y \in A$, $\|x - y\| < \delta$ implies

$$\|f(x) - f(y)\| = \|f(x - y)\| < \varepsilon,$$

and f is continuous at y. Conversely, let f be continuous at y, and let y and ε determine an appropriate δ. Suppose $\|z\| < \delta$; then

$$\|(y + z) - y\| = \|z\| < \delta,$$

and it follows that

$$\|f(z)\| = \|f(y + z - y)\| = \|f(y + z) - f(y)\| < \varepsilon. \quad \blacksquare$$

The gist of this theorem is that *a linear function continuous somewhere is continuous everywhere.* This is similar to the case of additive set functions (Section 40), only the proof is even simpler here.

Corollary 44.1.1. *If f is a sequence of continuous linear functions from a Banach space A to a Banach space B and if $f_0(x) = \lim_n f_n(x)$ for each $x \in A$, then f_0 is continuous and linear.*

Proof. Linearity is trivial:

$$f_0(ax + by) = \lim_n f_n(ax + by) = \lim_n [a f_n(x) + b f_n(y)]$$
$$= a \lim_n f_n(x) + b \lim_n f_n(y) = a f_0(x) + b f_0(y).$$

Since A is complete, it follows by 6.5.1 that f_0 is continuous somewhere; so by 44.1 it is continuous on all of A. $\blacksquare$

Theorem 44.2 (Banach-Steinhaus). *Let Φ be a set of continuous linear functions from a Banach space A to a Banach space B such that for each $x \in A$,*

$$\{f(x) \mid f \in \Phi\}$$

is a bounded set in B. Then there exists a number M such that

$$\|f(x)\| \leq M\|x\|$$

for all $x \in A$ and all $f \in \Phi$.

Proof. For each n, let

$$E_n = \{x \mid \|f(x)\| \leq n \quad \text{for all } f \in \Phi\}.$$

The hypothesis that Φ is pointwise bounded says that each $x \in A$ is in some E_n, or

$$\bigcup_{n=1}^{\infty} E_n = A.$$

Now each E_n may be described as

$$E_n = \bigcap_{f \in \Phi} \{x \mid \|f(x)\| \leq n\}$$

and since each f is continuous this gives E_n as an intersection of closed sets, hence closed. By the Baire category theorem (4.9) the complete space A cannot be a countable union of closed sets each with vacuous interior. Thus, some E_k contains a neighborhood $N(y, \varepsilon)$. For any $x \in A$, set

$$z = y + \frac{\varepsilon x}{2\|x\|};$$

then $z \in N(y, \varepsilon) \subset E_k$ and we have $\|f(z)\| \leq k$ and $\|f(y)\| \leq k$ for all $f \in \Phi$. However,

$$x = \frac{2\|x\|}{\varepsilon}(z - y);$$

so

$$\|f(x)\| \leq \frac{2\|x\|}{\varepsilon}(\|f(z)\| + \|f(y)\|) \leq \frac{4k}{\varepsilon}\|x\|. \quad \blacksquare$$

Corollary 44.2.1. *A linear function f from a Banach space A to a Banach space B is continuous if and only if there exists a number M such that*

$$\|f(x)\| \leq M\|x\|$$

for each $x \in A$.

Proof. If such an M exists, then f is obviously continuous at 0, hence continuous by 44.1. For the converse, let Φ in 44.2 consist of the single function f. $\blacksquare$

Let $C(A, B)$ be the space of all continuous linear functions from A to B. There is an obvious way to make this a linear space, and 44.2.1 indicates a way to make it a Banach space. The norm of f will be defined as the infimum of all numbers M that qualify in 44.2.1, but there is a more graceful way to say it. For $\|x\| = 0$, $M = 0$ will do, and for all other x there is a $y \, (= x/\|x\|)$ of norm 1 such that $\|f(y)\| \leq M$. Thus for $f \in C(A, B)$, we define

$$\|f\| = \sup\{\|f(x)\| \mid x \in A, \|x\| = 1\}.$$

Theorem 44.3. *If A and B are Banach spaces, then $C(A, B)$ is a Banach space.*

Proof. We leave verification of the norm postulates as exercises and prove only the completeness of $C(A, B)$. Let f be a Cauchy sequence in $C(A, B)$; then for each $x \in A$ and each m and n,

$$\|f_m(x) - f_n(x)\| = \|(f_m - f_n)(x)\| \leq \|f_m - f_n\| \|x\|;$$

so $f_0(x)$ is a Cauchy sequence in B. Since B is complete, we can set $f_0(x) = \lim_n f_n(x)$ for each $x \in A$; and it follows by 44.1.1 that $f_0 \in C(A, B)$. Given $\varepsilon > 0$, there exists N (because f is a Cauchy sequence) such that if $m > N$ and $n > N$, then $\|f_m - f_n\| < \varepsilon$. Thus, for all $x \in A$ with $\|x\| = 1$,

$$\|f_m(x) - f_n(x)\| \leq \|f_m - f_n\| < \varepsilon$$

provided $m > N$ and $n > N$. Letting $m \to \infty$, we thus have

$$\|f_0(x) - f_n(x)\| \leq \varepsilon$$

provided $\|x\| = 1$ and $n > N$. Therefore, for $n > N$,

$$\|f_0 - f_n\| = \sup_{\|x\|=1} \|f_0(x) - f_n(x)\| \leq \varepsilon;$$

and $\|f_0 - f_n\| \to 0$ as required. ∎

The ideas introduced so far are particularly interesting in the case in which B (the range space) is the real number system. In this case special terminology and notation are employed. In functional analysis the word *functional* is used as a noun to mean function into the real number system. Thus if B is the reals, what we have called $C(A, B)$ is the space of continuous linear functionals on A. This is commonly denoted by A^* and called the *conjugate space* to A. By 44.3, A^* is also a Banach space and so generates A^{**}, etc.

An important relation among spaces in this hierarchy of conjugates is introduced by the following consideration. Let A be a Banach space and let $x \in A$. Define a functional $\hat{x}$ (pronounced x-hat) on A^* by setting

$$\hat{x}(f) = f(x)$$

for every $f \in A^*$.

Theorem 44.4. *If A is a Banach space and $x \in A$, then $\hat{x} \in A^{**}$.*

Proof. For $f, g \in A^*$,

$$\hat{x}(af + bg) = (af + bg)(x) = af(x) + bg(x) = a\hat{x}(f) + b\hat{x}(g);$$

so $\hat{x}$ is linear on A^*. For any $f \in A^*$,

$$|\hat{x}(f)| = |f(x)| \leq \|x\| \|f\| \to 0$$

as $\|f\| \to 0$; so $\hat{x}$ is continuous at the zero functional and thus continuous by 44.1. ∎

The function H (for "hat") from A into A^{**} defined by

$$H(x) = \hat{x}$$

is easily seen to be linear because for any $f \in A^*$

$$\widehat{(ax + by)}(f) = f(ax + by) = a f(x) + b f(y) = a\hat{x}(f) + b\hat{y}(f).$$

In Section 45 it will be shown that H is continuous and that if A has the T_1 property H is one-to-one. In any case, H is called the *natural embedding* of A into A^{**} and A is called a *reflexive* Banach space if the range of H is all of A^{**}. It will be shown in Section 46 that some Banach spaces are reflexive while others are not.

EXERCISES

a. Prove the equivalence of the three statements: (1) $\|x\| > 0$ if $x \neq 0$, (2) $\{\theta\}$ is closed, (3) every singleton set is closed.

b. In general topology the T_1 property is frequently stated as follows. Given points x and y with $x \neq y$, each point has a neighborhood that does not contain the other. Prove that this is equivalent to the statements in Exercise a.

c. Verify that the norm of a continuous linear function, as defined in this section, satisfies postulates N–I through N–IV.

d. Prove that if f is a continuous linear function from one Banach space to another, then $f^{-1}(\overline{\{\theta\}}) \supset \overline{\{\theta\}}$.

e. Show that a linear space is a group with θ as the identity and that a linear function from one linear space to another is a homomorphism between the corresponding groups.

f. The kernel of a homomorphism is the inverse image of the identity. Suppose f is a continuous linear function from a Banach space to a T_1-Banach space. What can be said about the kernel of the group homomorphism f generates? See Exercise d.

g. The set of all ordered pairs (x, y) of numbers forms a linear space in an obvious way. Show that every linear functional f on this space is generated by an ordered pair (a, b) of numbers; specifically f assumes the form $f(x, y) = ax + by$. [*Hint:* Values of f on $(1, 0)$ and $(0, 1)$ determine f uniquely.]

h. In the space of Exercise g define a norm by setting $\|(x, y)\| = |x|$. Show that the resulting space is Banach but not T_1.

i. Show that on the space of Exercise h the functional generated by $(1, 0)$ is continuous while that generated by $(0, 1)$ is not continuous.

j. Let S be any linear space consisting of functions on $[0, 1]$ and let $t \in [0, 1]$. The number t generates a functional f_t on S called an *evaluation functional* defined by

$$f_t(x) = x(t).$$

Show that all evaluation functionals are linear.

k. Show that on C and M all evaluation functionals are continuous while on L_p they are all discontinuous.

l. In the proof of 44.3 the fact that $f_0 \in C(A, B)$ was deduced from 44.1.1. Give an alternative proof of this step as follows. From the fact that f is a Cauchy sequence show that f_0 is bounded on the unit sphere in A and hence continuous by 44.2.

m. Show that if A is any Banach space, then A^* is a T_1-Banach space.

n. Show that if A is reflexive, then A^* is reflexive. [*Hint:* For $x \in A$ and $F \in A^{***}$, $f(x) = F(\hat{x})$ defines $f \in A^*$ so that $F = \hat{f}$.]

o. Show that 44.1.1 is also a direct corollary of 44.2. Note that either with the proof given or with that suggested here, 44.1.1 depends on the Baire category theorem.

p. Give a direct proof of 44.2.1 that does not involve the Baire category theorem. [*Hint:* Suppose $\|f(x_n)\| \geq n \|x_n\|$; set $y_n = x_n/(n \|x_n\|)$ and look at $f(y_n)$.]

45 THE HAHN-BANACH THEOREM

Let A be a linear space and let $B \subset A$; B is a *linear subspace* of A if for $x, y \in B$ and all numbers a and b, $ax + by \in B$. If A is a Banach space, a *closed linear subspace* is a linear subspace which is also a closed set.

Let us think of the xy-plane as a Banach space and append a z-axis. Then the graph of a linear functional is (Exercise 44.g) the graph of an equation of the form $z = ax + by$; it is a plane passing through the origin. In the xy-plane the nontrivial linear subspaces are the lines through the origin. Now it is geometrically obvious that given a line L through the origin in the xy-plane and a point p of the plane not on L, there exists in 3-space a plane passing through L and passing a height 1 above the point p. In Banach space language this says that given a closed linear subspace and a point outside it there is a continuous linear functional which is zero on the subspace and one at the outside point. One of the principal results of this section is that this result, trivial for a simplified model, is true of Banach spaces in general.

Theorem 45.1 (Hahn-Banach). *Let A be a linear space and let B be a linear subspace of A. Let p be a functional on A such that $p(x + y) \leq p(x) + p(y)$ and $p(ax) = ap(x)$ for $a > 0$. Let f be a linear functional on B such that $f(x) \leq p(x)$ for all $x \in B$. Then, there exists a linear functional g on A such that $f \subset g$ and $g(x) \leq p(x)$ for every $x \in A$.*

Proof. Let $\mathcal{H}$ be the set of all linear functionals h with the following properties: (i) The domain of h is a linear space E with $B \subset E \subset A$. (ii) $h(x) \leq p(x)$ for all $x \in E$. (iii) $f \subset h$.

Lemma 45.1.1. *Let E be a linear space with $B \subset E \subset A$; let $z \in A - E$; let $E' = \{x + az \mid x \in E, a \text{ real}\}$. If there exists $h \in \mathcal{H}$ with domain E, then there exists $h' \in \mathcal{H}$ with domain E'.*

Proof of Lemma. Let $x, y \in E$; then
$$h(x) - h(y) = h(x - y) \le p(x - y) = p[(x + z) + (-y - z)]$$
$$\le p(x + z) + p(-y - z);$$
so
$$-p(-y - z) - h(y) \le p(x + z) - h(x).$$
Thus
$$\sup_{y \in E} [-p(-y - z) - h(y)] \le p(x + z) - h(x)$$
for every $x \in E$, and it follows that
$$\sup_{y \in E} [-p(-y - z) - h(y)] \le \inf_{x \in E} [p(x + z) - h(x)].$$
Therefore, there is a number t such that
$$-p(-y - z) - h(y) \le t \le p(y + z) - h(y) \quad (1)$$
for every $y \in E$. Next we note that the representation $x + az$ of points of E' is unique because if $x_1 + a_1 z = x_2 + a_2 z$, then $(a_1 - a_2) z = x_2 - x_1 \in E$, and since $z \in -E$ it follows that $a_1 = a_2$, whence $x_1 = x_2$. Thus, the formula
$$h'(x + az) = h(x) + at$$
defines an extension of h to E'. Clearly, h' is linear and $h' \supset h \supset f$; so the lemma is proved if we verify that h' is dominated by p. To this end we substitute $y = x/a$ in (1):
$$-p\left(-\frac{x}{a} - z\right) - h\left(\frac{x}{a}\right) \le t \le p\left(\frac{x}{a} + z\right) - h\left(\frac{x}{a}\right). \quad (2)$$
If $a > 0$, we multiply the right-hand inequality in (2) by a to get
$$at \le p(x + az) - h(x). \quad (3)$$
If $a < 0$, we multiply the left-hand inequality in (2) by $-a$ to get
$$-p(x + az) + h(x) \le -at. \quad (4)$$
From either (3) or (4) it follows that
$$h'(x + az) = h(x) + at \le p(x + az),$$
and the lemma is proved.

To complete the proof of the theorem, let $\mathscr{H}$ be partially ordered by inclusion. That is, define $h_1 \le h_2$ to mean $h_1 \subset h_2$ (h_2 is an extension of h_1). $\mathscr{H}$ is not empty because $f \in \mathscr{H}$. Let $\mathscr{L}$ be a linearly ordered subset of $\mathscr{H}$ and let
$$l = \bigcup_{h \in \mathscr{L}} h.$$

Clearly, l is an upper bound for $\mathscr{L}$, and it is readily verified that $l \in \mathscr{H}$. For example, let x and y be in the domain of l. Then, for some h_1 and h_2 in $\mathscr{L}$, x is in the domain of h_1 and y is in the domain of h_2. However, $\mathscr{L}$ is linearly ordered; so one of these h's contains the other; say $h_1 \subset h_2$. Then, x and y are both in the domain of h_2 and

$$l(ax + by) = h_2(ax + by) = ah_2(x) + bh_2(y) = al(x) + bl(y).$$

Thus, l is linear, and the other requirements that $l \in \mathscr{H}$ are checked in a similar manner. Zorn's lemma now yields a maximal element g of $\mathscr{H}$. This g is the required extension of f provided its domain is A. However, if the domain of g were not A, 45.1.1 would give a strictly larger element of $\mathscr{H}$, contradicting the fact that g is maximal. ∎

The Hahn-Banach theorem itself says nothing directly about continuity. Nevertheless, its principal application deals with extensions of continuous linear functionals. Indeed, the following theorem is frequently referred to as the Hahn-Banach theorem.

Theorem 45.2. *Let E be a linear subspace of a Banach space A and let f be a continuous linear functional on E. Then, there exists a continuous linear extension g of f to all of A such that $\|g\| = \|f\|$.*

Proof. Set $p(x) = \|f\| \, \|x\|$ for all $x \in A$. Theorem 45.1 yields a linear extension g of f, but for all $x \in A$

$$g(x) \leq p(x) = \|f\| \, \|x\|$$

and

$$-g(x) = g(-x) \leq p(-x) = \|f\| \, \|-x\| = \|f\| \, \|x\|;$$

so

$$|g(x)| \leq \|f\| \, \|x\|$$

and g is continuous with $\|g\| \leq \|f\|$. However, $g = f$ on E; so $\|g\| \geq \|f\|$. ∎

We promised a sweeping generalization of the fact that there is always a plane through a line and a point. This is furnished by the following.

Theorem 45.3. *Let E be a linear subspace of a Banach space A and let $y \in A$ with $\rho(y, E) = d > 0$. Then there exists a continuous linear functional f on A such that*

45.3.1 $f(x) = 0$ *for every $x \in E$.*

45.3.2 $f(y) = 1$.

45.3.3 $\|f\| = 1/d$.

Proof. Let $H = \{z + ay \mid z \in E, a \text{ real}\}$. As shown in the proof of 45.1.1, H is a linear subspace of A and the representation of points of H is unique. Thus the formula

$$g(z + ay) = a$$

defines a functional g on H. Clearly, g is linear, $g(z) = 0$ for $z \in E$, and $g(y) = 1$. Now

$$\|z + ay\| = |a| \left\| \frac{z}{a} + y \right\| \geq |a| d = |g(z + ay)| d;$$

that is, $|g(x)| \leq \|x\|/d$ for every $x \in H$. Thus by 44.2, g is continuous and $\|g\| \leq 1/d$. On the other hand, by the definition of $\rho(y, E)$, there is a sequence z of points in E such that

$$\lim_n \|z_n - y\| = d;$$

so for each n

$$\|g\| \, \|z_n - y\| \geq |g(z_n - y)| = |g(y)| = 1$$

and it follows that $\|g\| \geq 1/d$. Thus g has the required properties except that its domain is H rather than A. The required extension satisfying 14.3.1 through 14.3.3 is furnished by 14.2. ∎

Corollary 45.3.4. *If A is a Banach space and x, $y \in A$ with $\|x - y\| > 0$, then there is an $f \in A^*$ such that $f(x) \neq f(y)$.*

Proof. Let the E of 45.3 be $\{\theta\}$ and the y of 45.3 be $x - y$. ∎

EXERCISES

a. Show that 45.3.2 and 45.3.3 may be replaced by $f(y) = d$ and $\|f\| = 1$.

b. Show that for each x_0 in a Banach space A there is an $f_0 \in A^*$ such that

$$\|f_0\| = 1 \quad \text{and} \quad f_0(x_0) = \|x_0\|.$$

c. Show that if $f(x) = 0$ for every $f \in A^*$, then $\|x\| = 0$. Show also that if $f(x) = 0$ for every $x \in A$, then $\|f\| = 0$. Note: One of these proofs is trivial; the other requires the Hahn-Banach theorem.

d. Prove that $\|\hat{x}\| = \|x\|$. [*Hint*: $\|\hat{x}\| \leq \|x\|$ by basic definitions; for the reverse inequality use Exercise b.]

e. Let H be the natural embedding of A into A^{**}. Show that $H^{-1}(\theta) = \overline{\{\theta\}}$.

f. Deduce from Exercises d and e that H is continuous and that if A is T_1, then H is one-to-one.

g. Show that for every Banach space A the range of the natural embedding is closed in A^{**}.

h. Prove that if A^* is reflexive, then A is reflexive. *Hint*: Let $\hat{A}$ be the range of the natural embedding of A into A^{**} and suppose $f \in A^{**} - \hat{A}$. There exists $F \in A^{***}$ such that $F(f) \neq 0$ and $F = 0$ on $\hat{A}$. Why? Show that $F \notin A^*$.

i. Show that if A is a reflexive Banach space, then there is a norm-preserving linear function from A onto the conjugate of a Banach space.

j. Prove that if A is a reflexive Banach space and $f \in A^*$, then f assumes a maximum value on $\{x \mid \|x\| = 1\}$. *Hint:* Suppose $f(x) < \|f\|$ for $\|x\| = 1$; by Exercise b there is $F \in A^{**}$ such that $\|F\| = 1$ and $F(f) = \|f\|$. Can $F = \hat{x}$?

k. Let z be a sequence of nonnegative real numbers such that

$$\sum_{i=1}^{\infty} z_i = 1.$$

Show that

$$f(x) = \sum_{i=1}^{\infty} x_i z_i$$

defines a continuous linear functional on c_0 such that $\|f\| \leq 1$.

l. Show that $\|f\| = 1$ in Exercise k. *Hint:* Let

$$x_i^n = \begin{cases} 1 & \text{for } i \leq n, \\ 0 & \text{for } i \geq n; \end{cases}$$

then $x^n \in c_0$ and

$$f(x^n) = \sum_{i=1}^{n} z_i.$$

m. For the f of Exercise k, show that if $\|x\| = 1$, then $f(x) < 1$. Conclude from Exercise j that c_0 is not reflexive.

n. The following "proof" of the Hahn-Banach theorem does not use Zorn's lemma. Suppose f cannot be extended to all of A; let z be a point to which it cannot be extended; 45.1.1 gives an extension to z and thus a contradiction. What is the fallacy in this argument?

o. Let S be a set in a Banach space A such that for each $f \in A^*$, $\{f(x) \mid x \in S\}$ is a bounded set of numbers. Prove that S is norm bounded; that is, there exists M such that $\|x\| \leq M$ for every $x \in S$. *Hint:* Apply the Banach-Steinhaus theorem (44.2) to A^* with the Φ of that theorem being $\{\hat{x} \mid x \in S\}$.

p. Prove that if $\Phi \subset A^*$ and $\{\hat{x}(f) \mid f \in \Phi\}$ is bounded for each $x \in A$, then $\{F(f) \mid f \in \Phi\}$ is bounded for each $F \in A^{**}$.

We turn now from general consequences of the Hahn-Banach theorem to some interesting results obtained by applying the theorem to get extensions of specific functionals on specific spaces.

Theorem 45.4. *To every bounded sequence x of real numbers there corresponds a number*

$$\operatorname*{Lim}_{n} x_n$$

(called the generalized limit of x). This generalized limit has the following properties:

45.4.1 $\operatorname*{Lim}_{n} (ax_n + by_n) = a \operatorname*{Lim}_{n} x_n + b \operatorname*{Lim}_{n} y_n$.

45.4.2 *If $x_n \geq 0$ for every n, then $\operatorname*{Lim}_{n} x_n \geq 0$.*

45.4.3 $\operatorname*{Lim}_{n} x_{n+1} = \operatorname*{Lim}_{n} x_{n}$.

45.4.4 If $x_n = 1$ for every n, then $\operatorname*{Lim}_{n} x_n = 1$.

Proof. According to 45.4.1, the generalized limit is to be a linear functional over the space of all bounded sequences. In order to obtain such a functional all we need is a subadditive functional, homogeneous for positive multipliers; the Hahn-Banach theorem will do the rest. We define p by saying that for any bounded sequence x, $p(x)$ is the infimum (for all finite sets $i_1, i_2, \ldots, i_k$ of positive integers) of the numbers

$$\overline{\lim_{n}} \frac{1}{k} \sum_{j=1}^{k} x_{n+i_j}.$$

Clearly, $p(ax) = ap(x)$ for $a \geq 0$. To see that p is subadditive, we note that given bounded sequences x and y and given $\varepsilon > 0$, there are sets $i_1, i_2, \ldots, i_k$ and $m_1, m_2, \ldots, m_r$ such that

$$\overline{\lim_{n}} \frac{1}{k} \sum_{j=1}^{k} x_{n+i_j} \leq p(x) + \frac{\varepsilon}{2} \quad \text{and} \quad \overline{\lim_{n}} \frac{1}{r} \sum_{s=1}^{r} y_{n+m_s} \leq p(y) + \frac{\varepsilon}{2}.$$

Now

$$\frac{1}{kr} \sum_{j=1}^{k} \sum_{s=1}^{r} (x_{n+i_j+m_s} + y_{n+i_j+m_s})$$

$$= \frac{1}{r} \sum_{s=1}^{r} \left[\frac{1}{k} \sum_{j=1}^{k} x_{(n+m_s)+i_j} \right] + \frac{1}{k} \sum_{j=1}^{k} \left[\frac{1}{r} \sum_{s=1}^{r} y_{(n+i_j)+m_s} \right],$$

and the limit superior of a sequence of averages of terms from a sequence is no greater than the limit superior of the sequence; so

$$p(x + y) \leq \overline{\lim_{n}} \frac{1}{kr} \sum_{j=1}^{k} \sum_{s=1}^{r} (x_{n+i_j+m_s} + y_{n+i_j+m_s})$$

$$\leq \overline{\lim_{n}} \frac{1}{k} \sum_{j=1}^{k} x_{n+i_j} + \overline{\lim_{n}} \frac{1}{r} \sum_{s=1}^{r} y_{n+m_s}$$

$$\leq p(x) + p(y) + \varepsilon.$$

Since ε is arbitrary, we have p subadditive. The Hahn-Banach theorem now gives us a linear functional f on the space of all bounded sequences. We set

$$\operatorname*{Lim}_{n} x_n = f(x),$$

and 45.4.1 is automatically satisfied.

To prove 45.4.2, we note that if $x_n \geq 0$ for every n, then $p(-x) \leq 0$; so since $f(-x) \leq p(-x)$,
$$f(x) = -f(-x) \geq -p(-x) \geq 0.$$
To prove 45.4.3, we let y be defined by
$$y_n = x_{n+1} - x_n;$$
then for any positive integer k,
$$p(y) \leq \overline{\lim_n} \frac{1}{k} \sum_{i=1}^{k} y_{n+i} = \overline{\lim_n} \frac{1}{k} (x_{n+k+1} - x_{n+1}) \leq \frac{1}{k} \left(\sup_n x_n - \inf_n x_n \right).$$
Since this holds for every k,
$$f(y) \leq p(y) \leq 0.$$
However, the same argument will show that $p(-y) \leq 0$; so
$$f(y) = -f(-y) \geq -p(-y) \geq 0.$$
Thus,
$$0 = f(y) = \operatorname{Lim}_n (x_{n+1} - x_n) = \operatorname{Lim}_n x_{n+1} - \operatorname{Lim}_n x_n.$$
Finally, if $x_n = 1$ for every n, then $p(x) = 1$ and $p(-x) = -1$; so
$$1 \geq f(x) = -f(-x) \geq -p(-x) = 1. \blacksquare$$

EXERCISES

q. Show that the generalized limit of 45.4 has the property that
$$\underline{\lim_n} \, x_n \leq \operatorname{Lim}_n x_n \leq \overline{\lim_n} \, x_n;$$
thus it is equal to the ordinary limit whenever the latter exists.

r. Show that
$$p(x) = \overline{\lim_n} \, x_n$$
defines a subadditive functional p on the space of all bounded sequences.

s. Use the functional p of Exercise r to define a generalized limit for bounded sequences, and prove 45.4.1, 45.4.2, and 45.4.4 for this notion of limit.

t. Show that the generalized limit of Exercise s may be constructed in such a way that 45.4.3 fails.

u. Let $x \in M$ and let x' be the periodic extension of x to $(-\infty, \infty)$; that is, $x'(t + n) = x(t)$ for $n = \ldots, -1, 0, 1, 2, \ldots$. Define p on M by setting $p(x)$ equal to the infimum over all finite sets $\{a_i\}$ of real numbers of
$$\sup_{-\infty < t < \infty} \frac{1}{k} \sum_{i=1}^{k} x'(t + a_i).$$
Show that this p qualifies as a dominant functional in the Hahn-Banach theorem.

v. Let p be as in Exercise u and let $x \in M$ be Lebesgue measurable. Show that

$$\int_{[0,1]} x \leq p(x).$$

Hint: Use the periodic x' of Exercise u, and show that

$$\int_{[0,1]} x = \frac{1}{k} \sum_{i=1}^{k} \int_{[\alpha_i, \alpha_i+1]} x' \leq \sup_{-\infty < t < \infty} \frac{1}{k} \sum_{i=1}^{k} x'(t + a_i).$$

w. For $x \in M$ and x Lebesgue measurable, set

$$f(x) = \int_{[0,1]} x.$$

Show that f is linear on a linear subspace of M. Use Exercise v and the Hahn-Banach theorem to obtain a *generalized integral* $I(x)$ for every $x \in M$. Show that this generalized integral has the following properties:

1) If x is measurable, $I(x) = \int_{[0,1]} x$.

2) If $u = t + a \pmod{1}$ and $y(u) = x(t)$, then $I(y) = I(x)$.

3) $I(ax + by) = aI(x) + bI(y)$.

x. Restrict the generalized integral of Exercise w to characteristic functions and show that the result is an extension of Lebesgue measure to the class of all subsets of $[0, 1]$ which is finitely additive and invariant under translation of sets modulo 1. Note: The discussion in Section 18 shows that there is no example of this sort if we require that the extension be completely additive.

46 REPRESENTATION OF LINEAR FUNCTIONALS

In this section we look at some specific Banach spaces and determine the structure of their conjugate spaces. The spaces we look at are function spaces and for each space we examine we shall find a general formula for continuous linear functionals. In general, each such formula will indicate that the conjugate space is a continuous linear image of another function space.

We begin with the technically simplest example, which shows that c_0^* looks exactly like l_1.

Theorem 46.1. *To each $f \in c_0$ there corresponds a $z \in l_1$ such that*

46.1.1 $$f(x) = \sum_{i=1}^{\infty} x_i z_i$$

for every $x \in c_0$; furthermore,

46.1.2 $$\|f\| = \sum_{i=1}^{\infty} |z_i|.$$

Proof. Let $x^j \in c_0$ be the sequence such that

$$x_i^j = \begin{cases} 1 & \text{for } i = j, \\ 0 & \text{otherwise}; \end{cases}$$

and define a real sequence z by setting

$$z_j = f(x^j).$$

Then, clearly,

$$\sum_{i=1}^{\infty} x_i^j z_i = z_j = f(x^j);$$

so the representation 46.1.1 for f is established for $x = x^j$. The next step is to extend the formula to all of c_0. Let $y \in c_0$ and let $w^n \in c_0$ be defined by

$$w_i^n = \begin{cases} y_i & \text{for } i \leq n, \\ 0 & \text{for } i > n. \end{cases}$$

Now

$$w^n = \sum_{i=1}^{n} y_i x^i,$$

so

$$f(w^n) = \sum_{i=1}^{n} y_i f(x^i) = \sum_{i=1}^{n} y_i z_i.$$

However, $w_n \to y$ [unif.], which is to say $\|w^n - y\| \to 0$; and since f is continuous, $f(w^n) \to f(y)$. Thus,

$$f(y) = \lim_n f(w^n) = \lim_n \sum_{i=1}^{n} y_i z_i = \sum_{i=1}^{\infty} y_i z_i,$$

and 46.1.1 is established for all $y \in c_0$.

Now, let $u^j = x^j \operatorname{sgn} z_j$; then for each n

$$\sum_{j=1}^{n} |z_j| = \sum_{j=1}^{n} f(u^j) = f\left(\sum_{j=1}^{n} u^j\right) \leq \|f\|$$

because

$$\left\|\sum_{j=1}^{n} u^j\right\| = 1;$$

so $z \in l_1$ and

$$\sum_{j=1}^{\infty} |z_j| \leq \|f\|.$$

On the other hand, for any $x \in c_0$,

$$|f(x)| = \left|\sum_{i=1}^{\infty} x_i z_i\right| \leq \sum_{i=1}^{\infty} |x_i| |z_i| \leq \sup_i |x_i| \sum_{i=1}^{\infty} |z_i| = \|x\| \sum_{i=1}^{\infty} |z_i|;$$

and 46.1.2 is established. ∎

The outline of this proof is fairly standard for such theorems. We fix the sequence z by looking at f on the points x^j. Then, using linearity and continuity of f, we extend the representation of f to all c_0. Choosing points, $\sum \mathscr{U}^j$, on which the representation approaches $\|z\|$, we establish that $z \in l_1$ and get $\|z\| \leq \|f\|$. The reverse inequality follows from intrinsic properties of the representation formula.

We said in introducing 46.1 that it makes c_0^* "look like" l_1. A more precise statement is as follows.

Theorem 46.2. *There exists a function ϕ from l_1 onto c_0^* which is linear, norm-preserving and one-to-one.*

Outline of proof. Clearly, ϕ is defined by saying that

$$\phi(z) = f$$

if and only if

$$f(x) = \sum_{i=1}^{\infty} x_i z_i$$

for every $x \in c_0$. That the domain of ϕ is l_1 (that every $z \in l_1$ generates an $f \in c_0^*$) is implicit in the last sentence of the proof of 46.1. Computations showing linearity of ϕ are left to the reader. The difficult step is to show that ϕ is onto, and this is supplied by 46.1. If $z \neq 0$ in l_1, then for some j, $z_j \neq 0$ and $z_j x_j^j = \phi(z)(x^j) \neq 0$; so ϕ is one-to-one. ∎

Theorem 46.3. *Suppose $\mu(\Omega) < \infty$. For every $f \in L_p^*$ there is a $z \in L_q$ such that*

46.3.1
$$f(x) = \int_\Omega x\, z$$

for every $x \in L_p$; furthermore,

46.3.2
$$\|f\| = \left[\int_\Omega |z|^q\right]^{1/q}.$$

Proof. Let f be any linear functional on L_p. For each measurable set $E \subset \Omega$, we have $C_E \in L_p$. We now define a set function σ by setting

$$\sigma(E) = f(C_E).$$

It is obvious from the additivity of f that σ is finitely additive. Let E be a sequence of disjoint measurable sets in Ω; let

$$E_0 = \bigcup_{n=1}^{\infty} E_n;$$

and for each k, let

$$A_k = \bigcup_{n=1}^{k} E_n.$$

Then, $\mu\{t \mid C_{A_k}(t) \neq C_{E_0}(t)\} \to 0$ as $k \to \infty$; so
$$\lim_k C_{A_k} = C_{E_0} \text{ [meas.]}.$$
Since, for each k, $C_{A_k} \leq C_{E_0} \in L_p$, we have (note Exercise 41.m) that
$$\lim_k C_{A_k} = C_{E_0} \text{ [mean}^p\text{]};$$
therefore, by the continuity of f,
$$\lim_k \sum_{n=1}^{k} \sigma(E_n) = \lim_k \sigma(A_k) = \lim_k f(C_{A_k}) = f(C_{E_0}) = \sigma(E_0).$$
That is, σ is completely additive. If $\mu(E) = 0$, then as an element of L_p, $C_E = \theta$; so $\sigma(E) = f(C_E) = 0$. Thus, by 27.1, σ is absolutely continuous.

The Radon-Nikodym theorem (27.3) now tells us that there is a function z such that for every measurable $E \subset \Omega$,
$$f(C_E) = \sigma(E) = \int_E z = \int_\Omega C_E z.$$
This determines the function z and gives f the desired representation on the points C_E of L_p. The extension of this representation to the simple functions is obvious. If x is a monotone sequence of simple functions converging pointwise to $x_0 \in L_p$, then
$$\lim_n x_n = x_0 \text{ [meas.]}$$
by 38.3. For each n, $x_n \leq x_0 \in L_p$; so by Exercise 41.m,
$$\lim_n x_n = x_0 \text{ [mean}^p\text{]}.$$
Again using continuity of f, we have
$$f(x_0) = \lim_n f(x_n) = \lim_n \int_\Omega x_n z = \int_\Omega x_0 z,$$
the existence of $\int_\Omega x_0 z$ and the last equality coming from 26.1. Thus, the desired representation of f applies over the whole of L_p.

We still must show that $z \in L_q$ and that $\|f\| = \|z\|$. To this end, we define a sequence x as follows:
$$x_n(t) = \begin{cases} |z(t)|^{q-1} \operatorname{sgn} z(t) & \text{for} \quad |z(t)|^{q-1} \leq n, \\ n \operatorname{sgn} z(t) & \text{for} \quad |z(t)|^{q-1} > n. \end{cases}$$
Each x_n is bounded and measurable, therefore, in L_p; so we have from the definition of $\|f\|$ that for each n,
$$\left| \int_\Omega x_n z \right| = |f(x_n)| \leq \|f\| \, \|x_n\| = \|f\| \left(\int_\Omega |x_n|^p \right)^{1/p}.$$

However, from the definition of the sequence x, we have
$$x_n z = |x_n|\, |z| \geq |x_n|\, |x_n|^{1/(q-1)} = |x_n|^p.$$
Thus
$$\int_\Omega |x_n|^p \leq \|f\| \left(\int_\Omega |x_n|^p\right)^{1/p}$$
so
$$\left(\int_\Omega |x_n|^p\right)^{1-(1/p)} = \left(\int_\Omega |x_n|^p\right)^{1/q} \leq \|f\|.$$
It is easily checked from the definition of the sequence x that
$$\lim_n |x_n|^p = |z|^q \text{ [a. e.]};$$
and since the sequence $|x|$ is monotone, it follows from 26.1 that $z \in L_q$ and
$$\left(\int_\Omega |z|^q\right)^{1/q} \leq \|f\|.$$
Finally, by Hölder's inequality (41.3), we have for each $y \in L_p$,
$$|f(y)| = \left|\int_\Omega yz\right| \leq \int_\Omega |yz| \leq \|y\| \left(\int_\Omega |z|^q\right)^{1/q};$$
therefore
$$\|f\| \leq \left(\int_\Omega |z|^q\right)^{1/q}. \quad \blacksquare$$

Theorem 46.4. *For $\mu(\Omega) < \infty$, there exists a function ϕ from L_q onto L_p^* which is linear and norm-preserving; furthermore, $\phi^{-1}(\theta) = \overline{\{\theta\}}$.*

Outline of proof. Same as 46.2. Note that in this case ϕ is not one-to-one, but what it collapses into single points is sets of functions that are equal a. e. $\quad \blacksquare$

Our next theorem characterizes continuous linear functionals on C as Riemann-Stieltjes integrals. A function z on $[0, 1]$ is of bounded variation if
$$\sum_{i=1}^n |z(t_i) - z(t_{i-1})|$$
is bounded over all partitions, $0 = t_0 < t_1 < \cdots < t_n = 1$, of $[0, 1]$. The least upper bound of all such sums is the total variation of z denoted by $V(z)$. If z is of bounded variation and x is continuous, then
$$\int_0^1 x\, dz$$
exists and is equal to the limit of
$$\sum_{i=1}^n x(t_i)\, [z(t_i) - z(t_{i-1})]$$
for any sequence of partitions with mesh tending to 0.

Theorem 46.5 (F. Riesz). *To each of $f \in C^*$ there corresponds a function z of bounded variation such that*

46.5.1
$$f(x) = \int_0^1 x \, dz$$

for every $x \in C$; furthermore,

46.5.2
$$\|f\| = V(z).$$

Proof. By 45.2 there is an extension g of f to the space M such that $\|g\| = \|f\|$. Consider the points $y_t \in M$ defined by

$$y_t(u) = \begin{cases} 1 & \text{for } 0 \le u \le t, \\ 0 & \text{for } t < u \le 1. \end{cases}$$

We define z on $[0, 1]$ by setting

$$z(t) = g(y_t).$$

Next, we take a partition of $[0, 1]$, set

$$\varepsilon_i = \mathrm{sgn}\,[z(t_i) - z(t_{i-1})],$$

and note that

$$\left\|\sum_{i=1}^n \varepsilon_i(y_{t_i} - y_{t_{i-1}})\right\| = \sup_{0 \le u \le 1} \sum_{i=1}^n \varepsilon_i[y_{t_i}(u) - y_{t_{i-1}}(u)] = 1$$

because the functions $y_{t_i} - y_{t_{i-1}}$ are characteristic functions of disjoint intervals and $|\varepsilon_i| = 1$. Therefore

$$\sum_{i=1}^n |z(t_i) - z(t_{i-1})| = \sum_{i=1}^n \varepsilon_i[z(t_i) - z(t_{i-1})]$$

$$= \sum_{i=1}^n \varepsilon_i[g(y_{t_i}) - g(y_{t_{i-1}})]$$

$$= g\left[\sum_{i=1}^n \varepsilon_i(y_{t_i} - y_{t_{i-1}})\right]$$

$$\le \|g\| = \|f\|;$$

so z is of bounded variation and

$$V(z) \le \|f\|. \tag{1}$$

Now let $x \in C$; we define functions $w_n \in M$ by setting

$$w_n(t) = \sum_{i=1}^n x\left(\frac{i}{n}\right)[y_{i/n}(t) - y_{(i-1)/n}(t)].$$

Then w is a sequence of step functions converging uniformly to x; so

$$\lim_n g(w_n) = g(x) = f(x).$$

However,
$$g(w_n) = \sum_{i=1}^{n} x\left(\frac{i}{n}\right)\left[z\left(\frac{i}{n}\right) - z\left(\frac{i-1}{n}\right)\right];$$
so
$$\lim_n g(w_n) = \int_0^1 x\, dz,$$
and 46.5.1 is established. Finally,
$$|f(x)| = \left|\int_0^1 x\, dz\right| \leq \sup_{0 \leq t \leq 1} |x(t)|\, V(z) = \|x\|\, V(z);$$
and this, together with (1), establishes 46.5.2. ∎

Characterization of C^* as a function space is discussed in Exercises p through t below. We conclude this section by looking at the conjugate to a nonseparable space. This will involve the integral of a bounded function with respect to a finitely additive set function. The definition of such an integral is relatively simple. Let x be a bounded function on Ω; let σ be a finitely additive function on the class of all subsets; and assume that σ is of bounded variation. Partition the *range* of x into n equal subintervals $[h_{i-1}, h_i)$ and let $E_i = \{t \mid h_{i-1} \leq x(t) < h_i\}$. Define
$$y_n = \sum_{i=1}^n h_{i-1} C_{E_i}$$
and set
$$\int_\Omega y_n\, d\sigma = \sum_{i=1}^n h_{i-1}\, \sigma(E_i).$$
The sequence y converges uniformly to x, and
$$\left|\int_\Omega (y_m - y_n)\, d\sigma\right| \leq \sup_{t \in \Omega} |y_m(t) - y_n(t)| V(\sigma);$$
so the integrals of the y_n form a Cauchy sequence, and we define
$$\int_\Omega x\, d\sigma = \lim_n \int_\Omega y_n\, d\sigma.$$

Theorem 46.6. *To each $f \in m^*$ there corresponds a finitely additive set function σ on the class of all subsets of ω (the set of positive integers) such that*

46.6.1
$$f(x) = \int_\omega x\, d\sigma$$
for every $x \in m$; furthermore, σ is of bounded variation and

46.6.2
$$\|f\| = V(\sigma).$$

Proof. We define σ by setting
$$\sigma(E) = f(C_E)$$
for each $E \subset \omega$. If $E_1 \cap E_2 = \phi$, then $C_{E_1 \cup E_2} = C_{E_1} + C_{E_2}$; so
$$\sigma(E_1 \cup E_2) = f(C_{E_1 \cup E_2}) = f(C_{E_1} + C_{E_2}) = f(C_{E_1}) + f(C_{E_2}) = \sigma(E_1) + \sigma(E_2)$$
and σ is finitely additive. Let $x \in m$ and let h_i and y_n be defined as in the above definition of the integral. We have
$$f(y_n) = f\left(\sum_{i=1}^n h_{i-1} C_{E_i}\right) = \sum_{i=1}^n h_{i-1} \sigma(E_i) = \int_\omega y_n \, d\sigma.$$
Since $y_n \to x$ uniformly, $\|y_n - x\| \to 0$; so $f(y_n) \to f(x)$, and
$$f(x) = \lim_n f(y_n) = \lim_n \int_\omega y_n \, d\sigma = \int_\omega x \, d\sigma.$$
Thus 46.6.1 is established.

Given $\varepsilon > 0$, let $E_1 \subset \omega$ be such that
$$\sigma(E_1) > \overline{V}(\sigma) - \varepsilon,$$
and let $E_2 \subset \omega$ be such that
$$\sigma(E_2) < \underline{V}(\sigma) + \varepsilon.$$
Let $z \in m$ be defined by
$$z_i = \begin{cases} 1 & \text{for } i \in E_1, \\ -1 & \text{for } i \in E_2, \\ 0 & \text{otherwise.} \end{cases}$$
Then $\|z\| = 1$ and
$$\|f\| \geq |f(z)| = \left|\int_\omega z \, d\sigma\right| = \sigma(E_1) - \sigma(E_2) > \overline{V}(\sigma) - \underline{V}(\sigma) - 2\varepsilon;$$
thus
$$\|f\| \geq \overline{V}(\sigma) - \underline{V}(\sigma) = V(\sigma).$$
Finally, for any $x \in m$,
$$|f(x)| = \left|\int_\omega x \, d\sigma\right| \leq \sup_i |x_i| \, V(\sigma) = \|x\| \, V(\sigma);$$
so
$$\|f\| \leq V(\sigma). \quad \blacksquare$$

EXERCISES

a. Let $x^j \in c$ be as in the proof of 46.1; let $u \in c$ be the unit constant sequence ($u_i \equiv 1$). Show that if $y \in c$ with $\lim_i y_i = a$, then
$$y = au + \sum_{j=1}^\infty (y_j - a) x^j,$$
and for each $f \in c^*$,
$$f(y) = af(u) + \sum_{j=1}^\infty (y_j - a) f(x^j).$$

b. Prove that to each $f \in c^*$ there corresponds $z \in l_1$ such that for each $y \in c$,
$$f(y) = z_1 \lim_i y_i + \sum_{i=1}^{\infty} y_i z_{i+1}.$$
   Hint: Referring to Exercise a let
$$z_{j+1} = f(x^j), \quad z_1 = f(u) - \sum_{j=2}^{\infty} z_j.$$
c. Prove that with f and z related as in Exercise b,
$$\|f\| = \sum_{i=1}^{\infty} |z_i|.$$
d. Prove that there is a one-to-one, linear, norm-preserving function from l_1 onto c^*.
e. Let $x \in c_0$ with $\|x\| = 1$. Show that there exist $y, z \in c_0$ such that $y \ne x \ne z$, $\|y\| = \|z\| = 1$, and $x = (y + z)/2$. [*Hint:* Take i such that $|x_i| \le 1/2$; let $y_i = x_i + 1/2$, $z_i = x_i - 1/2$; otherwise let $y_j = z_j = x_j$.]
f. Show that $u \in c$ (as defined in Exercise a) is not half the sum of two distinct points, each of norm 1.
g. Use 46.2 and Exercise d to show that there is a one-to-one, linear, norm-preserving function from c_0^* onto c^*. Then, use Exercises e and f to show that there is no such function from c_0 onto c. Informally, quite different spaces may have the "same" conjugate space.
h. Prove that there is a one-to-one, linear, norm-preserving function from l_1^* onto m.
   Hint: $f \in l_1^*$ generates $z \in m$ such that
$$f(x) = \sum_{i=1}^{\infty} x_i z_i.$$
i. From the fact that c is separable and m is not, conclude that none of the spaces c_0, c, l_1, m is reflexive.
j. Prove that l_p^* is equivalent to l_q. Note: The proof of 46.3 will furnish hints, but 46.4 does not cover this because l_p is an L_p with $\mu(\Omega) = \infty$.
k. Show that there is a norm-preserving linear function ϕ from L_∞ onto L_1^* such that $\phi^{-1}(\theta) = \overline{\{\theta\}}$.
l. Show that there exists an $f \in L_\infty^*$ such that $\|f\| = 1$ and such that if $x \in C \subset L_\infty$, then $f(x) = x(1)$. [*Hint:* Show that $g(x) = x(1)$ defines a $g \in C^*$; then extend g to L_∞ by 45.2.]
m. Show that the f of Exercise l cannot assume the form
$$f(x) = \int_{[0,1]} xz$$
   where $z \in L_1$; thus L_1 is not reflexive. *Hint:* Let $x_n(t) = t^n$. Show that $f(x_n) = 1$ for all n but $\int_{[0,1]} x_n z \to 0$.
n. Let BV be the space of all functions x on $[0, 1]$ such that $x(0) = 0$ and such that x is of bounded variation. Set $\|x\| = V(x)$. Show that BV is a normed linear space.
o. Prove that if $\|x_n\| \to 0$ in BV, then $x_n(t) \to \theta(t)$ for every $t \in [0, 1]$.

p. Prove that if $V(x_n) \le K$ and $x_n(t) \to x_0(t)$ for every $t \in [0, 1]$, then $V(x_0) \le K$. [*Hint:* Total variation is a supremum of sums based on partitions of $[0, 1]$; each such sum generated by x_n is $\le K$; show that each such sum generated by x_0 is $\le K$.]

q. Using Exercises o and p, give a direct proof that BV is complete.

r. Give another proof that BV is complete based on the fact that C^* is known to be complete.

s. Show that every $f \in M^*$ has the form

$$f(x) = \int_{[0,1]} x \, d\sigma,$$

where σ is a finitely additive set function on the class of all subsets of $[0, 1]$.

t. Show that σ in Exercise s is of bounded variation and that $\|f\| = V(\sigma)$.

u. Let z be a sequence of real numbers such that $\lim_i x_i = 0$ implies

$$\sum_{i=1}^{\infty} z_i x_i$$

is convergent; then

$$\sum_{i=1}^{\infty} |z_i| < \infty.$$

This does not follow from 46.1. Why?

v. Prove the assertion of Exercise u. *Hint:* z is the pointwise limit on c_0 of the functionals z^n defined by

$$z^n(x) = \sum_{i=1}^{n} z_i x_i.$$

Use 44.1.1.

w. Prove that if z is a sequence of real numbers such that

$$\sum_{i=1}^{\infty} |x_i| < \infty$$

implies

$$\sum_{i=1}^{\infty} x_i z_i$$

is convergent, then z is bounded.

47 HAMEL BASES

A finite set $\{x_1, x_2, \ldots, x_n\}$ in a linear space is called *linearly independent* if

$$\sum_{i=1}^{n} a_i x_i = 0$$

implies $a_i = 0$ ($i = 1, 2, \ldots, n$). More generally, any set H in a linear space is called linearly independent if every finite subset of H is linearly independent in the sense just defined.

If S is any set in a linear space, the *linear hull* of S is the set of all finite linear combinations of elements of S; that is,

$$LH(S) = \left\{ \sum_{i=1}^{n} a_i x_i \mid x_i \in S, a_i \text{ real}, n \text{ a positive integer} \right\}.$$

If H is linearly independent and $z \in LH(H)$, then the expansion of z in terms of elements of H is unique because if

$$\sum_{i=1}^{n} a_i x_i = z = \sum_{j=1}^{m} b_j y_j,$$

then

$$\sum_{i=1}^{n} a_i x_i - \sum_{j=1}^{m} b_j y_j = 0.$$

Now, use linear independence. If $x_i = y_j$, then $a_i - b_j = 0$; otherwise, the a's and b's must themselves be zero.

Let A be a linear space and $H \subset A$; H is a *Hamel basis* for A if H is linearly independent and $LH(H) = A$. Another way of saying this is that every element of A has a unique expansion as a finite linear combination of elements of H.

Theorem 47.1. *Every linear space has a Hamel basis.*

Proof. Let A be a linear space and let $\mathscr{I}$ be the class of all linearly independent subsets of A. The subset relation $\subset$ defines a partial order on $\mathscr{I}$; let $\mathscr{L}$ be a linearly ordered subclass of $\mathscr{I}$ and set

$$J = \bigcup_{L \in \mathscr{L}} L.$$

Let $\{x_1, x_2, \ldots, x_n\} \subset J$; for each i, $x_i \in L_i \in \mathscr{L}$. However, $\mathscr{L}$ is linearly ordered; so for some j, $L_i \subset L_j$ ($i = 1, 2, \ldots, n$). Therefore,

$$\sum_{i=1}^{n} a_i x_i = 0$$

implies $a_i = 0$ ($i = 1, 2, \ldots, n$) because L_j is linearly independent. Thus we have shown that J is linearly independent and so $\mathscr{L}$ has an upper bound in $\mathscr{I}$. Thus, by Zorn's lemma, $\mathscr{I}$ has a maximal element H. Suppose $LH(H) \neq A$ and let $z \in -LH(H)$; this says

$$z - \sum_{i=1}^{n} a_i x_i \neq 0$$

for all finite sets $\{x_1, x_2, \ldots, x_n\} \subset H$. However, it follows from this that $H \cup \{z\}$ is linearly independent, contradicting maximality of H in $\mathscr{I}$. ∎

Suppose H is a Hamel basis for A and let g be any real valued function on H. Extend g to a function f on A by setting

$$f\left(\sum_{i=1}^{n} a_i x_i \right) = \sum_{i=1}^{n} a_i g(x_i)$$

for $x_i \in H$. Thus, f is a linear functional on A. Note that g does not have to satisfy any linearity condition because no point of H is a linear combination of other points of H.

Using this procedure we can construct discontinuous linear functionals. Suppose a Banach space has an infinite Hamel basis H. Let $x_1, x_2, x_3, \ldots$ be a sequence of distinct points of H and set

$$g(x_n) = n\|x_n\|;$$

if there are other points of H, let $g = 0$ there. The linear extension of g to the entire space is discontinuous by 44.2.1. If the basis is finite this construction does not work; and indeed it can be shown that, except for a triviality, linearity implies continuity on finite dimensional spaces. The triviality has to do with the T_1 property. If we have a basis element x with $\|x\| = 0$, we can define a linear f with $f(x) \neq 0$ and this f is clearly discontinuous. However, if the space is T_1 or if we limit our attention to linear functionals that vanish on $\overline{\{\theta\}}$, then the finite dimensional spaces are very well behaved. Some of this theory is indicated in the exercises that follow.

EXERCISES

a. Show that if A has a Hamel basis with n elements, then no set of $n + 1$ elements of A can be linearly independent. *Hint:* Any system

$$\sum_{k=1}^{n+1} a_{ki} b_k = 0 \quad (i = 1, 2, \ldots, n) \tag{1}$$

of n homogeneous linear equations in $n + 1$ unknowns b_k has a solution with not all b_k zero. Show that if

$$y_k = \sum_{i=1}^{n} a_{ki} x_i \quad (k = 1, 2, \ldots, n + 1)$$

and if the b_k are solutions of (1), then

$$\sum_{k=1}^{n+1} b_k y_k = 0.$$

b. Show that if A has a finite Hamel basis, then every Hamel basis for A has the same number of elements. (This number is called the *dimension* of A.)

c. Show that a linear functional on any linear space A is uniquely determined by its values on a Hamel basis for A.

d. Show that if A and B are linear spaces, each of dimension n, and $B \subset A$, then $B = A$. [*Hint:* If H is a Hamel basis for B and $x \notin B$, then $H \cup \{x\}$ is linearly independent.]

e. Let A be a linear space and let A' be the set of all linear functionals on A. (No norms here; so no considerations of continuity.) Show that if A has dimension n, then A'

has dimension n. *Hint:* Let $x_1, x_2, \ldots, x_n$ be a Hamel basis in A; set

$$f_i(x_j) = \begin{cases} 1 & \text{for } i = j, \\ 0 & \text{for } i \neq j; \end{cases}$$

extend the f_i to linear functionals on A and show that these form a Hamel basis in A'.

f. Show that if A has dimension n and if there is a one-to-one linear function from A onto B, then B has dimension n.

g. Show that if A is a finite dimensional Banach space, then $\dim(A^*) \leq \dim(A)$.

h. Show that every finite dimensional T_1-Banach space is reflexive. [*Hint:* By Exercise g, $\dim(A^{**}) \leq \dim(A)$; by Exercise 45.$f$ the natural embedding is one-to-one; use Exercises f and d.]

i. Show that on a finite dimensional T_1-Banach space every linear functional is continuous. [*Hint:* Show $\dim(A^*) = \dim(A')$.]

j. Show that in any T_1-Banach space every finite dimensional subspace is closed. [*Hint:* See Exercise h.]

k. Prove that a Hamel basis for a T_1-Banach space is either finite or noncountable. [*Hint:* Use Exercise j to show that finite dimensional proper subspaces are nowhere dense; apply the Baire category theorem.]

l. Let $E \subset c_0$ be the set of real sequences x such that $x_i \neq 0$ for only a finite number of integers i. Show that E is a linear subspace of c_0 and has a countably infinite Hamel basis.

m. Let A be any linear space and let E be any linearly independent set in A. Prove that there is a Hamel basis H for A such that $E \subset H$. [*Hint:* Modify the proof of 47.1, letting $\mathscr{I}$ be the class of linearly independent sets that contain E.]

n. For $0 < t < 1$ define $x^t \in l_2$ by setting $x_i^t = t^i$. Show that the set $\{x^t \mid 0 < t < 1\}$ is linearly independent. [*Hint:* A polynomial function on $(0, 1)$ is zero only if its coefficients are.]

o. Show that l_2 and c_0 each has a Hamel basis with the cardinal of the continuum. [*Hint:* Note Exercise n; $l_2 \subset c_0 \subset s$ and s has the cardinal of the continuum.]

p. Show that there is a linear isomorphism between l_2 and c_0. [*Hint:* Take the one-to-one correspondence (Exercise o) between their Hamel bases and extend it linearly.]

48 WEAK AND WEAK* SEQUENTIAL CONVERGENCE

A sequence x in a Banach space A *converges weakly* to $x_0 \in A$ if

$$\lim_n f(x_n) = f(x_0)$$

for each $f \in A^*$. A sequence f in A^* *converges weakly** to $f_0 \in A^*$ if

$$\lim_n f_n(x) = f_0(x)$$

for every $x \in A$.

For sequences f in A^* both these concepts are defined:

Weak convergence: $\lim_n F(f_n) = F(f_0)$ for every $F \in A^{**}$.

Weak* convergence: $\lim_n \hat{x}(f_n) = \hat{x}(f_0)$ for every $x \in A$.

When the definitions are stated this way it is clear that in any conjugate space weak convergence implies weak* convergence and that in a reflexive space the notions are equivalent.

Theorem 48.1. *In order that a sequence x in a Banach space A converges weakly to $x_0 \in A$, the following two conditions are necessary and sufficient:*

48.1.1 *There is a number M such that $\|x_n\| \leq M$ for all n.*

48.1.2 $\lim_n f(x_n) = f(x_0)$ *for every $f \in D$ where $LH(D)$ is dense in A^*.*

Proof. Necessity of 48.1.2 is obvious; A^* is dense in itself. Given weak convergence we have that

$$\{\hat{x}_n(f) \mid n = 0, 1, 2, \ldots\} = \{f(x_n) \mid n = 0, 1, 2, \ldots\}$$

is bounded for each $f \in A^*$; so by the Banach-Steinhaus theorem (44.2)

$$\{\|\hat{x}_n\| \mid n = 0, 1, 2, \ldots\}$$

is bounded, and 48.1.1 follows because $\|x_n\| = \|\hat{x}_n\|$.

Conversely, suppose we are given 48.1.1 and 48.1.2. Let $g \in A^*$ and let $\varepsilon > 0$. Since $LH(D)$ is dense in A^* there is an $f \in LH(D)$ such that

$$\|f - g\| < \frac{\varepsilon}{3M}.$$

Now, f is a finite linear combination of functionals from D. The sequence x converges with respect to each of these and therefore with respect to f. Thus there is an n_0 such that

$$|f(x_n) - f(x_0)| < \frac{\varepsilon}{3}$$

for all $n > n_0$. Thus, for $n > n_0$,

$$|g(x_n) - g(x_0)| \leq |g(x_n) - f(x_n)| + |f(x_n) - f(x_0)| + |f(x_0) - g(x_0)|$$

$$\leq \|g - f\| \|x_n\| + |f(x_n) - f(x_0)| + \|f - g\| \|x_0\|$$

$$\leq \|g - f\| M + |f(x_n) - f(x_0)| + \|f - g\| M$$

$$< \frac{\varepsilon}{3M} M + \frac{\varepsilon}{3} + \frac{\varepsilon}{3M} M = \varepsilon. \quad \blacksquare$$

Theorem 48.2. *In order that a sequence f in a conjugate space A^* converges weakly* to $f_0 \in A^*$ the following two conditions are necessary and sufficient:*

48.2.1 *There is a number M such that $\|f_n\| \leq M$ for all n.*

48.2.2 $\lim_n f_n(x) = f_0(x)$ *for every $x \in D$ where $LH(D)$ is dense in A.*

Proof. Same as 48.1 with points and functionals interchanged. ∎

A set $T \subset A$ is called *total* over A^* if $f(x) = 0$ for every $x \in T$ implies $f = 0$.

Theorem 48.3. *In order that a sequence f in a conjugate space A^* converges weakly* to $f_0 \in A^*$ the following two conditions are necessary and sufficient:*

48.3.1 *There is a number M such that $\|f_n\| \leq M$ for all n.*

48.3.2 $\lim_n f_n(x) = f_0(x)$ *for every $x \in T$ where T is total over A^*.*

Proof. The only point that requires proof is that 48.3.2 implies 48.2.2. Specifically, if T is total over A^*, then $LH(T)$ is dense in A. This follows by 45.3; if there is a point of A a positive distance from $LH(T)$, then there is a nonzero functional which vanishes on $LH(T)$. ∎

For weak convergence there is no parallel to 48.3. To prove such a parallel, we would need a dual to 45.3 which stated that if S is a linear subspace of A^* and g is a positive distance from S, then there is an $x \in A$ such that $f(x) = 0$ for all $f \in S$ and $g(x) \neq 0$. This assertion is not true; see Exercise *o* below. To put it another way, the Hahn-Banach theorem yields functionals that separate closed linear subspaces of A from outside points, but there is no dual theorem yielding points of A that separate closed linear subspaces of A^* from outside functionals. Of course, a closed linear subspace of A^* is separated from an outside functional by a functional $F \in A^{**}$ (45.3 again), but the F involved may not be an $\hat{x}$ for any $x \in A$.

Another interesting consequence of 45.3 is the following.

Theorem 48.4. *If $\lim_n x_n = x_0$ weakly, then there exists a sequence y of linear combinations of the x_n such that $\lim_n \|y_n - x_0\| = 0$.*

Proof. Let X be the range of the sequence x. If there is no sequence y as specified, then x_0 is a positive distance from $LH(X)$ and by 45.3 there is a functional f vanishing on $LH(X)$—hence with $f(x_n) = 0$ for each n—but with $f(x_0) \neq 0$. This contradicts the hypothesis that $\lim_n x_n = x_0$ weakly. ∎

Corollary 48.4.1. *If B is a closed linear subspace of a Banach space, then B is weakly sequentially closed (if $x_n \in B$ and $\lim_n x_n = x_0$, then $x_0 \in B$).*

Note: There is no dual to 48.4.1 in which weakly is replaced by weakly*. See Exercise *n* below.

EXERCISES

a. Prove that the following conditions are necessary and sufficient that $\lim_n x^n = y$ weakly in c_0. (1) There exists M such that $|x_i^n| \leq M$ for all n and all i. (2) $\lim_n x_i^n = y_i$ for each i. *Hint:* Define $z^j \in l_1$ by
$$z_i^j = \begin{cases} 1 & \text{for } j = i, \\ 0 & \text{for } j \neq i. \end{cases}$$
Linear combinations of the z^j form a dense set in l_1.

b. Prove that the following conditions are necessary and sufficient that $\lim_n x^n = y$ weakly in c. (1) There exists M such that $|x_i^n| \leq M$ for all n and all i. (2) $\lim_n x_i^n = y_i$ for each i. (3) $\lim_n \lim_i x_i^n = \lim_i y_i$.

c. Prove that the following conditions are necessary and sufficient that $\lim_n x_n = x_0$ weakly in L_1. (1) There exists M such that
$$\int_{[0,1]} |x_n| \leq M$$
for all n. (2) For every measurable $E \subset [0, 1]$,
$$\lim_n \int_E x_n = \int_E x_0.$$
[Hint: Every essentially bounded function can be approximated essentially uniformly by simple functions, which is to say that the linear hull of the characteristic functions is dense in L_∞.*]*

d. Compare Exercise c and Theorem 40.8. What is the difference between weak and strong convergence in L_1?

e. Prove that the following conditions are necessary and sufficient that $\lim_n x^n = y$ weakly in l_1. (1) There exists M such that
$$\sum_{i=1}^\infty |x_i^n| \leq M$$
for all n. (2) For every set E of positive integers
$$\lim_n \sum_{i \in E} x_i^n = \sum_{i \in E} y_i.$$

f. Let Ω be the set of positive integers and introduce a measure by setting $\mu(\{n\}) = 1/2^n$. Map each $x \in l_1$ onto a function z on Ω by setting $z_i = 2^i x_i$. Define $\|z\| = \int_\Omega |z|\, d\mu$. Show that the mapping described here is linear, one-to-one and norm-preserving from l_1 onto the L_1 space determined by Ω and μ.

g. Apply Exercises e and f and Theorem 40.8 to show that weak and strong convergence are equivalent in l_1. *[Hint:* In Exercise f, $\mu(\Omega) < \infty$; so pointwise convergence implies convergence in measure.*]*

h. Prove that in L_p ($1 < p \leq \infty$) necessary and sufficient conditions that $\lim_n x_n = x_0$ weakly* are: (1) there exists M such that $\|x_n\| \leq M$ for all n; (2) for every measurable $E \subset [0, 1]$,
$$\lim_n \int_E x_n = \int_E x_0.$$

i. Prove that if $p < \infty$ in Exercise h, the given conditions are equivalent to weak convergence.

j. Prove that in l_p ($1 < p < \infty$) necessary and sufficient conditions that $\lim_n x^n = y$ weakly* are: (1) there exists M such that
$$\sum_{i=1}^{\infty} |x_i^n|^p \leq M$$
for all n; (2) for each i
$$\lim_n x_i^n = y_i.$$
These conditions are also necessary and sufficient for weak convergence. Why?

k. Weak* convergence in l_1 is ambiguous because l_1 is equivalent to both c_0^* and c^*. In either case, the existence of M such that
$$\sum_{i=1}^{\infty} |x_i^n| \leq M$$
for all n is necessary. Show that this, together with the following, is sufficient that $\lim_n x^n = y$ weakly* in l_1.

Over c_0: $\lim_n x_i^n = y_i$ for each i.

Over c: $\lim_n x_i^n = y_i$ for each $i > 1$ and $\lim_n \sum_{i=1}^{\infty} x_i^n = \sum_{i=1}^{\infty} y_i$.

l. Let x be the sequence in l_1 defined by
$$x_i^n = \begin{cases} 1 & \text{if } i = n, \\ 0 & \text{if } i \neq n. \end{cases}$$
Show that in c_0^*, $\lim_n x^n = 0$ weakly* but not weakly.

m. Let x be as in Exercise l. Show that in c^*, $\lim_n x^n = x^1$ weakly*.

n. With x as in Exercises l and m, let $X \subset c^*$ be the linear hull of $\{x_2, x_3, x_4, \ldots\}$. Show that X is strongly closed (equivalent to l_1) but not weakly* closed (Exercise m).

o. Show that the subspace $X \subset c^*$ of Exercise n is total over c but that $\rho(x^1, X) = 1$. (Note: This vetoes the idea of a dual to 45.3.)

p. Prove that the following conditions are necessary and sufficient that $\lim_n x^n = y$ weakly* in m. (1) There exists M such that $|x_i^n| \leq M$ for all i and all n. (2) $\lim_n x_i^n = y_i$ for each i.

q. Prove that the following conditions are necessary and sufficient that $\lim_n x_n = x_0$ weakly in C. (1) There exists M such that $|x_n(t)| \leq M$ for all n and all $t \in [0, 1]$. (2) $\lim_n x_n(t) = x_0(t)$ for each $t \in (0, 1]$. [Hint: The conditions are uniform boundedness and pointwise convergence; prove convergence of the Riemann-Stieltjes integrals directly as in the proof of the Lebesgue dominated convergence theorem (28.1).]

r. Define functions $z_u \in BV$ on $[0, 1]$ by
$$z_u(t) = \begin{cases} 0 & \text{for } 0 < t < u, \\ 1 & \text{for } u \leq t < 1. \end{cases}$$

Show that condition (2) of Exercise q is

$$\lim_n \int_0^1 x_n \, dz_u = \int_0^1 x_0 \, dz_u$$

for every $u \in (0, 1]$. Let $U = \{z_u \mid 0 < u \le 1\}$; show that U is total over C but that $LH(U)$ is not dense in BV. [*Hint:* Let $w(t) = t$ and let y be any step function; show that $V(w - y) \ge 1$.]

s. Prove that if x is a uniformly bounded, pointwise convergent sequence of continuous functions on $[0, 1]$, then there is a sequence of linear combinations of the x's that is uniformly convergent.

t. Use the Banach-Steinhaus theorem (or the theorems of Moore and Osgood) to show that any conjugate space is weakly* sequentially complete.

u. Prove that L_1 is weakly sequentially complete. [*Hint:* Show that if x is a weak Cauchy sequence in L_1, then $\sigma(E) = \lim_n \int_E x_n$ for every measurable $E \subset [0, 1]$ defines a completely additive, absolutely continuous function σ; use the Radon-Nikodym theorem to get $\sigma(E) = \int_E x_0$ and show that x_0 is the desired weak limit of x.]

v. Show that neither c_0, c, nor C is weakly sequentially complete.

49 WEAK* TOPOLOGIES

Let A be a Banach space and let R be the real number system; for each $x \in A$, let $R_x = R$. Then

$$R^A = \underset{x \in A}{\times} R_x$$

is the set of all real valued functions on A, and the product topology in R^A is the topology of pointwise convergence on A. Now, A^* is the subset of R^A consisting of the continuous linear functions and in A^* pointwise convergence on A is what we have called weak* convergence. Thus in the language of general topology the weak* topology in A^* would be described as that induced on A^* by the product topology in R^A.

It is possible that a product topology be generated by a metric. For example, if I is the set of positive integers, then $s = R^I$ with the product topology; yet this topology of pointwise convergence on I is generated by a metric:

$$\rho(x, y) = \sum_{i=1}^\infty \frac{|x_i - y_i|}{2^i(1 + |x_i - y_i|)}.$$

In a metric space the closure of a set E is obtained by appending to E the set of all limits of convergent sequences from E, and the closure of $\bar{E}$ is just $\bar{E}$. Much of the impetus for the development of the theory of product topologies came from the discovery that "closing" a subspace of A^* by appending the weak* sequential limits does not necessarily yield a "closed" set.

Example 49.1. There is a linear subspace E of c_0^* such that if

$$E_s = \left\{ v \mid v = \lim_n z_n \text{ weakly*}; z_n \in E \right\},$$

then there is a functional $w \in c_0^* - E_s$ with

$$w = \lim_n v_n \text{ weakly*}$$

where each $v_n \in E_s$.

Proof. The set of all ordered pairs (i, k) of positive integers is countable, which is to say that there is a one-to-one function from $I \times I$ onto I. Let N be such a function; then clearly

$$\lim_i N(i, k) = \lim_k N(i, k) = \infty. \qquad (1)$$

For each (i, k) we define $z^{(i,k)} \in l_1 = c_0^*$ by setting

$$z_{2n-1}^{(i,k)} = \begin{cases} 1/2^n & \text{for } n \leq i, \\ 0 & \text{for } n > i, \end{cases}$$

$$z_{2n}^{(i,k)} = \begin{cases} i & \text{for } n = N(i, k), \\ 0 & \text{for } n \neq N(i, k). \end{cases}$$

Let E be the linear hull of the set of all $z^{(i,k)}$. Now define $z^{(i)} \in l_1$ by setting

$$z_{2n-1}^{(i)} = \begin{cases} 1/2^n & \text{for } n \leq i, \\ 0 & \text{for } n > i, \end{cases}$$

$$z_{2n}^{(i)} = 0 \qquad \text{for all } n.$$

Since weak* convergence in l_1 is pointwise convergence, it follows by (1) that for each i

$$\lim_k z^{(i,k)} = z^{(i)} \text{ weakly*};$$

thus $z^{(i)} \in E_s$. Again looking at pointwise limits, we see that $\lim_i z^{(i)} = w$ weakly* where

$$w_{2n-1} = \frac{1}{2^n} \quad \text{for all } n,$$

$$w_{2n} = 0 \quad \text{for all } n.$$

Clearly, $w \in l_1$; we complete the proof by showing that $w \notin E_s$. Suppose

$$w = \lim_j u^{(j)} \text{ weakly*} \qquad (2)$$

where $u^{(j)} \in E$. Each $u^{(j)}$ is a linear combination of the $z^{(i,k)}$, and we write

$$u^{(j)} = \sum_{i=1}^{\infty} \sum_{k=1}^{\infty} a_{i,k}^{(j)} z^{(i,k)}$$

with the understanding that all but a finite number of the coefficients $a_{i,k}^{(j)}$ are zero. From the definition of the $z^{(i,k)}$ we have

$$u_{2n-1}^{(j)} = \frac{1}{2^n} \sum_{i=n}^{\infty} \sum_{k=1}^{\infty} a_{i,k}^{(j)}, \qquad (3)$$

$$u_{2n}^{(j)} = i\, a_{i,k}^{(j)} \text{ where } n = N(i,k). \qquad (4)$$

So, from (2) and (3) and the definition of w, we must have

$$\lim_j \sum_{i=n}^{\infty} \sum_{k=1}^{\infty} a_{i,k}^{(j)} = 1 \qquad (5)$$

for every n. However, by (4)

$$\|u^{(j)}\| = \sum_{n=1}^{\infty} |u_n^{(j)}| \geq \sum_{n=1}^{\infty} |u_{2n}^{(j)}| = \sum_{i=1}^{\infty} i \sum_{k=1}^{\infty} |a_{i,k}^{(j)}| \geq n \sum_{i=n}^{\infty} \sum_{h=1}^{\infty} a_{i,k}^{(j)}$$

for every n; so by (5)

$$\overline{\lim_j} \|u^{(j)}\| \geq n$$

for every n, and u cannot be weakly* convergent because it violates 48.2.1. ∎

In the light of this example sequences must be replaced by nets in the general study of weak* topology. We give here only a few sample theorems from such a study.

Theorem 49.2. *If A is an infinite-dimensional, T_1-Banach space, then A^* is not weakly* complete.*

Proof. Let H be a Hamel basis for A and let $\mathscr{E}$ be the class of finite subsets of H, partially ordered by inclusion. Then $\mathscr{E}$ is also a directed set because if E_1 and E_2 are in $\mathscr{E}$, then $E_1 \cup E_2 \in \mathscr{E}$ and is greater than either E_1 or E_2. Since H is infinite, we can define a function on g_0 on H such that $g_0(x)/\|x\|$ is unbounded, and g_0 then has a linear extension g to all of A; by 44.2.1, $g \notin A^*$. For $E \in \mathscr{E}$, we define $f_E \in A^*$ by setting $f_E = g$ on $LH(E)$ and extending it (Hahn-Banach theorem) to a continuous linear functional on A. For each $x \in A$ there is an $E_0 \in \mathscr{E}$ such that $x \in E_0$; thus if $E \in \mathscr{E}$ and $E \supset E_0$, $f_E(x) = g(x)$. Therefore, $\lim_E f_E = g$ weakly* but $g \neq A^*$. ∎

Contrasting 49.2 with 44.1.1 shows that A^* loses something when we shift from sequences to nets. However, the general theory of product topologies yields one very interesting positive result for A^*.

Theorem 49.3 (Alaoglu). *Let A be any Banach space and let*

$$D = \{f \mid f \in A^*, \|f\| \leq 1\};$$

then D is weakly compact.*

Proof. For each $x \in A$ let K_x be the closed interval in the real number system

$$K_x = [-\|x\|, \|x\|].$$

Let

$$K = \underset{x \in A}{\times} K_x;$$

then by Tychonoff's theorem, K is compact in the product topology of R^A. Thus, the theorem is proved if we show that D is closed in K. Let f be a net in D; for each $x \in A$, $f(x)$ is a net in R with $|f_\alpha(x)| \leq \|x\|$; so if

$$\lim_\alpha f_\alpha = f_0 \text{ weakly*},$$

then $|f_0(x)| \leq \|x\|$ for every $x \in A$. So $f_0 \in D$ provided f_0 is linear, but

$$f_0(ax + by) = \lim_\alpha f_\alpha(ax+by) = a \lim_\alpha f_\alpha(x) + b \lim_\alpha f_\alpha(y) = af_0(x) + bf_0(y). \quad \blacksquare$$

We conclude this brief introduction to the theory of weak* topology by abandoning the idea of normed linear space altogether and letting the weak* scheme generate the only concept of limit in the structure. In this case every linear space becomes "reflexive" in a sense that we shall describe herewith.

Let A be a linear space. If we assume A has no norm, then continuity of functionals is not (as yet) defined; so A^* is meaningless. However, we have defined A' as the set of all linear functionals on A, and this depends only on the linear structure in A. Now at this stage two ideas are definable:

i) The natural embedding from A to A''. This is the mapping H such that $H(x) = \hat{x}$ where $\hat{x}(f) = f(x)$ for all $f \in A'$. That is, H is from A into $R^{A'}$, but it is readily verified that $\hat{x}$ is linear on A'; so H is from A into A''.

ii) Weak* convergence in A'. A net f in A' converges weakly* to a function f_0 on A provided $\lim_\alpha f_\alpha(x) = f_0(x)$ for every $x \in A$. If this is the case, it is readily verified that $f_0 \in A'$ (f_0 is linear on A).

Now the reflexivity idea we referred to is the following. Given A and A' as above, let $(A')^w$ be the subspace of A'' consisting of those linear functionals on A' that are weakly* continuous; then the natural embedding H is from A onto $(A')^w$.

The first step in the proof of this is of some interest in itself in that it offers a partial dual to the Hahn-Banach theorem. As shown in Section 45, the Hahn-Banach theorem leads to the result that given a closed linear subspace of A and a point outside the subspace, there is a functional in A^* which is zero on the subspace and not zero on the outside point. The complete dual is not true (Section 48); there are closed linear subspaces in A^* and outside functionals such that for no point in A is everything in the subspace zero and the outside functional not zero. However, we have a positive result if the subspace of A^* is finite dimensional.

Theorem 49.4. Let $g_1, g_2, \ldots, g_n$ be a finite set of linear functionals on a linear space A. Let $f \in A'$ be such that $g_i(x) = 0$ ($i = 1, 2, \ldots, n$) implies $f(x) = 0$. Then f is a linear combination of the g_i.

Proof. Let G be the function from A into Euclidean n-space defined by

$$G(x) = \{g_1(x), g_2(x), \ldots, g_n(x)\}.$$

By hypothesis, $G(x) = G(y)$ implies $f(x) = f(y)$; so the condition

$$H \circ G = f$$

defines a function H from Euclidean n-space into the real number system. However, H is linear because f and G are; so H assumes the form

$$H(u_1, u_2, \ldots, u_n) = \sum_{i=1}^{n} a_i u_i,$$

which is to say that

$$f = \sum_{i=1}^{n} a_i g_i. \quad \blacksquare$$

Corollary 49.4.1. Let A be a linear space and let E be a finite-dimensional linear subspace of A'. If $f \in A' - E$, then there is an $x \in A$ such that $g(x) = 0$ for all $g \in E$ but $f(x) \neq 0$.

Proof. This is just the contrapositive of 49.4. $\blacksquare$

The real weakness of 49.4.1 as compared to the Hahn-Banach theorem is not that 49.4.1 is restricted to finite-dimensional subspaces but that in the case of normed spaces it gives no clue to $\|x\|$, given $\rho(f, E)$; contrast the situation in 45.3.

Theorem 49.5. Let A be any linear space; then $F \in A''$ is weakly* continuous if and only if $F = \hat{x}$ for some $x \in A$.

Proof. Suppose $F = \hat{x}$; if $f_\alpha \to 0$ weakly*, then $F(f_\alpha) = f_\alpha(x) \to 0$; so F is weakly* continuous. Conversely, suppose F is weakly* continuous; then there is a weak* neighborhood N of 0 such that $f \in N$ implies $|F(f)| < 1$. However (see Section 8), the nature of weak* (product topology) neighborhoods is such that $f \in N$ means

$$|f(x_i)| < \varepsilon_i \ (i = 1, 2, \ldots, n)$$

for some finite set $\{x_1, x_2, \ldots, x_n\} \subset A$ and some set $\{\varepsilon_1, \varepsilon_2, \ldots, \varepsilon_n\}$ of real numbers. Now

$$|f(x_i)| < \frac{\varepsilon_i}{k} \ (i = 1, 2, \ldots, n)$$

implies

$$|kf(x_i)| < \varepsilon_i \ (i = 1, 2, \ldots, n)$$

implies
$$|kF(f)| = |F(kf)| < 1$$
implies
$$|F(f)| < \frac{1}{k}.$$
Therefore,
$$\hat{x}_i(f) = f(x_i) = 0 \ (i = 1, 2, \ldots, n)$$
implies
$$F(f) = 0;$$
and by 49.4 this implies that
$$F = \sum_{i=1}^{n} a_i \hat{x}_i. \quad \blacksquare$$

EXERCISES

a. Let A be an infinite-dimensional T_1-Banach space. Show that for each finite-dimensional subspace E of A there is an $f_E \in A^*$ such that $f_E(x) = 0$ for $x \in E$ and $\|f_E\| = \dim(E)$.

b. Show that in Exercise a the finite-dimensional subspaces of A form a directed set under inclusion and that $\lim_E f_E = 0$ weakly* but $\lim_E \|f_E\| = \infty$.

c. Norm boundedness of weakly* convergent sequences is provable (see proof of 48.1) from the Banach-Steinhaus theorem. However, the Banach-Steinhaus theorem is not restricted to countable families of functions. Why does this not preclude the situation in Exercise b?

d. Prove a dual to Exercise a: If A is T_1 and infinite-dimensional and E is a finite dimensional subspace of A^* there is an $x_E \in A$ such that $f(x_E) = 0$ for every $f \in E$ and $\|x_E\| = \dim(E)$. [Hint: 49.4.1 will do it.]

e. Prove a dual to Exercise b: If A is T_1 and infinite-dimensional, there is a net x in A such that $x_\alpha \to 0$ weakly but $\|x_\alpha\| \to \infty$.

f. Let E be a linear subspace of A' and let $f \in A'$ be such that for every finite-dimensional linear subspace X of A there exists $g \in E$ such that $g = f$ on X. Prove that f is in the weak* closure of E.

g. Let E be a linear subspace of A' which is total over A; let $\{x_1, x_2, \ldots, x_n\}$ be a linearly independent set in A; let $\{a_1, a_2, \ldots, a_n\}$ be any set of n numbers. Show that there exist $g_i \in E \ (i = 1, 2, \ldots, n)$ such that
$$g_i(x_j) = \begin{cases} a_i & \text{for } i = j, \\ 0 & \text{for } i \neq j. \end{cases}$$
[Hint: Show that $\{\hat{x}_1, \hat{x}_2, \ldots, \hat{x}_n\}$ is a linearly independent set of functionals on E because E is total over A; so $\hat{x}_1 \notin LH\{\hat{x}_2, \ldots, \hat{x}_n\}$ and g_1 is given by 49.4.1.]

h. Prove that if E is a linear subspace of A' which is total over A, then E is weakly* dense in A'. Hint: Let X be a finite-dimensional subspace of A with a Hamel basis $\{x_1, x_2, \ldots, x_n\}$ and let $f \in A'$. Let the a_i in Exercise g be $f(x_i)$; obtain the g_i and let their sum be the g of Exercise f.

i. Prove that for any Banach space A, $\{\hat{x} \mid x \in A\}$ is weakly* dense in A^{**}.

j. Let A be a linear space and let E be a linear subspace of A'; then E has a weak* topology; let E^w be the space of weakly* continuous functionals on E. Show that $H(x)(f) = f(x)$ for every $f \in E$ defines a linear function H from A onto E^w. [*Hint:* If $E = A$, this is 49.5; show that the proof of 49.5 still applies.]

k. Show that if, in Exercise j, E is total over A, then H is one-to-one.

l. Using the notation of Exercise j, show that for any linear space A, $(A')^w$ is total over A', hence weakly* dense in A''. [*Hint:* Use 49.5.]

m. Let A be an arbitrary linear space (no topology); then A' has a weak* topology and so $(A')^w$ is definable. Beginning with A', let "conjugate space" mean space of weakly* continuous linear functionals. Show that A', $(A')^w$, $((A')^w)^w$, etc. are all reflexive in that the "natural" map of each of these spaces is one-to-one onto its second conjugate.

n. Show that in Exercise m, "natural" map may mean $\hat{x}(f) = f(x)$ for every $f \in E$ where (if the map is from B to $(B^w)^w$) E may be any linear subspace of B' such that $B^w \subset E \subset B'$.

o. Given f linear and weakly* continuous over $(A')^w$, show that there is a unique linear and weakly* continuous extension g of f to A''. *Hint:* By Exercise n there is a one-to-one map from A' onto each of the spaces $((A')^w)^w$ and $((A')')^w$. Alternate proof: Use Exercise l.

50 THE CLOSED-GRAPH THEOREM

If f is a function from the reals to the reals, elementary analytic geometry says that the graph of f is the subset of the plane consisting of all points (x, y) such that $y = f(x)$. The modern definition of a function would say that f is this set of ordered pairs, thus making the word graph obsolete. In any case, let A and B be metric spaces and f a function from A to B. Now $A \times B$ has an induced metric and so we can look at topological properties of

$$\{(x, f(x)) \mid x \in A\} \tag{1}$$

and it does not make much difference whether we call this set f or the graph of f. If the set (1) is closed in $A \times B$, we say that f has a closed graph. Suppose the graph of f is closed, $x_n \to x_0$, and $f(x_n) \to y_0$; then we must have (x_0, y_0) in the graph, which is to say $y_0 = f(x_0)$. Thus, it is clear that a continuous function from one metric space to another has a closed graph. It is easily seen that the converse is not true. Define f on the real number system by

$$f(x) = \begin{cases} 1/x & \text{for } x \neq 0, \\ 0 & \text{for } x = 0; \end{cases}$$

the graph of f is closed but f is not continuous. Roughly speaking, the closed-graph theorem asserts that for linear functions a closed graph implies continuity. For a precise statement we need to look a little more critically at the structure of the spaces involved.

We could prove a closed-graph theorem for linear functions from one Banach space to another, but this would exclude an important application. Instead, we want to weaken the postulate system for norms by replacing N–IV by weaker conditions. A functional p on a linear space A is called a *paranorm* provided it satisfies the following postulates:

P–I. $p(\theta) = 0$;

P–II. $p(x) \geq 0$ for all $x \in A$;

P–III. $p(x + y) \leq p(x) + p(y)$ for all $x, y \in A$;

P–IV. $p(-x) = p(x)$ for all $x \in A$;

P–V. If $t_n \to t$ and $p(x_n - y) \to 0$, then $p(t_n x_n - ty) \to 0$;

It is easily verified that the norm postulate N–IV implies P–IV and P–V; so a norm is a paranorm. There are two prime examples that show that the converse is not true, and we are changing to paranorms in this section just to admit these two examples.

Example 1. The space s consisting of all sequences of real numbers. Define

$$p(x) = \sum_{i=1}^{\infty} \frac{|x_i|}{2^i(1 + |x_i|)};$$

then p is a paranorm but not a norm.

Example 2. The space S consisting of all measurable functions on $[0, 1]$. Define

$$p(x) = \int_{[0, 1]} \frac{|x|}{1 + |x|};$$

again p is a paranorm but not a norm.

A paranorm p generates a pseudo-metric ρ if we set

$$\rho(x, y) = p(x - y),$$

and we still have the feature that a linear function is continuous if and only if it is continuous at θ. (See the proof of 44.1; N–IV is not used.) The T_1 property is still an issue and, as in Section 44, we are going to proceed without it. We will define a *Fréchet space* to be a complete paranormed linear space and note in passing that standard usage reserves this title for the cases in which $p(x) = 0$ implies $x = \theta$.

At first this seems to be asking for trouble because if the range space is not T_1 we lose a lot of closed graphs. If $x_n \to x_0$, $f(x_n) \to y_0$ and $f(x_n) \to z_0$ with $z_0 \neq y_0$, then the graph of f is not closed because (x_0, y_0) and (x_0, z_0) are both accumulation points of the graph but by the definition of a function they cannot both be in it.

There is a natural generalization of the notion of graph that circumvents this difficulty and precisely fits the cases we want to study. Let f be a function from a Fréchet space A to a Fréchet space B; the *augmented graph* of f is the subset of $A \times B$ denoted by $G(f)$ and defined by

$$G(f) = \{(x, y) \mid p[y - f(x)] = 0, x \in A\}.$$

Theorem 50.1. *If f is a continuous function from one Fréchet space to another, then $G(f)$ is closed.*

Proof. Let $p(x_n - x_0) \to 0$; by continuity $p[f(x_n) - f(x_0)] \to 0$. Thus, if $p(y_n - y_0) \to 0$, where $(x_n, y_n) \in G(f)$, we have

$$p[f(x_0) - y_0] \leq p[f(x_0) - f(x_n)] + p[f(x_n) - y_n] + p(y_n - y_0)$$
$$= p[f(x_0) - f(x_n)] + p(y_n - y_0) \to 0,$$

and it follows that $(x_0, y_0) \in G(f)$. ∎

Let p_1 and p_2 be two paranorms on A; p_1 is *stronger* than p_2 if $p_1(x_n) \to 0$ implies $p_2(x_n) \to 0$. Paranorms p_1 and p_2 are *equivalent* if each is stronger than the other.

Theorem 50.2. *Let A have paranorms p_1 and p_2 with p_1 stronger than p_2 and let B have paranorms q_1 and q_2 with q_1 stronger than q_2. If f is continuous from (A, p_2) to (B, q_1), then f is continuous from (A, p_1) to (B, q_2).*

Proof. $p_1(x_n - x_0) \to 0$ implies $p_2(x_n - x_0) \to 0$ implies $q_1[f(x_n) - f(x_0)] \to 0$ implies $q_2[f(x_n) - f(x_0)] \to 0$. The second implication comes from the given continuity of f; the other two come from the given relative strengths of the paranorms. ∎

As can be seen from the proof, 50.2 is a relatively trivial result; if we strengthen convergence in the domain and/or weaken it in the range, we preserve continuity. In general, a change in the wrong direction may destroy continuity. However, the next theorem shows that in situations where continuity is equivalent to closed graphs, one type of "wrong way" change in topology is allowable.

Theorem 50.3. *Let A have paranorms p_1 and p_2 with p_1 stronger than p_2; let B have paranorms q_1 and q_2 with q_1 stronger than q_2; let $\overline{\{\theta\}}$ be the same in (B, q_1) and (B, q_2). If $G(f)$ is closed in $(A, p_2) \times (B, q_2)$, then $G(f)$ is closed in $(A, p_1) \times (B, q_1)$.*

Proof. Since $\overline{\{\theta\}}$ is the same in (B, q_1) and (B, q_2), $G(f)$ is the same subset of $A \times B$ for each choice of paranorms. Suppose $(x_n, y_n) \in G(f)$ with $p_1(x_n - x_0) \to 0$ and $q_1(y_n - y_0) \to 0$; then $p_2(x_n - x_0) \to 0$ and $q_2(y_n - y_0) \to 0$, and it follows that $(x_0, y_0) \in G(f)$. ∎

Theorem 50.4. *Let A have paranorms p_1 and p_2 with p_1 stronger than p_2. If (A, p_1) and (A, p_2) are both complete and if $\overline{\{\theta\}}$ is the same in both spaces, then p_1 and p_2 are equivalent.*

Proof. As a notational device for this proof only, for $E \subset A$ let $E^{(i)}$ denote the p_i-closure of E $(i = 1, 2)$. Let N be any p_1-neighborhood of θ and set $kN = \{kx \mid x \in N\}$. If

$$y \in A - \bigcup_{k=1}^{\infty} kN,$$

then $y/k \notin N$ for all k, contradicting P–V; so

$$\bigcup_{k=1}^{\infty} kN = A.$$

Now (A, p_2) is complete; so by the Baire category theorem not all kN can be nowhere dense with respect to p_2. Thus, for some k_0, $(k_0 N)^{(2)}$ contains a p_2-neighborhood $M = \{z \mid p_2(y - z) < \varepsilon\}$. Next, note that

$$k_0 p_2\left(\frac{y-z}{k_0}\right) = p_2\left(\frac{y-z}{k_0}\right) + \cdots + p_2\left(\frac{y-z}{k_0}\right) \geq p_2(y - z);$$

so

$$p_2\left(\frac{y-z}{k_0}\right) < \frac{\varepsilon}{k_0} \text{ implies } p_2(y - z) < \varepsilon$$

and it follows that

$$M_0 = \left\{w \mid p_2\left(\frac{y}{k_0} - w\right) < \frac{\varepsilon}{k_0}\right\} \subset N^{(2)}.$$

By P–IV (applied to p_1) it follows that $-M_0 = \{x \mid -x \in M_0\} \subset N^{(2)}$. However, if x_1 and x_2 are in N, then so is $(x_1 + x_2)/2$, and this is easily extended to $N^{(2)}$; so if

$$p_2(u) < \frac{\varepsilon}{k_0}, \tag{2}$$

we can write

$$u = \frac{1}{2}\left[\frac{y}{k_0} + u + \left(-\frac{y}{k_0} - u\right)\right]$$

where $(y/k_0) + u \in M_0 \subset N^{(2)}$ and $-(y/k_0) - u \in -M_0 \subset N^{(2)}$. Thus, (2) implies $u \in N^{(2)}$; that is, the p_2-closure of any p_1-neighborhood of θ contains a p_2-neighborhood of θ.

This is the first step. The proof is completed if we show that N (rather than $N^{(2)}$) contains a p_2-neighborhood of θ because then x_n in the p_2-neighborhood implies $x_n \in N$—p_2-convergence implies p_1-convergence. To establish that any N contains a p_2-neighborhood let $N = \{x \mid p_1(x) < \delta\}$ and let

$$N_n = \left\{x \mid p_1(x) < \frac{\delta}{2^n}\right\}.$$

We have shown that each $N_n^{(2)}$ contains a neighborhood
$$H_n = \{x \mid p_2(x) < \varepsilon_n\}$$
and we may assume that $\varepsilon_n \to 0$. Our plan is to show that $H_1 \subset N$. Accordingly, let $y \in H_1 \subset N_1^{(2)}$; there is then an $x_1 \in N_1$ such that
$$p_2(y - x) < \varepsilon_2.$$
This puts
$$y - x_1 \in H_2 \subset N_2^{(2)},$$
and so there is an $x_2 \in N_2$ such that
$$p_2(y - x_1 - x_2) < \varepsilon_3.$$
Continuing in this way, we obtain a sequence x with $x_i \in N_i$ and with
$$\sum_{i=1}^{\infty} x_i$$
convergent (p_2) to y. However, by the definition of N_i, we have $p_1(x_i) < \delta/2^i$; so
$$\sum_{i=1}^{\infty} x_i$$
is also convergent (p_1), and since (A, p_1) is complete the series converges (p_1) to z. Moreover,
$$p_1(z) \le \sum_{i=1}^{\infty} p_1(x_i) < \sum_{i=1}^{\infty} \frac{\delta}{2^i} = \delta;$$
so $z \in N$. Since p_1 is stronger than p_2, the series also converges (p_2) to z, and it follows that $p_2(y - z) = 0$. However,
$$\{\theta\}^{(1)} = \{\theta\}^{(2)}$$
by hypothesis; so $p_1(y - z) = 0$ and $y \in N$. ∎

Theorem 50.5 (closed-graph theorem). *Let A and B be Fréchet spaces and f a linear function from A to B such that $G(f)$ is closed; then f is continuous.*

Proof. Let p and q be the paranorms in A and B, respectively. We define a new paranorm p_1 in A by setting
$$p_1(x) = p(x) + q[f(x)].$$
Clearly, p_1 satisfies the first four postulates for a paranorm. To verify P-V, let $t_n \to t$ and $p_1(x_n - y) \to 0$. Then, $p(x_n - y) \to 0$ and $q[f(x_n - y)] \to 0$; so applying P-V to p and q and using linearity of f, we have
$$p_1(t_n x_n - ty) = p(t_n x_n - ty) + q[t_n f(x_n) - tf(y)] \to 0.$$

Next, suppose $p(x) = 0$ and $q[f(x)] \neq 0$. Define a sequence y by setting $y_n = x$ for all n; then in $A \times B$,

$$[y_n, f(y_n)] \to [\theta, f(x)] \notin G(f),$$

contradicting the hypothesis that $G(f)$ is closed. Therefore $p(x) = 0$ implies $p_1(x) = 0$. Clearly, $p_1 \geq p$; so $p_1(x) = 0$ implies $p(x) = 0$, and indeed p_1 is stronger than p. Finally, let x be a Cauchy sequence (p_1); clearly x is a Cauchy sequence (p) and $f(x)$ is a Cauchy sequence (q). Since A and B are complete we have $x_0 \in A$ and $y_0 \in B$ such that $p(x_n - x_0) \to 0$ and $q[f(x_n) - y_0] \to 0$. However, since $G(f)$ is closed, $q[f(x_0) - y_0] = 0$ and we have

$$p_1(x_n - x_0) = p(x_n - x_0) + q[f(x_n) - f(x_0)]$$
$$= p(x_n - x_0) + q[f(x_n) - y_0] \to 0;$$

so (A, p_1) is complete. It now follows by 50.4 that p and p_1 are equivalent. Since $p_1(x) \geq q[f(x)]$, it is clear that f is p_1-continuous at θ and hence everywhere. Thus, f is p-continuous. ∎

Let Ω be $[0, 1]$ with Lebesgue measure; then we have the following inclusion relations among the standard function spaces considered simply as point sets (consider $p > 1$):

$$C \subset L_\infty \subset L_p \subset L_1 \subset S. \tag{3}$$

Now, note the types of convergence that the "standard" paranorms impose on these spaces:

C: uniform
L_∞: essentially uniform
L_p: mean^p
L_1: mean
S: measure

Since we are on a space of finite measure, each convergence type in this list implies those below it. This is to say that, reading from left to right in (3), we encounter weaker paranorms. There is another interesting feature to the list (3). The L_∞ norm restricted to C is equivalent to the C norm, but this is not true at any other stage in the list. In parallel to this observation note that C is a closed subspace of L_∞, while at no other stage in the list is the subspace closed as a subset of the larger space.

Our final project is to show that any nested set of Fréchet spaces must be related in this way. The smaller the space, the stronger the paranorm. Paranorms are equivalent if and only if the smaller space is closed in the larger. Given the paranorm on the largest space, all the others are uniquely determined by the conditions that they be stronger than the big one and that the smaller space be complete.

Since we are operating without the T_1-property, we need to explain one item of terminology. Let A and B be Fréchet spaces with perhaps different paranorms but with $A \subset B$. If we say that $\overline{\{\theta\}}$ is the same in A and B we mean that if $\overline{\{\theta\}} = B_0$ in B, then in A, $\overline{\{\theta\}} = A \cap B_0$. For example, $\overline{\{\theta\}}$ is the same in C and L_∞ because in L_∞ it is the set of functions a. e. zero and the only one of these in C is the identically zero function.

Theorem 50.6. *Let H be a Fréchet space with paranorm p. Let $Y \subset H$ be a Fréchet space with paranorm q such that* (i) $\overline{\{\theta\}}$ *is the same in H and Y and* (ii) q *is stronger than p on Y. Let X be any Fréchet space and let f be a linear function from X to Y. Then, f is q-continuous if and only if it is p-continuous.*

Proof. If f is q-continuous, then it is p-continuous by 50.2 because p is weaker than q. Conversely, if f is p-continuous, then $G(f)$ is closed in $X \times (Y, p)$ by 50.1. So, by 50.3, $G(f)$ is closed in $X \times (Y, q)$ and by 50.5 it is q-continuous. ∎

Corollary 50.6.1. *Let H, Y, and X be Fréchet spaces with paranorms p, q, and r respectively. Suppose* (i) $H \supset Y \supset X$; (ii) $\overline{\{\theta\}}$ *is the same in H, Y, and X;* (iii) q *is stronger than p on Y and r is stronger than p on X. Then r is stronger than q on X.*

Proof. Define f from X to Y by setting $f(x) = x$ for $x \in X$. Since r is stronger than p we have $f(x_n) = x_n \to x_0 = f(x_0)(p)$ provided $x_n \to x_0(r)$; that is, f is continuous from (X, r) to (Y, p). By 50.6 f is then continuous from (X, r) to (Y, q), which is to say that $x_n \to x_0(r)$ implies $x_n = f(x_n) \to f(x_0) = x_0(q)$; r is stronger than q. ∎

Corollary 50.6.2. *Let H be a Fréchet space with paranorm p and let $Y \subset H$ be a Fréchet space with each of two paranorms q and r such that* (i) $\overline{\{\theta\}}$ *is the same in (H, p), (Y, q) and (Y, r);* (ii) q *and r are each stronger than p on Y. Then, q and r are equivalent.*

Proof. Set $X = Y$ in 50.6.1. ∎

Corollary 50.6.3. *In 50.6.1, q and r are equivalent on X if and only if X is closed in (Y, q).*

Proof. If X is closed in (Y, q), then (X, q) is a Fréchet space (it is complete); so by 50.6.2, q and r are equivalent on X. Conversely, if q and r are equivalent, then (X, q) is complete because (X, r) is and so X is closed in (Y, q). ∎

So we end with a note on modes of convergence. The "standard" modes of convergence in C and L_p ($1 \le p \le \infty$) are the "natural" ones in that they are the only modes of convergence stronger than convergence in measure with respect to which the respective spaces are complete.

EXERCISES

a. Show that if p_1 is strictly stronger than p_2 on A and $f(x) = x$, then f is not continuous from (A, p_2) to (A, p_1).

b. Conclude from Exercise a that there is no parallel to 50.3 for continuous functions in general. Thus the closed-graph theorem is essential in the proof of 50.6.

c. Let A and B be T_1-Fréchet spaces and let f be a one-to-one function from A to B with $G(f)$ closed. Show that $G(f^{-1})$ is closed. [*Hint:* How are $G(f)$ and $G(f^{-1})$ related in this case?]

d. Suppose that (A, p_1) and (A, p_2) in Exercise a both have the T_1-property. Show that $G(f)$ is closed. [*Hint:* f^{-1} is continuous by 50.2; use Exercise c.]

e. Define p_1 and p_2 on l_1 by

$$p_1(x) = \sup_i |x_i|, \qquad p_2(x) = \sum_{i=1}^{\infty} |x_i|.$$

Now, apply Exercises a and d. The identity map from (l_1, p_2) to (l_1, p_1) is linear and has a closed graph but is not continuous. Why does this not contradict the closed-graph theorem?

f. Define p_1 and p_2 on the plane by

$$p_1(x, y) = \sqrt{x^2 + y^2}, \qquad p_2(x, y) = |x|.$$

Referring to 50.4, show that all hypotheses but one are satisfied and the conclusion fails.

g. Give a streamlined version of the first half of the proof of 50.4 assuming p_1 and p_2 are norms rather than only paranorms.

h. Let f be a sequence of linear functionals on a Fréchet space A such that $\{f_i(x) \mid i = 1, 2, 3, \ldots\}$ is bounded for each $x \in A$. Define g on A by setting $g(x) = y$, where y is the real sequence such that $y_i = f_i(x)$. Show that g is a linear function from A to m.

i. Show that if the function g of Exercise h is continuous, then the sequence f is equicontinuous.

j. Show that if each f_i in Exercise h is continuous, then g is indeed continuous; so f is equicontinuous. [*Hint:* g is continuous from A to s (where convergence means pointwise convergence); so it has a closed graph in this setting; by 50.3 the graph remains closed with s replaced by m; now use 50.5.]

k. The overall result of Exercises h through j is that a pointwise bounded sequence of continuous linear functionals is equicontinuous. Give another proof of this using the Banach-Steinhaus theorem (44.2). Note: The Banach-Steinhaus theorem obviously imposes uniformity on pointwise behavior of continuous linear functions. When used as suggested in Exercise j, the closed-graph theorem has much the same effect.

l. On each of the sequence spaces l_p ($p \geq 1$), c_0, c, and m the coordinate functionals defined by $f_i(x) = x_i$ are continuous. Show that each of these spaces has the only norm it could have so as to be complete and have the coordinate functionals continuous. [*Hint:* s has the weakest paranorm for which all coordinate functionals are continuous; use 50.6.2.]

m. There is (Exercise 47.p) a linear isomorphism f from c_0 to l_2. Define p_1 on c_0 by $p_1(x) = \|x\|$ and p_2 by $p_2(x) = \|f(x)\|$. Show that (c_0, p_1) and (c_0, p_2) are both complete but p_1 and p_2 are not equivalent. [*Hint:* l_2 is reflexive; c_0 is not.]

n. Show that in Exercise m neither p_1 nor p_2 is stronger than the other and there is no T_1-paranorm on c_0 weaker than both.

REFERENCES FOR FURTHER STUDY

On Banach spaces in general:
  Banach [1]
  Hewitt and Stromberg [13]
  Riesz and Nagy [24]
  Wilansky [30]

On representation of linear functionals:
  Banach [1]
  Zaanen [31]
On weak and weak* convergence:
  Banach [1]
  Wilansky [30]

BIBLIOGRAPHY AND INDEXES

BIBLIOGRAPHY

1. BANACH, S., *Théorie des opérations linéaires*, Monografje Matematyczne, Tom I. Warsaw, 1932.
2. BARTLE, R. G., *The Elements of Integration*. New York: Wiley, 1966.
3. BERBERIAN, S. K., *Measure and Integration*. New York: Macmillan, 1965.
4. CARATHÉODORY, C., *Vorlesungen über reelle Funktionen*, 2. Auflage. Leipzig: Teubner, 1927.
5. CARATHÉODORY, C., *Algebraic Theory of Measure and Integration* (edited by Finsler, Rosenthal, and Steuerwald). New York: Chelsea, 1963.
6. DOOB, J. L., Probability in function spaces, *Bull. Amer. Math. Soc.*, **53** (1947), 15–30.
7. FRÉCHET, M., Sur l'intégrale d'une fonctionnelle étendue à un ensemble abstrait, *Bull. Soc. Math. France*, **43** (1915), 249–67.
8. GOFFMAN, C., *Real Functions*. New York: Rinehart, 1953.
9. HAHN, H. and ROSENTHAL, A., *Set Functions*. Albuquerque: The University of New Mexico Press, 1948.
10. HALMOS, P. R., The foundations of probability, *Amer. Math. Monthly*, **51** (1944), 403–510.
11. HALMOS, P. R., *Measure Theory*. New York: Van Nostrand, 1950.
12. HAUSDORFF, F., *Mengenlehre*, 3. Auflage. New York: Dover Publications, 1944.
13. HEWITT, E. and STROMBERG, K., *Real and Abstract Analysis*. Berlin: Springer-Verlag, 1965.
14. HOBSON, E. W., *The Theory of Functions of a Real Variable and the Theory of Fourier's Series* (2 vols.). Cambridge: Cambridge University Press; Vol. I, 3rd ed., 1927; Vol. II, 2nd ed., 1926.
15. HUREWICZ, W. and WALLMAN, H., *Dimension Theory*. Princeton: Princeton University Press, 1941.
16. KELLEY, J. L., *General Topology*, Princeton: Van Nostrand, 1955.
17. KESTELMAN, H., *Modern Theories of Integration*. Oxford: Oxford University Press, 1937.
18. KOLMOGOROFF, A. N., *Foundations of the Theory of Probability*. New York: Chelsea, 1950.
19. LEBESGUE, H., *Leçons sur l'intégration*, 2nd ed. Paris: Gauthier-Villars, 1928.

20. McShane, E. J. and Botts, T. A., *Real Analysis*. Princeton: Van Nostrand, 1959.
21. Munroe, M. E., *Introductory Real Analysis*. Reading: Addison-Wesley, 1965.
22. Natanson, J. P., *Theory of Functions of a Real Variable*. New York: Ungar, 1955.
23. Neumann, J. von, *Functional Operators*: Vol. 1, *Measures and Integrals*. Princeton: Princeton University Press, 1950.
24. Riesz, F. and Nagy, B., *Functional Analysis*. New York: Ungar, 1955.
25. Royden, H. L., *Real Analysis*, 2nd ed. New York: Macmillan, 1968.
26. Rudin, W., *Principles of Mathematical Analysis*. New York: McGraw-Hill, 1953.
27. Saks, S., *Theory of the Integral*, Monografje Matematyczne, Tom VII. Warsaw, 1937.
28. Titchmarsh, E. C., *The Theory of Functions*, 2nd ed. Oxford: Oxford University Press, 1939.
29. Vallée Poussin, Ch.-J. de la, *Cours d'analyse infinitésimale*, 2nd ed. Paris: Gauthier-Villars; Vol. I, 1909; Vol. II, 1912.
30. Wilansky, A., *Functional Analysis*. New York: Blaisdell, 1964.
31. Zaanen, A. C., *Integration*, 2nd ed. Amsterdam: North-Holland, 1967.

INDEX OF POSTULATES

Key letter	Entity being defined	Page
a	finitely additive set function	33
A	completely additive set function	31
c	finitely additive class	25
C	completely additive class	25
E	equivalence relation	9
G	group	86
H	Haar measure	87
L	linear space	232
M	I–III outer measure	42
	IV metric outer measure	57
N	norm	232
P	paranorm	269
PM	pseudo-metric space	11
PO	partially ordered set	5

INDEX OF SYMBOLS

[a.e.]	almost everywhere convergence, 104
[a.un.]	almost uniform convergence, 199
$\mathscr{B}$	class of Borel sets, 30
$\mathscr{C}$	sequential covering class, 47
C	space of continuous functions, 19
c	space of convergent sequences, 19
c_0	space of sequences convergent to zero, 19
C_Γ	characteristic function of the set Γ, 7
$\overline{D}\tau, \underline{D}\tau, D\tau$	strong derivates, derivative of τ, 169
$\overline{D}_{\mathscr{N}\mu}\tau, \underline{D}_{\mathscr{N}\mu}\tau, D_{\mathscr{N}\mu}\tau$	derivates, derivative with respect to sets, 192
$d(E)$	diameter of E, 12
$\dim(E)$	dimension of E, 85
$e(f)$	expectation of f, 158
$e_0(f), e^0(f)$	conditional expectations, 162
essup	essential supremum, 220
$\mathscr{F}$	class of closed sets, 28
$\mathscr{G}$	class of open sets, 28
$G(f)$	augmented graph of f, 270
inf	infimum, 3
L_p	space of pth power integrable functions, 215
L_∞	space of essentially bounded functions, 220
l_p	space of pth power convergent series, 220
l_2	Hilbert space, 19
$LH(S)$	linear hull of S, 255
$\lim_{x \to x_0} f(x)$	limit of a function, 15
$\lim_\alpha x_\alpha$	limit of a net, 21
$\overline{\lim}_n u_n, \underline{\lim}_n u_n$	limit superior, inferior of a sequence of numbers, 4
$\overline{\lim}_n \Gamma_n, \underline{\lim}_n \Gamma_n$	limit superior, inferior of a sequence of sets, 8
$\mathrm{Lim}_n x_n$	generalized limit of a bounded sequence of numbers, 242

283

$\mathscr{M}$	class of measurable sets, 44
M	space of bounded functions, 20
m	space of bounded sequences, 19
[mean]	convergence in the mean, 205
[mean^p]	convergence in the mean of order p, 218
[meas.]	convergence in measure, 200
$N(x, \varepsilon)$	ε-neighborhood of x, 12
$\underline{N}(x, \varepsilon)$	deleted ε-neighborhood of x, 12
$p(x)$	paranorm of x, 269
R_n	Euclidean n-space, 19
s	space of all sequences of real numbers, 19
S	space of all measurable functions, 220
sup	supremum, 3
[unif.]	uniform convergence, 199
$\overline{V}(\sigma, E), \underline{V}(\sigma, E), V(\sigma, E)$	upper, lower, total variation of σ on E, 36
δ	(as subscript) class of countable intersections, 27
θ	zero element in a function space, 198; in any linear space, 232
μ	measure function, 40; restriction of outer measure μ^*, 49
μ^*	outer measure function, 42
μ_*	inner measure function, 51
$\bar{\mu}$	completion of the measure μ, 54
μ_f	Lebesgue-Stieltjes measure induced by f, 69
$\mu^{*(p)}$	Hausdorff p-dimensional outer measure, 84
$\mu_0(E), \mu^0(E)$	conditional probabilities, 162
$\rho(x, y)$	distance between x and y, 11
σ	(as subscript) class of countable unions, 27
τ	function used to define outer measure, 47
ϕ	vacuous set, 1
Ω	space, 1
$\pm\infty$	infinite elements in extended number system, 4
$\in$	is an element of, 1
$\subset$	is a subset of, 1
$\cup, \cap, -$	union, intersection, complement, 5
$\{\alpha \mid P(\alpha)\}$	the set of all α for which $P(\alpha)$ is true, 1
$\Gamma \times \Delta$	product of Γ and Δ, 2
$\bigcup_{\Gamma \in \mathscr{E}} \Gamma, \bigcap_{\Gamma \in \mathscr{E}} \Gamma$	union, intersection of a class of sets, 6
$\bigcup_{n=1}^{\infty} \Gamma_n, \bigcap_{n=1}^{\infty} \Gamma_n$	union, intersection of a sequence of sets, 6
$\underset{a \in A}{\times} X_a$	product of a class of sets, 22

INDEX OF SYMBOLS

$\bar{A}$	closure of A, 12
A^0	interior of A, 13
A^*	conjugate of Banach space A, 236
f^+, f^-	positive, negative parts of f, 102
f^{-1}	inverse of f, 2
$f(\alpha,)$	function on Δ generated by f on $\Gamma \times \Delta$, 3
$f \circ g$	composite function, 2
$\int_E f$	integral over E of f, 115
$\int_a^b f(x)\, dx$	Riemann integral, 122
$\int_a^b x\, dz$	Riemann-Stieltjes integral, 249
$\int_Y \int_X f\, d\alpha\, d\beta$	iterated integral, 150
$\int_{X \times Y}' f\, d\mu_{(x)}$	pseudo-integral with respect to conditional probability, 163
$\|f\|$	norm of functional f, 235
$\|x\|$	norm of element x, 232
$\hat{x}$	natural image of x in second conjugate space, 236
$x \cdot y$	scalar product in Hilbert space, 224
$\bar{\tau}', \underline{\tau}', \tau'$	derivates, derivative of τ, 168
$\bar{\tau}'^*, \underline{\tau}'^*, \tau'^*$	regular derivates, derivative of τ, 168

SUBJECT INDEX

Absolutely continuous function of a real variable, 137–141, 144, 167
Absolutely continuous set function, 136–137, 142–144, 180–183
Accumulation point, 12, 31
Additive set function, 33
Almost everywhere, 104
Almost everywhere convergence, 104, 108, 146, 199–200, 206–208, 224
Almost uniform convergence, 199–200, 202, 206–208, 224
Approximately continuous function, 188–191
Axiom of choice, 5

Baire category theorem, 14
Baire classification of functions, 105, 111
Baire function, 97–99, 101
Banach space, 233–237
Banach-Steinhaus theorem, 234–235
Bessel's inequality, 228
Borel classification of sets, 29
Borel measure, 55
Borel sets, 30, 55, 59, 97–98, 100–101
Bounded set, 12
Bounded variation, 39, 71, 140, 167, 249–254

Cantor function, 139, 184
Cantor ternary set, 14, 67
Cardinal number, 9–11
Category of a set, 14
Cauchy sequence, 12–14, 198, 202–203, 208, 220–224, 236
Cauchy-Riemann integral, 135, 157
Characteristic function of a set, 7, 106
Class, 1

Closed set, 13, 21, 28–30, 65, 110
Closed-graph theorem, 272
Closure of a set, 12–13, 21, 262–264
Coefficient of linear correlation, 218
Collection, 1
Compact set, 14, 17, 22–23, 87–93, 264
Complement of a set, 6
Complete measure, 54
Complete pseudo-metric space, 12, 208–209, 220–224, 233, 236
Completely additive class, 25–26, 44
Completely additive set function, 31, 35–39, 41–42, 135, 136, 141–143, 179, 210
Completion of a measure, 54
Conditional expectation, 161–165
Conditional probability, 161–165
Conjugate space, 236, 245–254
Continuous function, 15–17, 19, 97–98, 100, 110, 138–139, 210, 270
Continuous linear function, 234–235, 240, 266, 271
Contracting sequence of sets, 8–9, 35
Convergence in measure, 147, 200–208, 212–214, 224, 274
Convergence in probability, 203
Convergent sequence of intervals, 168
Convergent sequence of set functions, 61, 183, 210–211
Convergent sequence of sets, 8–9, 41
Countable set, 10
Cross partial derivative, 169, 190
Cube, 168

Darboux sum, 122
Dense set, 13
Density of a set, 184
Derivative, 132, 144, 166–170, 176–183

287

288   SUBJECT INDEX

Derivative with respect to nets, 191–197
Diameter of a set, 12
Dimension of a linear space, 256
Dimension of a set, 85
Directed set, 21
Discrete probability distribution, 32
Disjoint sets, 6
Distance function, 11
Distribution function, 69–71, 75–78, 113
Domain of a function, 2

Egoraff's theorem, 108–110, 199–200
Element of a set, 1
Equicontinuous point functions, 18–19, 211–212, 234–235, 275
Equicontinuous set functions, 212–213
Equivalence class, 9, 11–12
Equivalence relation, 9
Essential supremum, 220
Essentially bounded functions, 220, 253
Euclidean n-space, 19, 72
Events, 32, 78
Expanding sequence of sets, 8–9, 35, 51
Expectation, 158–160
Extended real number system, 4
Extension of a function, 3, 55, 238–241

Fatou's lemma, 222
Finite dimensional space, 256–257
Finite dimensional subspaces, 266–267
Finite intersection property, 23, 89
Finite set, 9
Finitely additive class, 25
Finitely additive set function, 33–34, 251–252
Fourier expansion, 228
Frichet space, 269–274
Fubini's theorem, 155, 164, 186, 190
Function, 2
Functional, 236
Functional analysis, 231

Generalized limit of a bounded sequence, 242–244
Generalized rectangle, 149
Graph of a function, 268–270, 272
Greatest lower bound, 3
Group, 86

Haar measure, 86–93
Hahn decomposition theorem, 141
Hahn-Banach theorem, 238–241
Hamel basis, 255–256
Hausdorff measure, 84–85
Hausdorff space, 21
Hermite polynomial, 226
Hilbert space, 19, 224–230
Hölder's inequality, 216

Indefinite integral, 115, 167
Indefinitely fine sequence of sets, 196
Independent experiments, 79
Independent random variables, 113, 159, 218
Infimum, 3
Infinite set, 9–10
Inner measure, 51–53
Integrable function, 116, 120
Integral defined, 116, 121, 126–127, 133–134, 136, 205
Integration by parts, 211
Interior of a set, 13
Intersection of sets, 5–6
Interval, 69, 72, 77, 79, 168
Inverse function, 2, 16, 97–99
Iterated integral, 150–157

Joint distribution, 79, 113
Jordan decomposition theorem, 38

Laguerre polynomial, 226
Law of large numbers, 203–204, 220
Least upper bound, 3
Lebesgue decomposition theorem, 183
Lebesgue dominated convergence theorem, 146
Lebesgue integral, 121, 125, 167, 181, 245
Lebesgue measure, 49, 50, 67, 73, 93–95, 245
Lebesgue monotone convergence theorem, 132
Lebesgue-Stieltjes integrals, 130–131, 144, 211
Lebesgue-Stieltjes measures, 68–78, 113, 156, 167, 177–178, 180–183
Legendre polynomial, 226

Limit inferior of a sequence of numbers, 4
Limit inferior of a sequence of sets, 8, 40, 50
Limit superior of a sequence of numbers, 4
Limit superior of a sequence of sets, 8, 41
Limit under the integral sign, 145
Linear functional, 236, 238–240, 265–268
Linear hull, 255
Linear space, 232, 238–240, 254–257, 265–268
Linear subspace, 238–241, 259, 261, 263, 266
Linearly independent elements, 254
Locally compact space, 14, 87–93
Lower density, 184
Lower derivate, 168–170
Lower variation of a set function, 36–39, 142
Lusin's theorem, 110

Marginal probability measure, 160
Mass density function, 182
Mass distribution, 71–72, 76, 182
Maximal element, 5
Mean convergence, 145–149, 205–208, 212–214
Mean^p convergence, 218–219, 222–223, 227–230
Mean value theorem, 131
Measurable cover, 50, 62
Measurable function, 97–111, 188
Measurable kernel, 53
Measurable set, 43–46, 49, 52, 62–65, 187
Measure function, 40–41, 44–46, 49
Method I: construction of outer measures, 47–49, 53–56, 66, 69, 73, 80, 89, 149
Method II: construction of outer measures, 60–62, 84
Metric outer measure, 57–67, 70, 76, 80–81, 84, 98
Metric space, 11, 19–20, 57, 83, 196
Minimal completely additive class, 26, 30
Minkowski's inequality, 217
Monotone sequence of nets, 192
Monotone sequence of sets, 8–9
Monotone set function, 36
Moore-Osgood theorem, 18

Natural embedding, 237
Negative set, 142
Neighborhood, 12, 21, 22
Net (in differentiation theory), 192
Net (in general topology), 21
Noncountable set, 10–11
Nondecreasing set function, 36
Nonincreasing set function, 36
Nonmeasurable set, 93–95, 100–101
Norm, 232–233
Norm of a linear function, 235
Normed linear space, 232–238
Nowhere dense set, 14

Open set, 13, 16, 21, 28, 62, 64, 77, 97
Ordinate set, 127
Orthogonal elements of Hilbert space, 225
Orthonormal sequence, 225–230
Outer measure, 42–67

Parameter of regularity, 168
Paranorm, 269
Parswal's theorem, 229
Partially ordered set, 5
Perfect set, 14
Point, 1
Point of density, 184–187
Point of dispersion, 184–188
Pointwise convergence, 16–19, 20, 22, 105, 106, 210, 224, 234, 260–262
Positive set, 142
Probability of an event, 32, 78
Probability in function spaces, 82
Probability in sequence spaces, 78–83
Product measure, 76, 79, 82–83, 149, 156, 160
Product space, 2, 3, 16–19, 22–23, 88–89, 149–158, 160–165
Product topology, 22–23, 88–89, 262, 264–267
Pseudo-integral, 163–165
Pseudo-metric space, 11–19, 208, 218, 220, 233, 269

Radon-Nikodym theorem, 142–144
Random variable, 112–114, 158–165
Range of a function, 2

Reflexive Banach space, 237, 247–249, 257, 258, 265–267
Regular derivative, 168, 179
Regular measure, 65, 177, 192
Regular outer measure, 50–57
Regular sequence of intervals, 168, 171, 176
Regular sequence of sets, 193–197
Restriction of a function, 3, 44, 49, 55, 110
Riemann integral, 121–126, 221–222
Riemann sum, 123
Riemann-Stieltjes integral, 249
Riesz representation theorem, 250
Riesz-Fischer theorem, 222, 227–228

Sample mean, 112
Sampling, 112
Scalar product, 224–225
Schmidt orthogonalization process, 226, 229
Schwarz's inequality, 218
Separable space, 13, 20, 83, 197, 219, 224, 230
Sequence, 3
Sequential covering class, 47
Set, 1
Simple function, 106–107, 116–121
Singular function, 182–184
Space, 1
Strong derivative, 168–169, 184–191
Sturm-Liouville equation, 225
Subadditive set function, 40
Subsequence, 3
Subset, 1
Subuniform convergence, 17–19
Summable function, 133
Supremum, 3

T_1-space, 21, 233–234, 237, 241, 256–257, 267, 269, 274, 275
Topological group, 87
Topological space, 21
Total variation of a set function, 37, 142
Translation in a group, 86
Translation modulo 1, 94
Triangle postulate, 11, 215–218, 232, 269
Tychonoff theorem, 23, 89, 265

Uniform absolute continuity, 212–215
Uniform continuity, 17, 123, 139
Uniform convergence, 16–18, 20, 108–110, 149, 198–200, 206–208
Union of sets, 5–6
Upper bound, 3, 5
Upper density, 184
Upper derivate, 168–170
Upper variation of a set function, 36–39, 142

Vacuous set, 1
Variance, 160
Vitali covering, 171–176
Vitali's theorem, 171–192

Weak sequential convergence, 257–262
Weak* sequential convergence, 257–262
Weak topology, 267
Weak* topology, 262–268
Weakly* continuous functionals, 266, 268

Zero-one law, 83
Zorn's lemma, 5

Date: Sep 21 1979

QA312
.M74
1971

275770